Lecture Notes in Control and Information Sciences

Edited by M. Thoma and A. Wyner

Vol. 97: I. Lasiecka/R. Triggiani (Eds.)
Control Problems for Systems
Described by Partial Differential Equations
and Applications
Proceedings of the IFIP-WG 7.2
Working Conference
Gainesville, Florida, February 3-6, 1986
VIII, 400 pages, 1987.

Vol. 98: A. Aloneftis
Stochastic Adaptive Control
Results and Simulation
XII, 120 pages, 1987.

Vol. 99: S. P. Bhattacharyya
Robust Stabilization Against
Structured Perturbations
IX, 172 pages, 1987.

Vol. 100: J. P. Zolésio (Editor)
Boundary Control and Boundary Variations
Proceedings of the IFIP WG 7.2 Conference
Nice, France, June 10-13, 1987
IV, 398 pages, 1988.

Vol. 101: P. E. Crouch,
A. J. van der Schaft
Variational and Hamiltonian
Control Systems
IV, 121 pages, 1987.

Vol. 102: F. Kappel, K. Kunisch,
W. Schappacher (Eds.)
Distributed Parameter Systems
Proceedings of the 3rd International Conference
Vorau, Styria, July 6–12, 1986
VII, 343 pages, 1987.

Vol. 103: P. Varaiya, A. B. Kurzhanski (Eds.)
Discrete Event Systems:
Models and Applications
IIASA Conference
Sopron, Hungary, August 3-7, 1987
IX, 282 pages, 1988.

Vol. 104: J. S. Freudenberg/D. P. Looze
Frequency Domain Properties of Scalar
and Multivariable Feedback Systems
VIII, 281 pages, 1988.

Vol. 105: Ch. I. Byrnes/A. Kurzhanski (Eds.)
Modelling and Adaptive Control
Proceedings of the IIASA Conference
Sopron, Hungary, July 1986
V, 379 pages, 1988.

Vol. 106: R. R. Mohler (Editor)
Nonlinear Time Series and
Signal Processing
V, 143 pages, 1988.

Vol. 107: Y. T. Tsay, L.-S. Shieh, St. Barnett
Structural Analysis and Design
of Multivariable Systems
An Algebraic Approach
VIII, 208 pages, 1988.

Vol. 108: K. J. Reinschke
Multivariable Control
A Graph-theoretic Approach
274 pages, 1988.

Vol. 109: M. Vukobratović/R. Stojić
Modern Aircraft Flight Control
VI, 288 pages, 1988.

Vol. 110: B. J. Daiuto, T. T. Hartley,
S. P. Chicatelli
The Hyperbolic Map and Applications
to the Linear Quadratic Regulator
VI, 114 pages, 1989

Vol. 111: A. Bensoussan, J. L. Lions (Eds.)
Analysis and Optimization
of Systems
XIV, 1175 pages, 1988.

Vol. 112: Vojislav Kecman
State-Space Models of Lumped
and Distributed Systems
IX, 280 pages, 1988

Vol. 113: M. Iri, K. Yajima (Eds.)
System Modelling and Optimization
Proceedings of the 13th IFIP Conference
Tokyo, Japan, Aug. 31 – Sept. 4, 1987
IX, 787 pages, 1988.

Vol. 114: A. Bermúdez (Editor)
Control of Partial Differential Equations
Proceedings of the IFIP WG 7.2
Working Conference
Santiago de Compostela, Spain, July 6–9, 1987
IX, 318 pages, 1989

Vol. 115: H.J. Zwart
Geometric Theory for Infinite
Dimensional Systems
VIII, 156 pages, 1989.

Vol. 116: M.D. Mesarovic, Y. Takahara
Abstract Systems Theory
VIII, 439 pages, 1989

For information about Vols. 1-96 please contact your bookseller or Springer-Verlag

Lecture Notes in Control and Information Sciences

Edited by M. Thoma and A. Wyner

170

J. M. Skowronski, H. Flashner, R. S. Guttalu (Eds.)

Mechanics and Control

Proceedings of the 4th Workshop
on Control Mechanics, January 21–23, 1991
University of Southern California, USA

Springer-Verlag Berlin Heidelberg GmbH

Editors
Prof. Janislaw M. Skowronski
Prof. Henryk Flashner
Prof. Ramesh S. Guttalu

Dept. of Mechanical Engineering
University of Southern California
Los Angeles, CA 90089-1453
USA

ISBN 978-3-540-54954-3 ISBN 978-3-540-46606-2 (eBook)
DOI 10.1007/978-3-540-46606-2

Originally published by Springer-Verlag Berlin Heidelberg New York in 1992.

Typesetting: Camera ready by authors

61/3020-5 4 3 2 1 0 Printed on acid-free paper.

INTRODUCTION

The Workshop in Control Mechanics held annually at the University of Southern California since 1988 has already established an international tradition in promoting control of nonlinear mechanical systems subject to uncertainty, and promoting the use of advanced mechanics methods in control theory. The present proceedings cover our fourth meeting, with 17 papers contributing to the mentioned profile. The order in the proceedings is alphabetic, but the following groups of topics can be selected. A large number of papers deals with *feedback control of deterministic uncertain systems*: Benedé - Leitmann - Ryan, Corless - Da, Hui - Zac, Nwokah - Jayasuriya - Chait, Olas, Soldatos, Vincent - Lin; some of these papers *introduce observers*: Jabbari, Gawronski; or *adaptive control* and observers: Skowronski. *Stochastic* uncertainty is discussed by Blaquiére. Next large group of papers refers to control of *flexible mechanical systems approximated by lumped models*: Corless - Da, Pszczel - Payne, Skowronski. Control of *robotic manipulators* is discussed by Ahmad - Zribi and Reithmeier - Leitmann, with robotic *coordination control* studied by Ahmad - Zribi and Stonier, and in rather general set-up by Skowronski. Finally *flight control* problems are investigated by Ardema - Bowles - Whittaker and Gawronski, the latter discussing an interesting NASA problem in deep space antenna guidance.

The methods used in almost all works are those of the Liapunov formalism applied to untruncated nonlinearity, large amplitudes and global studies in the state space of multidimensional systems. We hope that the results published here will move a step further to the development in the above direction.

Thanks are in order to the Department of Mechanical Engineering, University of Southern California, in particular to its Head Professor Satwinder Sadhal and Administrative Officer Mrs Jacquette Givens for valuable help in organizing the workshop.

J.M. Skowronski
H. Flashner
R.S. Guttalu
Los Angeles, July 1991

CONTENTS

LYAPUNOV BASED CONTROL DESIGN FOR MULTIPLE ROBOTS HANDLING A COMMON OBJECT

Shaheen Ahmad and Mohamed Zribi

Real-Time Robot Control Laboratory
School of Electrical Engineering
Purdue University
West Lafayette, IN 47907

Abstract

In this paper, we address the problem of multiple robots manipulating a rigid object cooperatively. First we propose and prove several fundamental properties of the multi-robot system. These properties are then exploited to design a set point regulation controller. The proposed controller takes into account the dynamics of both the object and the manipulators. This controller enables us to control the position of the object and the internal forces acting on the object. An adaptive version of the proposed controller is then introduced; the adaptive controller is able to account for the uncertainty in the mass of the load. The convergence analysis of the position, the load mass estimate and the internal forces is also given.

1. Introduction

Recently considerable amount of research has focused on the problem of cooperative control and coordination of multiple robots. Cooperative robots can be used in hazardous or unsafe environments such as in space, in deep waters and in radioactive environments. Interest in multi-robot systems has arisen because several tasks require the use of two or more robots. Examples of such tasks include the joining and securing of large pipes for the construction of space structures, picking up and carrying heavy loads, and manipulating difficult to grasp objects. By using more than one robot the manipulation capability and the workspace of the robots are further increased. However multi-robot systems are more difficult to control than single robots. Additional problems arise as the parameters of the robots and the manipulated load may not be known exactly.

It is important to review some major developments in the single robot adaptive control literature because of their relevance to the multiple robots case.

Ortega and Spong [10] presented a summary of several recent papers on adaptive control for rigid robots. Craig et. al. [5] proposed an adaptive control law which is basically a modification of the a computed torque control scheme. Slotine and Li [13] exploited the strúctural properties of the rigid manipulator dynamics to design an adaptive controller capable of general trajectory tracking control. They tested their design on a two degree of freedom semi-direct drive robot. They experimentally showed that the parameters of the robot which were assumed to be unknown initially can be estimated within a very short time. They also showed that their controller has the same level of robustness to unmodeled dynamics as the PD controlled system. Better tracking performance than the PD or the computed torque schemes were also obtained.

Bayard and Wen [3] introduced a class of asymptotically stable adaptive control laws for robotic manipulators. In their analysis they used a parameterization of the dynamics of the robot, the parameterization is based on physical quantities. They also used the property that the robot dynamics are linear with respect to the unknown parameters of the manipulator. They proposed an energy like Lyapunov function that retains the nonlinear character and structure of the dynamics. The authors proved that their approach leads to asymptotically stable adaptive systems. Their approach does not require the convergence of the parameter estimates, invertibility of the mass matrix estimate or the measurement of joint accelerations.

Seraji [12] proposed a decentralized adaptive control algorithm for robot manipulators. Seraji's control scheme results in a PID controller and position velocity acceleration feedforward controller both with adjustable gains. Sadegh and Horowitz [11] presented a control scheme that uses the desired trajectory outputs to update the parameters and the controller. They also addressed the robustness properties of the proposed scheme. Middleton and Goodwin [8] developed the use of predictive adaptive control laws for rigid link manipulators. They used a computed torque controller in conjunction with linear estimation techniques to prove the global convergence of the adaptive system.

Few control schemes for cooperative multiple robots manipulating a common load have been proposed. Tarn et al. [16] presented a control method that is based on the linearization of the robot models with nonlinear feedback, they then use the theory of linear optimal control to design a robust controller. Yun et al. [17] used exact linearization and output decoupling to design a controller for two manipulators working in cooperation. Zheng and Luh [20] considered the dual arms system as a closed dynamic chain and then used master/slave method to design the control laws. Yoshikawa and Zheng [19] proposed a cooperative dynamic hybrid control method; this method takes into consideration both the manipulator dynamics and the object dynamics. Ahmad and Guo [2] looked at the control problems when the robots have compliant joints. Ahmad [1] has developed a nonlinear cancellation feedback control scheme for multiple robots with flexible joints. In this scheme, both internal force control and load position control are achieved, and the interaction forces, accelerations and jerk do not need to be measured

Hsu et al [6] developed a control algorithm for the coordinated manipulation of multifingered robot hand. Their control law guarantees the convergence of the load motion and internal forces to the desired values respectively. Walker et al. [18] developed an algorithm for the control of two robots handling a load of unknown mass. Their controller achieves global convergence of the load motion and *the internal forces* to their respective desired trajectories. Their controller basically amounts to an open loop controller for the internal forces and a closed loop controller for the positions, however they did not prove the convergence of the internal forces. The attractive feature of the proposed control law is that the computational algorithm is the same for the two manipulators.

Hu and Goldenberg [7] addressed the problem of multiple robots manipulating a load in contact with the environment. First they decomposed the multi-robot system into three subsystems, one for the position error another one for the contact forces and a third one for the internal forces. Then they used these subsystems errors and Popov's hyperstability theory to derive the update laws to estimate the unknown parameters. The proposed control law guarantees the global asymptotic convergence to the desired position, internal forces and contact forces trajectories. Carignan [4] used reduced order models for the tracking control of two manipulator arms handling a load. An adaptive controller was designed to track a desired trajectory using a reduced model of the system instead of the full order model of the system. Carignan's results were very encouraging, however a lot of work remains to be done on the use of reduced order models.

Zribi and Ahmad [22] - [23] proposed an adaptive controller for the multi-robot system manipulating a rigid object cooperatively. Their controller takes into account the dynamics of the manipulators and the load and does not require feedback of joint acceleration or the inversion of the inertia or the Jacobian matrices. They also considered the effects of bounded disturbances on the system, a control law which guarantees the convergence of the tracking error to a bounded set when disturbances are present is given. Zribi and Ahmad [21] also developed a multi-robot predictive adaptive control scheme which is able to ensure the tracking of the manipulators, the load positioning and the convergence of the internal forces to the desired internal forces, however measurements of the end effectors forces are required. The parameters of the load and the robots are updated by least squares estimation techniques.

In this paper we address the problem of controlling multiple robots handling an object cooperatively. In section 2 we develop the dynamic and kinematic models of the manipulators and the load. In section 3 we state and prove few properties of the multi-robot system. In section 4 we use these properties to develop a controller which guarantees the convergence of the load motion to the desired set point. In section 5 we give an adaptive version of the controller proposed previously.

2. Cooperative Multi-Robot System Model

2.1 Dynamics model

The general dynamic model for n cooperative multi-robot system has been investigated thoroughly in the literature, and is also described in the below for completeness, (see Figure 1 for a depiction of the multi-robot system). The dynamic equation of the ith manipulator in cooperative manipulation is given as:

$$D_i(q_i)\ddot{q}_i + C_i(q_i,\dot{q}_i)\dot{q}_i + G_i(q_i) + J_i(q_i)^T F_{e_i} = \tau_i \qquad i=1, \cdots n \tag{1}$$

where, $q_i \in R^{n_i}$ is the vector of joint displacements, and n_i is the number of joints of the ith robot. The inertia matrix of the ith robot is $D_i(q_i) \in R^{n_i \times n_i}$, this is a positive definite and symmetric matrix. The matrix of centrifugal and Coriolis forces is $C_i(q_i,\dot{q}_i) \in R^{n_i \times n_i}$; the vector of gravity forces is $G_i(q_i) \in R^{n_i}$, and the manipulator Jacobian is $J_i(q_i) \in R^{6 \times n_i}$. The control input torque for the ith robot is $\tau_i \in R^{n_i}$.

In the following we will assume that n_i is 6. The forces/moments applied by the ith manipulator on the object at the point of contact are F_{e_i} and they are assumed to be exactly measurable. The contact forces/moments $F_{e_i} \in R^6$ can be written in terms of the contact forces $f_i \in R^3$ and contact moments $\eta_i \in R^3$, (where 6 represents the dimension of the Cartesian work space), such that

$$F_{e_i} = \begin{bmatrix} f_{e_i}{}^T & \eta_{e_i}{}^T \end{bmatrix}^T \qquad i = 1,\dots n \tag{2}$$

If we assume that the object is rigidly grasped, then the equations of motion of the object are obtained from the Newton-Euler mechanics as,

$$M_1 \ddot{z} + M_1 g = \sum_{i=1}^{i=n} f_i \tag{3}$$

$$I\dot{\omega} + \omega \times (I\omega) = \sum_{i=1}^{i=n} (\eta_i + r_i \times f_i) \tag{4}$$

where the position of the center of mass of the object expressed in world coordinate frame is $z \in R^3$. The rotational velocity of the center of mass of the object in world coordinate frame is $\omega \in R^3$, and the gravity force vector for the object, expressed in world coordinate reference frame is $g \in R^3$. The mass matrix $M_1 \in R^{3\times3}$ is a diagonal matrix whose diagonal elements are the mass of the load; the matrix $I \in R^{3\times3}$ is the inertia matrix of the load. The position of the end effector of the ith manipulator with respect to the object center of mass, expressed in the world coordinate frame is $r_i \in R^3$.

Now if we define $B_i \in R^{6\times6}$, and $\Omega_i(r_i)$ such that

$$B_i = \begin{bmatrix} R_i^c & 0 \\ \Omega_i(r_i)R_i^c & R_i^c \end{bmatrix} \quad \text{and} \quad \Omega_i(r_i) = \begin{bmatrix} 0 & -r_{iz} & r_{iy} \\ r_{iz} & 0 & -r_{ix} \\ -r_{iy} & r_{ix} & 0 \end{bmatrix} \tag{5}$$

and $R_i^c \in R^{3\times3}$ is the rotation matrix that converts any vector expressed in i-th base coordinate frame into a vector in the center of mass of the object reference frame. The vector $r_i = [\, r_{ix}, r_{iy}, r_{iz}]^T$ represents the translational displacements from the center of mass of the object to the contact point of the object and the ith manipulator (see Figure 1). Equations (3) and (4) can be grouped into one equation such that

$$M\ddot{x} + N_2\dot{x} + G_l = \sum_{i=1}^{i=n} B_i^{-T} F_{e_i} \tag{6}$$

where,

$$M = \begin{bmatrix} M_1 & 0 \\ 0 & I \end{bmatrix}, \quad N_2\dot{x} = \begin{bmatrix} 0 \\ \omega \times (I\omega) \end{bmatrix} \quad \text{and} \quad G_l = \begin{bmatrix} M_1 g \\ 0 \end{bmatrix} \tag{7}$$

2.2 Kinematic Model

Occasionally, we might be interested in controlling the manipulators in some predefined Cartesian task space such that:

$$x_{e_i} = K_i(q_i) \qquad i = 1,...n \tag{8}$$

where $K_i(.) : R^6 \longrightarrow R^6$ is the transformation from the joint angle space of q_i to the task space containing x_{e_i}, and $x_{e_i} \in R^6$ is the position and orientation of the point of contact of the ith manipulator with the load, expressed in the world coordinate frame. Notice that the end effector velocity $\dot{x}_{e_i} = \begin{bmatrix} v_i^T & \omega_i^T \end{bmatrix}^T \in R^6$.

If we define the manipulator Jacobian $J_i(q_i)$ to be the differential map of q_i space to x_{e_i} space, then we can write

$$\dot{x}_{e_i} = J_i(q_i)\dot{q}_i \qquad i = 1,...n \tag{9}$$

We will also define $B_i : R^6 \longrightarrow R^6$ to be a transformation between the center of mass of the load and the contact point with the end effectors, expressed in the world coordinate frame, then we can write:

$$B_i\dot{x}_{e_i} = B_i \begin{bmatrix} v_i \\ \omega_i \end{bmatrix} = \begin{bmatrix} v \\ \omega \end{bmatrix} \qquad i = 1,...n \tag{10}$$

or,

$$\dot{x} = B_i J_i \dot{q}_i \tag{11}$$

The angular velocity of the point of contact of the ith manipulator, with the load, expressed in world coordinate frame is $v_i \in R^3$. The rotational velocity of the point of contact of the ith manipulator, with the load, expressed in world coordinate frame is $\omega_i \in R^3$. The angular velocity of the center of mass of the object expressed in world coordinate frame is $v \in R^3$; $x \in R^6$ is the position and orientation vector of the center of mass of the load. If we take the derivative with respect to time of equation (11), we get

$$\ddot{x} = (B_i J_i)'\dot{q}_i + B_i J_i \ddot{q}_i \tag{12}$$

where $(.)'$ denotes the time derivative of the entity.

We require each robot manipulator to follow a predefined trajectory q_{id} and $q_{id} = K^{-1}(x_{e_{id}})$, then the trajectory tracking error of the ith robot, $e_i \in R^6$, is:

$$e_i = q_i - q_{id} \qquad i = 1,...n \tag{13}$$

where the vector of desired joint displacements for the ith manipulator is $q_{id} \in R^6$. We will also define the object error as

$$e_o = x - x_d$$

where x_d correspond to the desired position/orientation of the object.

2.3 Motion and Internal Forces

The end effector force F_{e_i} can be decomposed into two parts, F_{eI_i} and F_{eM_i}, then we can write

$$F_{e_i} = F_{eM_i} + F_{eI_i}$$

where F_{eI_i} corresponds the internal force at the ith robot end effector, and the motion force F_{eM_i} corresponds to the force responsible for the motion of the object generated by the ith robot. The motion forces are responsible for the work exchange with the load. The internal forces result in zero motion of the load.

3. Properties of Multi-robot System

Few important properties of the inertia matrix, the centrifugal/Coriolis matrix and the contact and internal forces are required for the development; these properties are,

P1 :

$$\dot{q}_i^T(\dot{D}_i - 2C_i)\dot{q}_i = 0 \qquad i = 1,...,n \tag{14}$$

P2 :

$$\dot{x}^T(\dot{M} - 2N_2)\dot{x} = 0 \tag{15}$$

P3 :

$$-\sum_{i=1}^{i=n}\dot{q}_i^T J_i^T F_{e_i} + \dot{x}^T\sum_{i=1}^{i=n} B_i^{-T}F_{e_i} = 0 \tag{16}$$

P4 :

$$\sum_{i=1}^{i=n}\dot{q}_i^T J_i^T F_{eI_i} = 0 \tag{17}$$

P5 :

$$\sum_{i=1}^{i=n} B_i^{-T}F_{eI_i} = 0 \tag{18}$$

Proof:

The proof of property P1 is well documented in the literature (see Spong and Vidyasagar [15]). Property P3 is easily proved by using equation (11)

$$\begin{aligned}
-\sum_{i=1}^{i=n}\dot{q}_i^T J_i^T F_{e_i} + \dot{x}^T\sum_{i=1}^{i=n} B_i^{-T}F_{e_i} &= -\sum_{i=1}^{i=n}\dot{q}_i^T J_i^T F_{e_i} + \sum_{i=1}^{i=n}\dot{x}^T B_i^{-T}F_{e_i} \\
&= -\sum_{i=1}^{i=n}\dot{q}_i^T J_i^T F_{e_i} + \sum_{i=1}^{i=n}(B_i J_i \dot{q}_i)^T B_i^{-T}F_{e_i} \\
&= -\sum_{i=1}^{i=n}\dot{q}_i^T J_i^T F_{e_i} + \sum_{i=1}^{i=n}\dot{q}_i^T J_i^T B_i^T B_i^{-T}F_{e_i} = 0
\end{aligned} \tag{19}$$

Thus property P3 is proved. □

Now to prove property P2, we will define H to be the Hamiltonian of the multi-robot system

$$H = \tfrac{1}{2}\sum_{i=1}^{i=n}\dot{q}_i^T D_i\dot{q}_i + \sum_{i=1}^{i=n} P_{1i}(q_i) + \tfrac{1}{2}\dot{x}^T M\dot{x} + P_2(x) \tag{20}$$

where $\tfrac{1}{2}\dot{q}_i^T D_i\dot{q}_i$ represents the Kinetic energy of the ith robot; $P_{1i}(q)$ is the potential energy of the ith robot. The term $\tfrac{1}{2}\dot{x}^T M\dot{x}$ corresponds to the kinetic energy of the load, and $P_2(x)$ is the potential energy of the load. It has been shown [10] that the Hamiltonian is an exact expression of the total energy of the system, thus,

$$\frac{dH}{dt} = \sum_{i=1}^{i=n} \dot{q}_i^T \tau_i \tag{21}$$

This is true because the power input to the system is only through the actuators. Now taking the the derivative of the Hamiltonian with respect to time, we obtain

$$\frac{dH}{dt} = \sum_{i=1}^{i=n} \dot{q}_i^T D_i \ddot{q}_i + \tfrac{1}{2}\sum_{i=1}^{i=n} \dot{q}_i^T \dot{D}_i \dot{q}_i + \sum_{i=1}^{i=n} \dot{q}_i^T \frac{\partial P_{1i}(q_i)}{\partial q_i}$$
$$+ \dot{x}^T M \ddot{x} + \tfrac{1}{2}\dot{x}^T \dot{M} \dot{x} + \dot{x}^T \frac{\partial P_2(x)}{\partial x} \tag{22}$$

Now if we substitute equations (1) and (6) into equation (22), we can write

$$\frac{dH}{dt} = \sum_{i=1}^{i=n} \dot{q}_i^T (\tau_i - C_i \dot{q}_i - G_i - J_i^T F_{e_i}) + \tfrac{1}{2}\sum_{i=1}^{i=n} \dot{q}_i^T \dot{D}_i \dot{q}_i + \sum_{i=1}^{i=n} \dot{q}_i^T \frac{\partial P_{1i}(q_i)}{\partial q_i}$$
$$+ \dot{x}^T (-N_2 \dot{x} - G_l + \sum_{i=1}^{i=n} B_i^{-T} F_{e_i}) + \tfrac{1}{2}\dot{x}^T \dot{M} \dot{x} + \dot{x}^T \frac{\partial P_2(x)}{\partial x} \tag{23}$$

Now using the fact that

$$\frac{\partial P_{1i}(q_i)}{\partial q_i} = G_i \tag{24}$$

$$\frac{\partial P_2(x)}{\partial x} = G_l \tag{25}$$

We get

$$\frac{dH}{dt} = \sum_{i=1}^{i=n} \dot{q}_i^T \tau_i + \sum_{i=1}^{i=n} \dot{q}_i^T (\tfrac{1}{2}\dot{D}_i - C_i)\dot{q}_i + \dot{x}^T (\tfrac{1}{2}\dot{M} - N_2)\dot{x}$$
$$- \sum_{i=1}^{i=n} \dot{q}_i^T J_i^T F_{e_i} + \dot{x}^T \sum_{i=1}^{i=n} B_i^{-T} F_{e_i} \tag{26}$$

Using equation (21) and property P1 we can conclude that

$$\dot{x}^T (\tfrac{1}{2}\dot{M} - N_2)\dot{x} - \sum_{i=1}^{i=n} \dot{q}_i^T J_i^T F_{e_i} + \dot{x}^T \sum_{i=1}^{i=n} B_i^{-T} F_{e_i} = 0 \tag{27}$$

and therefore

$$\dot{x}^T (\tfrac{1}{2}\dot{M} - N_2)\dot{x} = 0 \tag{15}$$

because of property P3. Thus property P2 is proved. □

To prove property P4, we go back to the definition of internal forces. By definition the internal forces are the forces that do not contribute to the motion of the object (i.e. the net effect of the internal forces to the motion of the load is zero). So the net work done by the internal forces is zero. Thus we can write:

$$\sum_{i=1}^{i=n} \dot{x}_{e_i}^T F_{el_i} = 0 \tag{28}$$

Or by definition $\dot{x}_{e_i} = J_i \dot{q}_i$, hence we have

$$\sum_{i=1}^{i=n} \dot{q}_i^T J_i^T F_{eI_i} = 0 \tag{29}$$

which correspond exactly to property P4. □

To prove property P5, we state the fact that by definition the sum of the internal forces at the load frame of reference is equal to zero, thus,

$$\sum_{i=1}^{i=n} F_{I_i} = 0 \tag{30}$$

where F_{I_i} is the ith robot internal force expressed in the center of mass of the load frame (i.e. F_{I_i} is the projection of F_{eI_i} on the center of mass of the object). So we can write the above equation as:

$$\sum_{i=1}^{i=n} B_i^{-T} F_{eI_i} = 0 \tag{31}$$

which correspond to property P5. □

By using properties P3, P4 and P5, it can be easily shown that the following equation is true,

$$-\sum_{i=1}^{i=n} \dot{q}_i^T J_i^T F_{eM_i} + \dot{x}^T \sum_{i=1}^{i=n} B_i^{-T} F_{eM_i} = 0 \tag{32}$$

Lemma 1: There exists a passive mapping between the joint velocities and the joint input torques, such that

$$\sum_{i=1}^{i=n} < \dot{q}_i \mid \tau_i > = \sum_{i=1}^{i=n} \int_0^T \dot{q}_i^T \tau_i dt \geq -\gamma^2 \tag{33}$$

where gamma is a real number.

Proof:

From equation (21) we have

$$\sum_{i=1}^{i=n} \int_0^T \dot{q}_i^T \tau_i dt = \int_0^T \sum_{i=1}^{i=n} \dot{q}_i^T \tau_i dt = \int_0^T dH = H(T) - H(0) \geq -H(0) \tag{33a}$$

as the energy of the system H is always greater than zero.
Hence Lemma 1 is proved. □

4. Proposed Controller

In this section we will introduce a controller which guarantee trajectory tracking and internal force control. The properties described in the previous section will be utilized. We should note that the multi-robot system is redundantly actuated, for example if each robot has six degrees of freedom, n robots manipulating an object have a total of 6n actuators. In order to determine the extent at which each robot and its actuators are used to manipulate the load we will assume that an affine load strategy has been developed. The scalars α_i such that $\sum_{i=1}^{i=n} \alpha_i = 1$ are the load distribution scalars.

Theorem 1:

The below control law ensures asymptotic trajectory stabilization and internal force regulation (i.e. $x \rightarrow x_d$, $q \rightarrow q_d$ and $F_{el_i} \rightarrow F_{eld_i}$ as $t \rightarrow \infty$, where x_d and F_{eld_i} are the desired load position and desired internal forces).

$$\tau_i = G_i - \alpha_i J_i^T B_i^T K_p e_o + \alpha_i J_i^T B_i^T G_l - K_{d_i} \dot{q}_i + J_i^T F_{eld_i} \tag{34}$$

where the desired internal forces are chosen such that $\sum_{i=1}^{i=n} B_i^{-T} F_{eld_i} = 0$ and the load distribution scalar parameters are chosen such that $\sum_{i=1}^{i=n} \alpha_i = 1$.

$K_p \in R^{6\times6}$ and $K_{d_i} \in R^{6\times6}$ are positive definite matrices.

Proof:

Consider the following Lyapunov function candidate which is a modified energy function :

$$V = \sum_{i=1}^{i=n} \tfrac{1}{2} \dot{q}_i^T D_i \dot{q}_i + \tfrac{1}{2} e_o^T K_p e_o + \tfrac{1}{2} \dot{x}^T M \dot{x} \tag{36}$$

Now if we differentiate V with respect to time, we get

$$\dot{V} = \sum_{i=1}^{i=n} (\dot{q}_i^T D_i \ddot{q}_i + \tfrac{1}{2} \dot{q}_i^T \dot{D}_i \dot{q}_i) + \dot{e}_o^T K_p e_o + \dot{x}^T M \ddot{x} + \tfrac{1}{2} \dot{x}^T \dot{M} \dot{x} \tag{37}$$

Using the fact that $\dot{e}_o = \dot{x}$, and using properties P1 and P2, we get,

$$\dot{V} = \sum_{i=1}^{i=n} \dot{q}_i^T (D_i \ddot{q}_i + C_i \dot{q}_i) + \dot{x}^T K_p e_o + \dot{x}^T (M\ddot{x} + N_2 \dot{x}) \tag{38}$$

substituting the dynamics of the manipulators, equation (1), and the dynamics of the load, equation (6), into equation (38),

$$\dot{V} = \sum_{i=1}^{i=n} \dot{q}_i^T (\tau_i - G_i - J_i^T F_{e_i}) + \dot{x}^T (K_p e_o - G_l + \sum_{i=1}^{i=n} B_i^{-T} F_{e_i}) \tag{39}$$

Using property P3, and using the fact that $\sum_{i=1}^{i=n} \alpha_i = 1$, we can write

$$\dot{V} = \sum_{i=1}^{i=n} \dot{q}_i^T (\tau_i - G_i) + \sum_{i=1}^{i=n} \alpha_i \dot{x}^T (- G_l + K_p e_o) \tag{40}$$

Using the fact that $\dot{x} = B_i J_i \dot{q}_i$, we have

$$\begin{aligned} \dot{V} &= \sum_{i=1}^{i=n} \dot{q}_i^T (\tau_i - G_i) + \sum_{i=1}^{i=n} \alpha_i (B_i J_i \dot{q}_i)^T (-G_l + K_p e_o) \\ &= \sum_{i=1}^{i=n} \dot{q}_i^T (\tau_i - G_i + \alpha_i J_i^T B_i^T K_p e_o - \alpha_i J_i^T B_i^T G_l) \end{aligned} \tag{41}$$

Now substituting for τ_i from equation (34), we get

$$\dot{V} = \sum_{i=1}^{i=n} (- \dot{q}_i^T K_{d_i} \dot{q}_i + \dot{q}_i^T J_i^T F_{eld_i}) \tag{42}$$

By definition, the internal forces F_{eId_i} are chosen consistently such that $\sum_{i=1}^{i=n}\dot{q}_i^T J_i^T F_{eId_i} = \sum_{i=1}^{i=n}\dot{x}^T B_i^{-T} F_{eId_i} = \dot{x}^T \sum_{i=1}^{i=n} B_i^{-T} F_{eId_i} = 0$, hence we have

$$\dot{V} = -\sum_{i=1}^{i=n}\dot{q}_i^T K_{d_i}\dot{q}_i \leq 0 \tag{43}$$

Therefore we can conclude that $\dot{e}_i = \dot{q}_i \rightarrow 0$ as $t \rightarrow \infty$ because K_{d_i}'s are positive definite matrices.

Now we need to prove that $e_i \rightarrow 0$ as $t \rightarrow \infty$. We will start by proving that $e_o \rightarrow 0$ as $t \rightarrow \infty$. At the equilibrium point, we have $\dot{q}_i = 0$, $\ddot{q}_i = 0$, $\dot{x} = 0$ and $\ddot{x} = 0$, thus we can write equation (1) as

$$G_i + J_i^T F_{e_i} = \tau_i \quad i=1, \cdots n \tag{44}$$

Also recall that the controller at the equilibrium point reduces to

$$\tau_i = G_i - \alpha_i J_i^T B_i^T K_p e_o + \alpha_i J_i^T B_i^T G_l + J_i^T F_{eId_i} \tag{45}$$

Now equating equations (44) and (45), we obtain

$$J_i^T F_{e_i} = -\alpha_i J_i^T B_i^T K_p e_o + \alpha_i J_i^T B_i^T G_l + J_i^T F_{eId_i} \tag{46}$$

Solving for F_{e_i}, we get

$$F_{e_i} = -\alpha_i B_i^T K_p e_o + \alpha_i B_i^T G_l + F_{eId_i} \tag{47}$$

At the equilibrium point equation (6) can be written as,

$$G_l = \sum_{i=1}^{i=n} B_i^{-T} F_{e_i} \tag{48}$$

Now combining equations (47) and (48), we get

$$G_l = \sum_{i=1}^{i=n} B_i^{-T}(-\alpha_i B_i^T K_p e_o + \alpha_i B_i^T G_l + F_{eId_i}) \tag{49}$$

or,

$$G_l = G_l - K_p e_o + \sum_{i=1}^{i=n} B_i^{-T} F_{eId_i} \tag{50}$$

Using property P5 as applied to the desired internal forces, the above equation reduces to the following

$$K_p e_o = 0 \tag{51}$$

Because K_p is a symmetric positive definite matrix, we have $e_o = 0$ at the equilibrium. Thus we can conclude that $e_i = 0$ at the equilibrium because the of the rigid contact between the object and the manipulators.

Notice that at the equilibrium, equation (47) becomes

$$F_{e_i} = \alpha_i B_i^T G_l + F_{eId_i} \tag{47a}$$

the first term is used to compensate for the gravitational load of the carried object and the second term correspond to the desired internal forces.

Note the above controller is a direct consequenc of the passivity relationship given in Lemma 1.

Remark:

The above control law given in theorem 1 can also be implemented by the following independent joint control law,

$$\tau_i = G_i - K_{p_i} e_i + \alpha_i J_i^T B_i^T G_l - K_{d_i} \dot{q}_i + J_i^T F_{eld_i} \quad i=1,\ldots n \tag{52}$$

where F_{eld_i} and α_i are chosen such that $\sum_{i=1}^{i=n} B_i^{-T} F_{eld_i} = 0$ and $\sum_{i=1}^{i=n} \alpha_i = 1$.

Proof:

Consider the following Lyapunov function candidate:

$$V = \sum_{i=1}^{i=n} (\tfrac{1}{2} \dot{q}_i^T D_i \dot{q}_i + \tfrac{1}{2} e_i^T K_{p_i} e_i) + \tfrac{1}{2} \dot{x}^T M \dot{x} \tag{53}$$

It is easy to show that

$$\dot{V} = - \sum_{i=1}^{i=n} \dot{q}_i^T K_{d_i} \dot{q}_i \leq 0 \tag{54}$$

Thus $\dot{e}_i = \dot{q}_i \rightarrow 0$ as $t \rightarrow \infty$.

At the equilibrium point, we have $\dot{q}_i = 0$, $\ddot{q}_i = 0$, $\dot{x} = 0$ and $\ddot{x} = 0$, thus we can write equation (1) as

$$G_i + J_i^T F_{e_i} = \tau_i \qquad i=1, \cdots n \tag{55}$$

Also recall that the controller at the equilibrium point reduces to

$$\tau_i = G_i - K_{p_i} e_i + \alpha_i J_i^T B_i^T G_l + J_i^T F_{eld_i} \tag{56}$$

Now equating equations (55) and (56), we obtain

$$J_i^T F_{e_i} = - K_{p_i} e_i + \alpha_i J_i^T B_i^T G_l + J_i^T F_{eld_i} \tag{57}$$

Solving for F_{e_i}, we get

$$F_{e_i} = - J_i^{-T} K_{p_i} e_i + \alpha_i B_i^T G_l + F_{eld_i} \tag{58}$$

One way to prove that $e_i \rightarrow 0$ as $t \rightarrow \infty$ is by noting that at the equilibrium point F_{e_i} is given by equation (47a),

$$F_{e_i} = \alpha_i B_i^T G_l + F_{eld_i} \tag{47a}$$

hence equations (47a) and (58) imply that

$$J_i^{-T} K_{p_i} e_i = 0$$

but $J_i^{-T} K_{p_i}$ is not a singular matrix, thus $e_i = 0$ at the equilibrium. Another way of proving that $e_i \rightarrow 0$ is by noting that at the equilibrium point equation (6) can be written as,

$$G_l = \sum_{i=1}^{i=n} B_i^{-T} F_{e_i} \tag{59}$$

Now combining equations (58) and (59) to get

$$G_l = \sum_{i=1}^{i=n} B_i^{-T} (- J_i^{-T} K_{p_i} e_i + \alpha_i B_i^T G_l + F_{eld_i}) \tag{60}$$

or,

$$G_l = G_l - \sum_{i=1}^{i=n} B_i^{-T} J_i^{-T} K_{p_i} e_i + \sum_{i=1}^{i=n} B_i^{-T} F_{eld_i} \tag{61}$$

Using property P5 as applied to the desired forces, the above equation, reduces to the following

$$\sum_{i=1}^{i=n} B_i^{-T} J_i^{-T} K_{p_i} e_i = 0 \tag{62}$$

Now let A_i be such,

$$A_i = B_i^{-T} J_i^{-T} K_{p_i} \tag{63}$$

and let A be such that,

$$A = \left[A_1 \;\; A_2 \;\; \ldots\ldots A_n \right] \tag{64}$$

Also let e denote

$$e = \left[e_1^T \ldots\ldots e_i^T \ldots e_n^T \right]^T \tag{65}$$

Note that equation (62) can be written as,

$$Ae = 0 \tag{66}$$

Notice that the matrix A is a 6 by 6n matrix, whose rank is 6 (because we are assuming that B_i , J_i and K_{p_i} are not singular).

From the kinematics equations, we can write nonlinear equations such that,

$$x = f_1(q_1) = f_2(q_2) = \ldots\ldots = f_i(q_i) = \cdots = f_n(q_n) \tag{67}$$

or,

$$x = f_1(e_1 + q_{1d}) = f_2(e_2 + q_{2d}) = \ldots\ldots = f_i(e_i + q_{id}) = \cdots = f_n(e_n + q_{nd}) \tag{68}$$

Now using Taylor's Series, and assuming that e_i is small (i.e ignore high order terms), we get

$$\left(\frac{\partial f_i}{\partial q_i} \Big|_{q_{id}} \right) e_i = \left(\frac{\partial f_j}{\partial q_j} \Big|_{q_{jd}} \right) e_j \qquad i, j = 1, \ldots, n \tag{69}$$

Using equations (62) and (69), we can write

$$A_i' e_i = 0 \tag{70}$$

where A_i' is a 6 by 6 nonsingular matrix. Thus $e_i = 0$ at the equilibrium point.

If we don't assume that e_i is small to start with, we have

$$x_d = f_1(q_{1d}) = f_2(q_{2d}) = \ldots\ldots = f_i(q_{id}) = \cdots = f_n(q_{nd}) \tag{71}$$

It is obvious from equations (68) and (71) that $e_i = 0 \Rightarrow e_j = 0$ for (j=1,...n). Hence it is sufficient to prove that one of the $e_i = 0$, to prove that all of the e_i's are equal to zero.

We claim that for equation (66) and the Taylor's series expansion of equation (68) to hold true at the same time, we have to have one of the $e_i = 0$, and thus all of the $e_i = 0$.

5. Adaptive Controller

In this section we will describe an adaptive controller to compensate for the uncertainty of the mass of the object (i.e. the unknown gravity term).

Theorem 2:
The below controller ensures asymptotic trajectory stabilization and internal force regulation (i.e. $x \rightarrow x_d$, $q \rightarrow q_d$ and $F_{eI_i} \rightarrow F_{eId_i}$ as $t \rightarrow \infty$). Also the unknown gravity vector converges to its actual value.

$$\tau_i = G_i - \alpha_i J_i^T B_i^T K_p e_o + \alpha_i J_i^T B_i^T \hat{G}_l - K_{d_i} \dot{q}_i + J_i^T F_{eld_i} \tag{72}$$

where F_{eld_i} and α_i are chosen such that $\sum_{i=1}^{i=n} B_i^{-T} F_{eld_i} = 0$ and $\sum_{i=1}^{i=n} \alpha_i = 1$.

and the parameter adaptation law is given as,

$$\dot{e}_p = -\Gamma^{-1}\dot{x} \qquad \text{with } e_p(0) = \Gamma^{-1}(x_d(t_f) - x(0)) \tag{73}$$

where $e_p = \tilde{G}_l = \hat{G}_l - G_l$, $\hat{G}_l$ is the estimate of the gravity vector and Γ is a symmetric positive definite matrix. Note that t_f is the time at which the equilibrium is reached and $t_f \rightarrow \infty$.

Proof:

Consider the following Lyapunov function candidate:

$$V = \sum_{i=1}^{i=n} \tfrac{1}{2}\dot{q}_i^T D_i \dot{q}_i + \tfrac{1}{2} e_o^T K_p e_o + \tfrac{1}{2}\dot{x}^T M \dot{x} + \tfrac{1}{2} e_p^T \Gamma e_p \tag{74}$$

Now if we differentiate V with respect to time, we get

$$\dot{V} = \sum_{i=1}^{i=n} (\dot{q}_i^T D_i \ddot{q}_i + \tfrac{1}{2}\dot{q}_i^T \dot{D}_i \dot{q}_i) + \dot{e}_o^T K_p e_o + \dot{x}^T M \ddot{x} + \tfrac{1}{2}\dot{x}^T \dot{M}\dot{x} + e_p^T \Gamma \dot{e}_p \tag{75}$$

Now using the fact that $\dot{e}_o = \dot{x}$, and using property P1 and P2, we get,

$$\dot{V} = \sum_{i=1}^{i=n} \dot{q}_i^T (D_i \ddot{q}_i + C_i \dot{q}_i) + \dot{x}^T K_p e_o + \dot{x}^T (M\ddot{x} + N_2 \dot{x}) + e_p^T \Gamma \dot{e}_p \tag{76}$$

substituting the dynamics of the manipulators, equation (1), and the dynamics of the load, equation (6), into equation (76)

$$\dot{V} = \sum_{i=1}^{i=n} \dot{q}_i^T (\tau_i - G_i - J_i^T F_{e_i}) + \dot{x}^T K_p e_o + \dot{x}^T (-G_l + \sum_{i=1}^{i=n} B_i^{-T} F_{e_i}) + e_p^T \Gamma \dot{e}_p \tag{77}$$

Using property P3, and $e_p = \hat{G}_l - G_l$, we can write

$$\dot{V} = \sum_{i=1}^{i=n} \dot{q}_i^T (\tau_i - G_i) + \dot{x}^T K_p e_o - \dot{x}^T \hat{G}_l + \dot{x}^T e_p + e_p^T \Gamma \dot{e}_p \tag{78}$$

now using the fact that $\dot{x} = B_i J_i \dot{q}_i$, we have

$$\dot{V} = \sum_{i=1}^{i=n} \dot{q}_i^T(\tau_i - G_i) + \sum_{i=1}^{i=n} \alpha_i (B_i J_i \dot{q}_i)^T(-\hat{G}_l + K_p e_o) + \dot{x}^T e_p + e_p^T \Gamma \dot{e}_p$$

$$= \sum_{i=1}^{i=n} \dot{q}_i^T(\tau_i - G_i + \alpha_i J_i^T B_i^T K_p e_o - \alpha_i J_i^T B_i^T \hat{G}_l) + \dot{x}^T e_p + e_p^T \Gamma \dot{e}_p \tag{79}$$

Now substituting for τ_i from equation (72), we get

$$\dot{V} = \sum_{i=1}^{i=n}(- \dot{q}_i^T K_{d_i} \dot{q}_i + J_i^T F_{eld_i}) + \dot{x}^T e_p + e_p^T \Gamma \dot{e}_p \tag{80}$$

However F_{eld_i} were chosen such that $\sum_{i=1}^{i=n} B_i^{-T} F_{eld_i} = 0$, thus we have

$$\dot{V} = - \sum_{i=1}^{i=n} \dot{q}_i^T K_{d_i} \dot{q}_i + e_p^T(\dot{x} + \Gamma \dot{e}_p) \tag{81}$$

Using the adaptation law given by (73), we thus get

$$\dot{V} = - \sum_{i=1}^{i=n} \dot{q}_i^T K_{d_i} \dot{q}_i \tag{82}$$

Therefore we can conclude that $\dot{e}_i = \dot{q}_i \longrightarrow 0$ as $t \longrightarrow \infty$.
At the equilibrium point, we have $\dot{q}_i = 0$, $\ddot{q}_i = 0$, $\dot{x} = 0$ and $\ddot{x} = 0$, thus we can write equation (1) as

$$G_i + J_i^T F_{e_i} = \tau_i \qquad i=1, \cdots n \tag{83}$$

Also recall that the controller at the equilibrium point reduces to

$$\tau_i = G_i - \alpha_i J_i^T B_i^T K_p e_o + \alpha_i J_i^T B_i^T \hat{G}_l + J_i^T F_{eld_i} \tag{84}$$

Now equating equations (83) and (84), we obtain

$$J_i^T F_{e_i} = - \alpha_i J_i^T B_i^T K_p e_o + \alpha_i J_i^T B_i^T \hat{G}_l + J_i^T F_{eld_i} \tag{85}$$

Solving for F_{e_i}, we get

$$F_{e_i} = - \alpha_i B_i^T K_p e_o + \alpha_i B_i^T \hat{G}_l + F_{eld_i} \tag{86}$$

At the equilibrium point equation (6) can be written as,

$$G_l = \sum_{i=1}^{i=n} B_i^{-T} F_{e_i} \tag{87}$$

Now combining equations (86) and (87) to get

$$G_l = \sum_{i=1}^{i=n} B_i^{-T}(- \alpha_i B_i^T K_p e_o + \alpha_i B_i^T \hat{G}_l + F_{eld_i}) \tag{88}$$

or,

$$G_l = \hat{G}_l - K_p e_o + \sum_{i=1}^{i=n} B_i^{-T} F_{eld_i} \tag{89}$$

Using property P5 as applied to the desired forces, the above equation, reduces to the following

$$G_l = \hat{G}_l - K_p e_o \tag{90}$$

equation (90) can be written as,

$$K_p e_o(t_f) = e_p(t_f) \tag{91}$$

The solution to equation (73) (the adaptation law) is

$$e_p(t) = e_p(0) - \Gamma^{-1}x(t) + \Gamma^{-1}x(0) \tag{92}$$

at the equilibrium (i.e. $t = t_f$)

$$e_p(t_f) = e_p(0) - \Gamma^{-1}x(t_f) + \Gamma^{-1}x(0) \tag{93}$$

if $e_p(0)$ is chosen such that $e_p(0) = \Gamma^{-1}(x_d(t_f) - x(0))$ then equation (93) becomes

$$e_p(t_f) = -\Gamma^{-1}e_o(t_f) \tag{94}$$

Now combining equations (91) and (94) we get

$$K_p e_o(t_f) = -\Gamma^{-1}e_o(t_f) \tag{95}$$

or,

$$(K_p + \Gamma^{-1})e_o(t_f) = 0 \tag{96}$$

however $(K_p + \Gamma^{-1})$ is a positive definite matrix, hence equation (96) implies $e_o(t_f) = 0$ at the equilibrium. Thus we can conclude that $e_i = 0$ at the equilibrium because of the rigid grasp between the object and the manipulators. Therefore $F_{eI_i} \longrightarrow F_{eId_i}$ at the equilibrium and $e_p(t) \longrightarrow 0$ subject to the choice of $e_p(0)$. □

6. Conclusion

In this paper, we proposed a controller for the multi-robot system. The proposed controller takes into account the dynamics of the objects and manipulators. Few special properties such as the property of the linearity with respect to the parameters, were exploited during the derivation of the controller. The proposed controller guarantees global stability of the system in the sense of Lyapunov. The advantage of this controller is that it does not require measurements of joint accelerations and inversion of the inertia or Jacobian matrices or the end effectors forces. An adaptive version of the control law was also proposed. The adaptive controller enables us to estimate the mass of the load while ensuring asymptotic trajectory stabilization and convergence of the internal forces to their desired values.

References

[1] S. Ahmad, "Control of Cooperative Multiple Flexible Joint Robots," TR-EE-91-19, Purdue University, W. Lafayette, Indiana, 1991.

[2] S. Ahmad, and H. Guo "Dynamic Coordination of Dual-Arm Robotic Systems With Joint Flexibility," Proceedings of the 1988 IEEE International Conference on Robotics and Automation, Philadelphia, Pennsylvania, 1988, pp. 332 - 337.

[3] D. Bayard, and J. Wen, "New Class of Control Laws for Robotic Manipulators: Part 2. Adaptive case," Int. J. Control, Vol. 47, No. 5, 1985, pp. 1387-1406.

[4] C. Carignan, "Adaptive tracking for Complex Systems Using Reduced-Order Models," Proceedings of the 1990 IEEE International Conference on Robotics and Automation, Cincinnati, Ohio, 1990, pp. 2078-2083.

[5] J. Craig, P. Hsu, and S. Sastry, "Adaptive Control of Mechanical Manipulators," Proceedings of the 1986 IEEE International Conference on Robotics and Automation, San Francisco, CA, 1986, pp. 190-195.

[6] P. Hsu, Z. Li, and S. Sastry, "On Grasping and Coordinated Manipulation by a Multifingered Robot Hand," Proceedings of the 1988 IEEE International Conference on Robotics and Automation, Philadelphia, Pennsylvania, 1988, pp. 384-389.

[7] Y. Hu and A. Goldenberg, "An Adaptive Approach to Motion and Force Control of Multiple Coordinated Robot Arms," Proceedings of the 1989 IEEE International Conference on Robotics and Automation, Scottsdale, Arizona, 1989, pp. 1091-1096.

[8] R. Middleton and G. Goodwin, "Adaptive Computed Torque Control for rigid Link Manipulations," Systems & Control Letters, Vol. 10, 1988, pp. 9-16.

[9] R. Murray and S. Sastry, "Grasping and Manipulation using Multifingered Robots Hands," Proceedings of Symposia in Applied Mathematics, Vol. 41, 1990, pp. 91-127.

[10] R. Ortega and M. Spong, "Adaptive Motion Control of Rigid Robots: a Tutorial," Automatica, Vol. 25, No 6, 1989, pp. 877-888.

[11] N. Sadegh, and R. Horowitz, "Stability and Robustness Analysis of a Class of Adaptive Controllers for Robotic Manipulators," Int. J. of Robotics Research, Vol. 9, No 3, 1990, pp. 74-92.

[12] H. Seraji, "Decentralized Adaptive Control of manipulators: Theory, Simulation, and Experimentation," IEEE Trans. on Robotics and Automation, Vol. 5, No. 2, 1989, pp. 183-201.

[13] J-J. Slotine and W. Li, "Adaptive Manipulator Control: A Case Study," IEEE Trans. Automatic Control, Vol. AC-33, No. 11, 1988, pp. 995-1003.

[14] J-J. Slotine and W. Li, "Applied Nonlinear Control," Prentice Hall, New Jersy, 1991.

[15] M. Spong and M. Vidyasagar, "Robot Dynamics and Control," John Wiley & Sons, New York, 1989.

[16] T. Tarn, A. Bejczy, and X. Yun, "Design of Dynamic Control of Two Cooperating Robot Arms: Closed chain Formulation," Proceedings of the 1987 IEEE International Conference on Robotics and Automation, Raleigh, NC, 1987, pp. 7-13.

[17] X. Yun, T. Tarn, and A. Bejczy, "Dynamic Coordinated Control of Two Robot Manipulators," Proceedings of 28th Conference on Decision and Control, Tampa, FL., 1989, pp. 2476-2481.

[18] M. Walker, D. Kim and J. Dionise, "Adaptive Coordinated Motion Control of Two Manipulator Arms," Proceedings of the 1989 IEEE International Conference on Robotics and Automation, Scottsdale, Arizona, 1989, pp 1084-1090.

[19] T. Yoshikawa and X. Zheng, "Coordinated Dynamic Hybrid Position/Force Control for Multiple Robot Manipulators Handling One Constraint Object," Proceedings of the 1990 IEEE International Conference on Robotics and Automation, Cincinnati, Ohio, 1990, 1178-1183.

[20] Y. Zheng and J. Luh, "Joint Torques Control of Two Coordinated Moving Robots," Proceedings of the 1986 IEEE International Conference on Robotics and Automation, San Francisco, CA, 1986, 1375-1380.

[21] M. Zribi and S. Ahmad, "Predictive Adaptive Control of Multiple Robots in Cooperative Motion," TR-EE-91-20, Purdue University, W. Lafayette, Indiana, 1991.

[22] M. Zribi and S. Ahmad, "Robust Adaptive Control of Multiple Robots in Cooperative Motion Using σ Modification," Proceedings of the 1991 IEEE International Conference on Robotics and Automation, Sacramento, CA, 1991.

[23] M. Zribi and S. Ahmad, "Robust Adaptive Control of Multiple Robots in Cooperative Motion," Proceedings of the 1991 American Control Conference, Boston, MA, 1991.

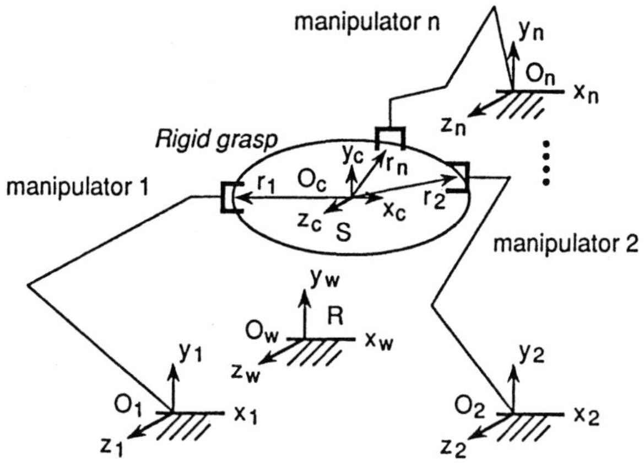

Figure 1. Multirobot system.

OPTIMAL TRAJECTORIES FOR HYPERSONIC LAUNCH VEHICLES

Mark D. Ardema, Jeffrey V. Bowles, Thomas Whittaker
NASA Ames Research Center, Moffett Field, CA 94035-1000

Abstract

In this paper, we derive a near-optimal guidance law for the ascent trajectory from Earth surface to Earth orbit of a hypersonic, dual-mode propulsion, lifting vehicle. Of interest are both the optimal flight path and the optimal operation of the propulsion system. The guidance law is developed from the energy-state approximation of the equations of motion. The performance objective is a weighted sum of fuel mass and volume, with the weighting factor selected to give minimum gross take-off weight for a specific payload mass and volume.

Introduction

Over the past six years, interest in hypersonic, air-breathing aircraft has revived. Missions being studied include space transportation and long distance commercial travel. The vehicles being considered have scramjet propulsion systems augmented by other propulsion modes for low-speed and, possibly, orbital flight. The National Aerospace Plane (NASP) Program has been initiated to demonstrate the capability of a hypersonic airplane to take-off from the surface of the Earth and achieve Earth orbit with a single stage.

One feature of a single-stage to orbit (SSTO) vehicle is its low payload-to-gross weight fraction. This means that vehicle performance is extremely sensitive to perturbations in vehicle design and operation. In particular, it is essential to "optimize" the flight path and the operation of the propulsion system to the extent possible in order to attain adequate mission performance, and to do this for every competing design under consideration. Although there are well-developed numerical methods for trajectory optimization of point-mass vehicle models, these methods are too expensive computationally and not robust enough to use at the conceptual design stage, in which many hundreds of vehicles must be evaluated and compared on a consistent basis.

Beginning in about 1965, NASA Ames Research Center has developed a hypersonic vehicle synthesis code, called HAVOC [1, 2]. Figure 1 shows the main elements of the code; the vehicle is defined by a geometry data base. Other program elements, such as aerodynamics, propulsion, and structures, use this data to make detailed computations of vehicle performance. An important feature of HAVOC is that it may be used iteratively to compute "closed" vehicles; that is, vehicles that meet pre-specified requirements of both payload mass and payload volume. Figure 2 shows a typical hypersonic vehicle configuration.

In the past, the trajectory module in HAVOC has been a path-following routine: a flight path is determined external to the synthesis code, perhaps by a numerical optimization, and then inserted into the module. This is undesirable because it is not easily responsive to changes in vehicle design and other characteristics. The purpose of the present paper is to develop and demonstrate a relatively simple near-optimal guidance law appropriate for use in a vehicle synthesis code such as HAVOC.

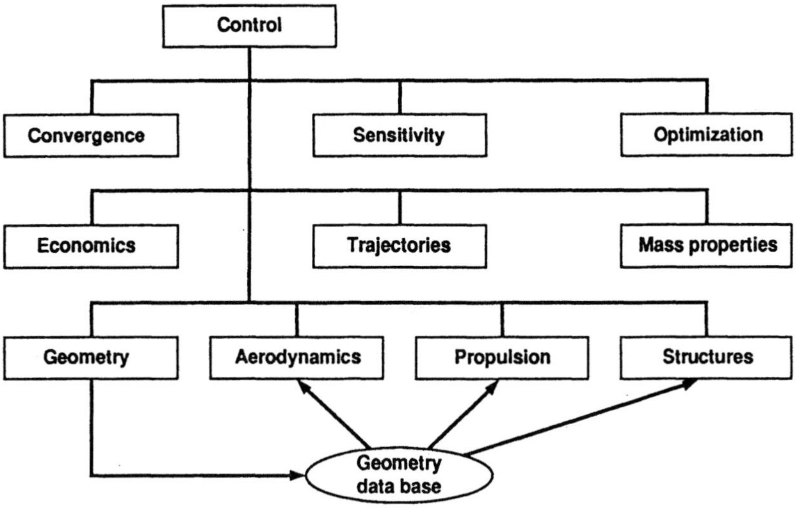

Figure 1. Hypersonic Aircraft Vehicle Optimization Code (HAVOC).

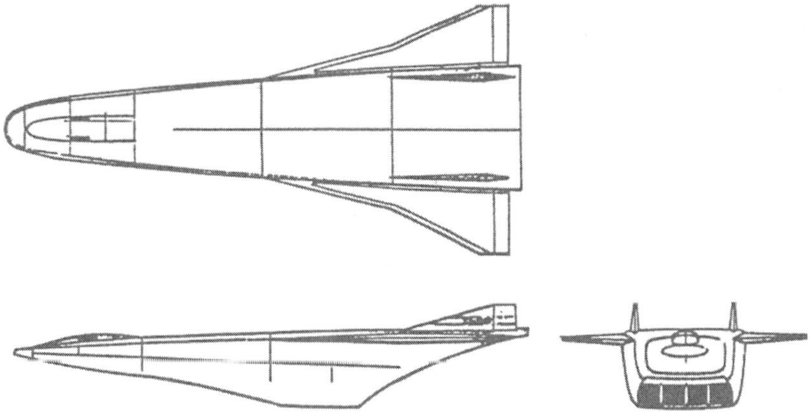

Figure 2. Hypersonic aircraft configuration.

The basis of the guidance law is the energy-state dynamic model. This dynamic model has been used successfully many times to obtain effective guidance laws for a wide variety of aircraft and missions [3, 4, 5, 6, 7]. The key idea is to introduce the total mechanical energy as a state variable, and then to neglect all other dynamics. When flight path optimization is done with this model, simple rules for the optimal path and for the optimal operation of the propulsion system are obtained. In the trajectory module in HAVOC, the guidance law is used to integrate the full point-mass equations of motion.

Reference [7] investigates SSTO optimal trajectories using singular perturbation methods. (For a review of the singular perturbation approach to aircraft trajectory optimization, see [8] - [11], for example.) The slow system in [7] is the energy-state model, and consequently the analysis of the slow system in [7] is similar to that of the present paper. The differences are in the cost functional ([7] considers minimum fuel mass only), the use of closed vehicles for comparisons in the present paper, and the very detailed vehicle modeling in the HAVOC program.

Guidance Law Development

The trajectory equations in HAVOC are those of a point-mass airplane with the following assumptions: (1) The aircraft flies in a great circle about a spherical, rotating Earth (terms in the square of the Earth's rotational speed are neglected), (2) the time rate of change of the flight path angle, $\dot{\gamma}$, is neglected, (3) the effect of side-slip on vehicle drag is ignored (side-slip is necessary to maintain great-circle flight over a rotating Earth), and (4) zero ambient winds. Under these assumptions, the equations of motion are

$$\begin{aligned} \dot{h} &= V \sin\gamma \\ \dot{V} &= \frac{T_V - D - mg \sin\gamma}{m} \\ \dot{m} &= -\beta \\ o &= \frac{V \cos\gamma}{R+h} - \frac{g\cos\gamma}{V} + \frac{T_\gamma + L}{mV}\cos\phi + 2\Omega_\gamma \\ o &= T_s - \frac{2W}{g}(\Omega_r V \cos\gamma - \Omega_T V \sin\gamma) \end{aligned} \tag{1}$$

In these equations, the state variables are h, the height above the surface of the Earth, V, the airspeed, and m, the vehicle mass; T_V, T_γ, and T_S are the components of thrust along the velocity vector, perpendicular to the velocity vector and in the great circle, and perpendicular to the great circle, respectively; D is drag; R is the radius of the Earth; g is the gravitational acceleration at the Earth's surface; and Ω_γ, Ω_r, and Ω_T are the Earth rotation (Coriolis) terms, which depend on instantaneous heading and latitude, as well as on speed and Earth rotation rate. The control variables are α, the angle of attack, and β, the engine fuel mass flow rate.

The energy-state approximation requires two additional assumptions. The first assumption is that α is small; more specifically, that the dependence of D and T_V on α can be neglected. This is a reasonable assumption because α is typically 2° and stays less than 5° for hypersonic flight paths, and thus the zero-lift drag is several times greater than the drag-due-to-lift. The second assumption is that the flight path angle is small enough to be neglected, again a very good approximation for hypersonic vehicle trajectories.

Now we introduce the total mechanical energy per unit weight:

$$E = h + \frac{1}{2g}V^2 \tag{2}$$

Differentiating and using (1) results in the new state equations

$$\begin{aligned} \dot{E} &= \frac{V}{mg}(T_V - D) = P \\ \dot{m} &= -\beta \end{aligned} \tag{3}$$

In these equations, the controls are V (or equivalently h) and β. If ϕ is the quantity to be minimized for a given energy gain and $\Phi = \dot{\phi}$, then

$$\phi_f - \phi_o = \int_{\phi_o}^{\phi_f} d\phi = \int_{t_o}^{t_f} \Phi dt = \int_{E_o}^{E_f} \frac{\Phi}{P} dE \qquad (4)$$

Consequently, the optimal controls $\underline{u}$ are to be determined from

$$\underline{u}^* = \arg \max_{\underline{u} \in U} \left(\frac{P}{\Phi}\right)_E \qquad (5)$$

where $\underline{u}^*$ is the vector of optimal controls and U is the set of admissible controls.

Two simple examples are minimum time and minimum fuel mass. For minimum time, $\phi = t$ and $\Phi = 1$ so that

$$\underline{u}^*_{\text{min. time}} = \arg \max_{\underline{u} \in U} (P)_E \qquad (6)$$

For minimum fuel mass, $\phi = -m$ and $\Phi = \beta$, so that

$$\underline{u}^*_{\text{min. fuel mass}} = \arg \max_{\underline{u} \in U} \left(\frac{P}{\beta}\right)_E \qquad (7)$$

For a SSTO mission with a hypersonic aircraft, neither of the above two cost functionals is appropriate. What is desired is a trajectory that gives the minimum gross-take-off-weight vehicle to put a given payload mass and volume in orbit. Because liquid hydrogen fueled aircraft have relatively low gross densities and correspondingly high surface area to gross weight ratios, they are very sensitive to perturbations in volume as well as in mass, and it is therefore necessary to minimize a weighted sum of fuel mass and volume. Thus, we introduce the cost functional

$$\phi = -[mK + V_f(1-K)] \qquad (8)$$

where V_f is the fuel volume and $K \in [0,1]$ is a weighting parameter to be chosen later. Another feature of SSTO aircraft that needs to be taken into account is that they have two (or more) independent propulsion modes. Thus, the total thrust (along the velocity vector) and fuel flow rates are

$$\begin{aligned} T_v &= f_v(\pi_a T_{M_a} \cos\varsigma_a + \pi_r T_{M_r} \cos\varsigma_r) \\ \beta &= C_a \pi_a T_{M_a} + C_r \pi_r T_{M_r} \end{aligned} \qquad (9)$$

where $\pi \in [0,1]$ is throttle setting, T_M is maximum thrust, ς is thrust offset angle, C is specific fuel consumption, and f_v is a parameter to account for the fact that the thrust and velocity vectors are not collinear. The subscripts "a" and "r" refer to the two propulsion modes, later to be identified as airbreather and rocket, respectively. The quantity to be maximized is now

$$\frac{P}{\Phi}=\frac{V(f_v\pi_a T_{M_a}\cos\varsigma_a+f_v\pi_r T_{M_r}\cos\varsigma_r-D)}{mg\left\{C_a\pi_a T_{M_a}\frac{[K(\rho_a-1)+1]}{\rho_a}+C_r\pi_r T_{M_r}\frac{[K(\rho_r-1)+1]}{\rho_r}\right\}} \tag{10}$$

The controls are now $\underline{u}=(\pi_a, \pi_r, V)^T$. The constraints on these controls will now be discussed. Figure 3 shows the constraints on the throttle settings. Part (a) of the figure shows equivalence ratio, e, as a function of Mach number, M, for the airbreathing engine. Equivalence ratio is defined as the ratio of actual fuel flow to the flow for stoichiometric combustion. Since, if there are no other considerations than combustion efficiency, it is not optimum to have $e > 1$, the limit $e \leq 1$ is imposed. At about Mach 6, at the start of the supersonic combustion ramjet mode, there is a pronounced reduction in the allowable engine equivalence ratio, due to engine operating limits of thermal choke/burner exit Mach number for the limited variable geometry scramjet engines. The net result is a sharp dip in the upper bound for the airbreathing engine, as shown in Figure 3(a). As the Mach number is increased in the scramjet mode, this constraint vanishes and the engine is allowed to again burn fuel stoichiometrically. Another factor that is taken account of is the need to circulate the liquid hydrogen fuel through the engine structure before combustion in order to keep the engine structure to an acceptable temperature. At approximately Mach 12, this fuel flow requirement exceeds the fuel flow rate for $e = 1$. One option always is to turn the airbreathing engine completely off. All of these constraints and requirements define the set of admissible controls, and result in the definitions of $\pi_a = 1$ and $\pi_a = 0$, shown on Figure 3(b). The bounds on the throttle setting of the rocket engine, Figure 3(b), are straightforward.

The bounds on V are shown on Figure 4. The four constraints are as follows: (1) a minimum dynamic pressure (q_{min}), (2) a maximum dynamic pressure (q_{max}), (3) a duct pressure limit, and (4) an aerodynamic heating limit on the airframe structure. These four constraints define V_m and V_M, the minimum and maximum values of V, as a function of M.

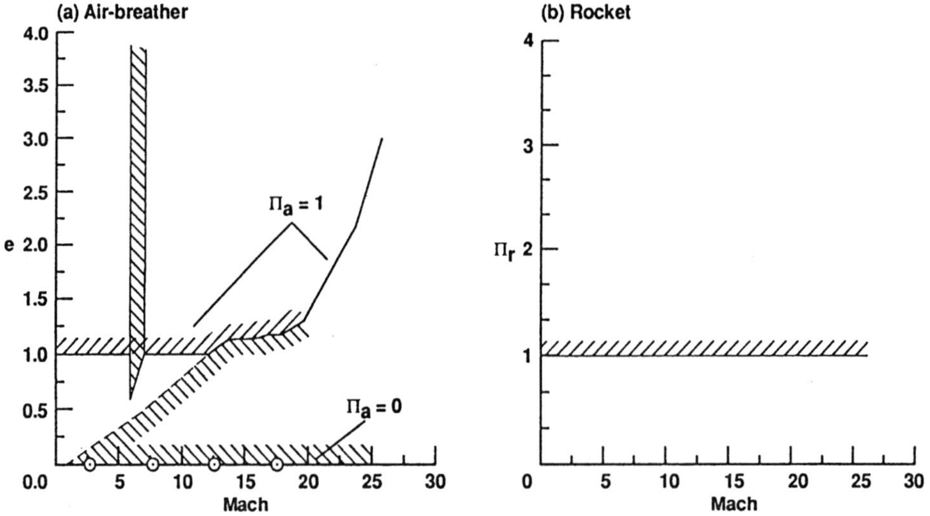

Figure 3. Constraints on throttle controls.

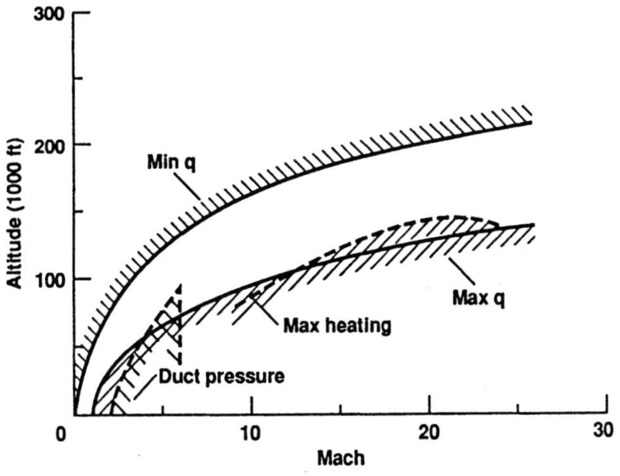

Figure 4. Constraints on velocity control.

It is well known that energy-state trajectories may have discontinuous jumps in altitude and velocity, requiring in principle infinite normal acceleration [3, 4, 12, 13]. This is especially prevalent in the transonic region, where the jumps are due to drag rise. To circumvent this, a maximum and a minimum load factor constraint is imposed. This leads to some non-optimality, but seems to give good results. Another way to handle these discontinuities would be to treat them as transition layers in a singular perturbation approach, as is done in [12, 13].

The maximization of (10) with respect to π_a, π_r, and V, subject to the control constraints depicted in Figures 3 and 4, gives the following guidance law for the optimal controls:

$$V^* = \underset{V_m \le V \le V_M}{\arg\max} \left(\frac{P}{\Phi}\right)_{\pi_a^*, \pi_r^*, E} \tag{11}$$

$$\text{If } \frac{C_a^* \rho_r}{C_r^* \rho_a} \ge \frac{K(\rho_r - 1) + 1}{K(\rho_a - 1) + 1}, \text{ then}$$

$$\pi_r = 1$$

$$\pi_a = \begin{cases} 0 \text{ if } 1 - \dfrac{D^*}{T_{r_M}^*} > \dfrac{C_r^* \rho_a [K(\rho_r - 1) + 1]}{C_a^* \rho_r [K(\rho_a - 1) + 1]} \\ 1 \text{ otherwise} \end{cases}$$

$$\text{If } \frac{C_a^* \rho_r}{C_r^* \rho_a} < \frac{K(\rho_r - 1) + 1}{K(\rho_a - 1) + 1}, \text{ then}$$

$$\pi_a = 1$$

$$\pi_r = \begin{cases} 0 \text{ if } 1 - \dfrac{D^*}{T_{a_M}^*} > \dfrac{C_a^* \rho_r [K(\rho_a - 1) + 1]}{C_r^* \rho_a [K(\rho_r - 1) + 1]} \\ 1 \text{ otherwise} \end{cases} \tag{12}$$

This guidance law is similar to that developed in [7].

Due to the great complexity of the engine, and especially to the variable engine geometry features, the variation of P/Φ with V is quite complex. A section of this function in the transonic region is shown in Figure 5. Because of this complexity, the following numerical procedure has been adopted:

(1) At a given energy, search across the admissible range of V, using (12) to evaluate π_a and π_r at each point, to find the value of V that maximizes P/Φ ; this is V*;

(2) use the value of V* and the corresponding π_a^* and π_r^* in the integration of the equations (1) to the next energy level.

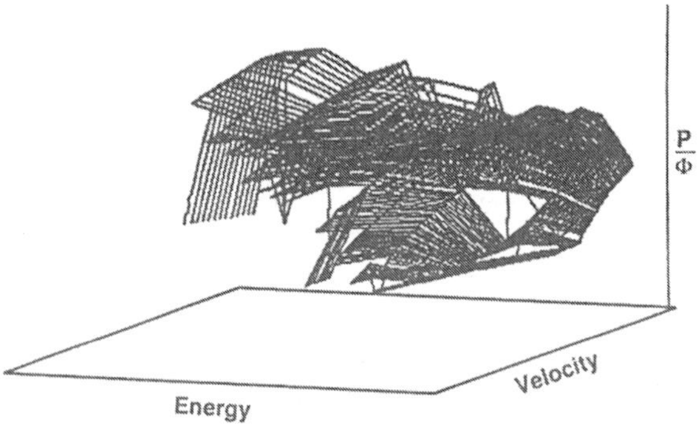

Figure 5. Cost functional in the transonic region.

The procedure is repeated starting at the energy level at take-off and ending at orbital energy. The resulting trajectory is near-optimal in the sense that the dynamic model used to get the optimal control is slightly different than dynamic model that is integrated to get the trajectories.

Numerical Results

The airplane of Figure 2 will be used in all numerical results. This baseline design and its corresponding trajectory have been refined by numerous simulations and are considered to be close to optimal. It is our purpose to start with this design and use the energy-state guidance law just developed to resize the vehicle. Specifically, the mass and volume of payload to orbit is held constant in all cases by iteratively exercising HAVOC with the guidance law used in the trajectory module.

The first case considered is K = 1, that is, minimizing fuel mass. Running the synthesis code for this case without resizing indeed gives a vehicle with less fuel mass than the baseline; however, because it gives a vehicle that uses the airbreathing engine relatively more and the rocket relatively less (because the density of liquid hydrogen is much less than that of liquid oxygen), there is insufficient payload volume. When the vehicle is resized to meet both payload mass and volume requirements, the vehicle gross take-off weight (W_{GTO}) actually increases relative to the baseline (Figure 6). Since the ultimate goal is the minimization of W_{GTO}, the K = 1 design is worse than the baseline.

This procedure was repeated for a range of weighting parameter values, Figure 6. It is seen that W_{GTO} is minimum for a K of about 0.95; this value is denoted K*. The K* vehicle gives a small but significant improvement in W_{GTO} relative to the baseline. As K is decreased below K*, the rocket is used more and more and the airbreather less, and the gross weight begins to rise rapidly.

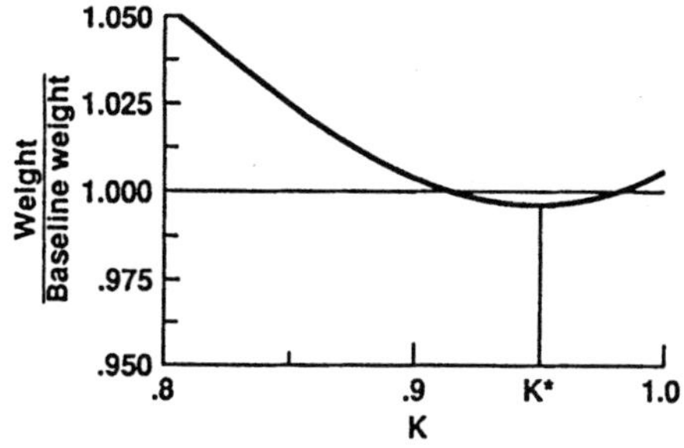

Figure 6. Variation of gross take-off weight with parameter K.

Figure 7 shows the throttle histories for both propulsion systems for two values of the weighting parameter. For K = 0.95 = K*, the airbreathing engine (scramjet above Mach 6) is full on ($\pi_a = 1$) until about 85% of orbital energy is achieved. The rocket is on briefly for take-off and again transonically, and comes on full again at about 50% of orbital energy. There is a significant portion of the trajectory, from Mach 17 to 22, during which the propulsion modes are at full throttle simultaneously. For K = 0.3, the rocket is used more and the airbreather less, relative to K = K*, as expected. The airbreather is now turned off at about 50% of orbital energy, whereas the rocket comes on at 25%. Again, the rocket is on briefly during transonic flight.

The optimal flight paths for three values of the weighting parameter are shown on Figure 8. (Since these are for constant payload mass and volume, the gross take-off weights are different for the three cases.) The K = 0.3, K = K*, and K = 1.0 trajectories are all similar from about Mach 2.5 to about Mach 16; during this interval the trajectories are not constrained by either the upper or lower dynamic pressure limits. At Mach 16, the K = 0.3 vehicle shuts off the scramjet, and the trajectory transitions to the minimum dynamic pressure (q_{min}) boundary. At Mach 23, the q_{min} constraint is removed and the vehicle begins a maximum load factor pull-up that ends at orbital energy. In all cases, there is considerable "chatter" in the trajectories in the low supersonic region, due to the cost functional behavior shown on Figure 5.

The K = K* and K = 1.0 cases are very similar to each other. They briefly touch the q_{min} boundary at about Mach 21, then decrease in altitude until Mach 22, when the airbreather is turned off, and then they commence a maximum load factor climb.

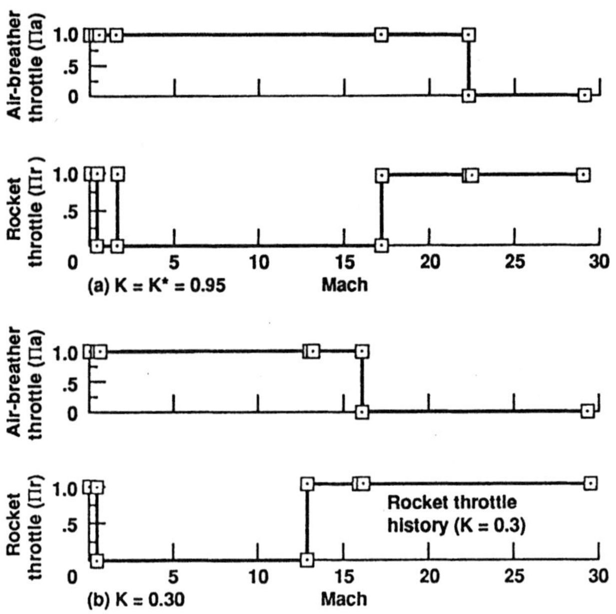

Figure 7. Optimal throttle schedules.

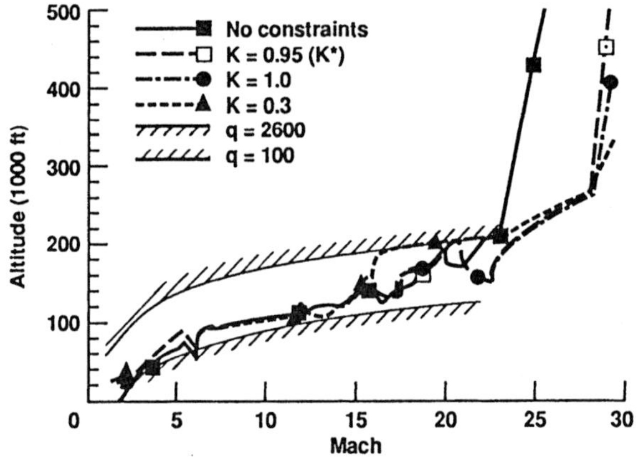

Figure 8. Optimal flight paths.

Also shown on Figure 8 is a case for which the lower path constraints (duct pressure) and the load factor constraint have been removed (the q_{min} constraint is retained) for K = K*. At several points in the hypersonic region, the no-constraint trajectory jumps at high load factors between different altitudes; this trajectory also makes a big jump in altitude when the q_{min} constraint ends.

The behavior in the transonic region is shown on Figure 9 for K = K*. Both constrained and unconstrained cases are considered. All trajectories begin at zero altitude and then rise at constant Mach number at a high subsonic speed. They then dive back down, along constant energy paths. The unconstrained trajectory goes back down to zero altitude, but the constrained trajectories encounter the duct pressure boundary and begin to climb up this boundary.

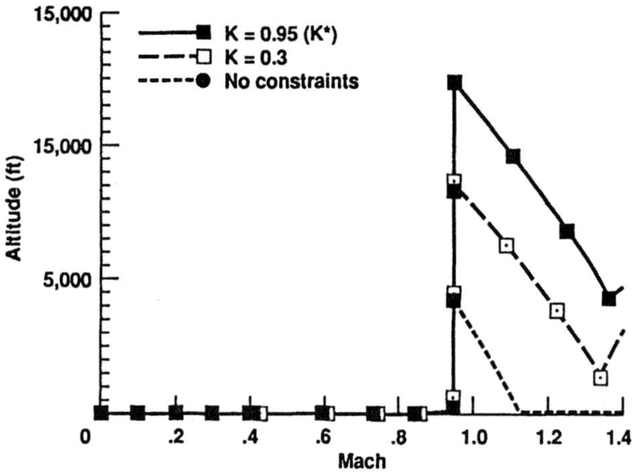

Figure 9. Optimal paths in the transonic region.

Concluding Remarks

A simple guidance law has been developed for use in a hypersonic aircraft synthesis computer code. The guidance law is based on energy-state modeling. Its use in a vehicle synthesis code allows rapid, consistent, and accurate assessment of hypersonic aircraft performance.

The synthesis code with the energy-state guidance law was used to study the single-stage-to-orbit (SSTO) mission performance of a dual-mode hypersonic vehicle. One of the propulsion modes is an airbreathing engine capable of either ramjet or scramjet operation, and the other mode is a rocket. The optimal (in the sense of energy-state modeling) flight paths and throttle histories of the two propulsion modes were determined for a variety of conditions.

It was found that propellant mass and volume must both be considered in the cost functional when seeking minimum gross take-off weight. Typical trajectories begin with ramjet propulsion to about Mach 6 and then scramjet propulsion to a hypersonic speed. The rocket is turned on briefly transonically but then stays off until high hypersonic speeds. There is an interval when both propulsion modes are on full. The flight paths exhibit the typical dive in the transonic region; in the hypersonic region, they tend to be unconstrained, at least while the airbreather is still on. At the end, there is a constant load factor climb with only the rocket on.

References

[1] Petersen, R. H., Gregory, T. J., and Smith C. L.: Some Comparisons of Turboramjet-Powered Hypersonic Aircraft for Cruise and Boost Missions. *J. Aircraft*, Vol. 3, No. 5, 1966.

[2] Bowles, J. V.: Ames Conceptual Studies Activities. Proceedings of the Second National Aerospace Plane Symposium. Applied Physics Laboratory, Laurel, MD, Nov. 1986.

[3] Bryson, A. E., Jr., Desai, M. N., and Hoffman, W. C.: Energy-State Approximation in Performance Optimization of Supersonic Aircraft. *J. Aircraft*, Vol. 6, No. 6, 1969.

[4] Kelley, H. J. Cliff, E. M., and Weston, A. R.: Energy State Revisited. *Optimal Control Applications and Methods*, Vol. 7, April 1986.

[5] Ardema, M. D.: Solution of the Minimum Time-to-Climb Problem by Matched Asymptotic Expansions. *AIAA J.*, Vol. 14, July 1976.

[6] Lee, H.Q. and Erzberger, H.: Algorithm for Fixed-Range Optimal Trajectories. NASA TP-1565, July 1980.

[7] Van Buren, M. A. and Mease, K. D.: Aerospace Plane Guidance Using Geometric Control Theory. Proceedings of the 1990 American Control Conference, May 23-25, 1990, San Diego, CA.

[8] Kelley, H. J.: Aircraft Maneuver Optimization by Reduced-Order Approximation, *Control and Dynamic Systems*, Vol. 10, edited by C.T. Leondes, Academic Press, New York, 1973.

[9] Ardema, M. D.: An Introduction to Singular Perturbations in Nonlinear Optimal Control. In *Singular Perturbations in Systems and Control*, edited by M. D. Ardema, Courses and Lectures No. 280, International Center for Mechanical Sciences, Udine, Italy, 1983.

[10] Calise, A. J.: Singular Perturbation Methods for Variational Problems in Aircraft Flight. *IEEE Transactions Automatic Control*, Vol. AC-21, No. 3, 1976.

[11] Shinar, J.: Remarks on Singular Perturbation Technique Applied in Nonlinear Optimal Control. Proceedings of 2nd IFAC Workshop on Control Applications on Nonlinear Programming and Optimization, 1980.

[12] Weston, A. R., Cliff, E. M., and Kelley, H. J.: Altitude Transitions in Energy Climbs. *Automatica*, Vol. 19, No. 2, February 1983.

[13] Ardema, M. D. and Yang, L.: Interior Transition Layers in Flight-Path Optimization. *J. Guidance, Control, Dynamics*, Vol. 11, No. 1, Jan.-Feb. 1988.

REDUCED-ORDER OUTPUT FEEDBACK CONTROL OF A CLASS OF UNCERTAIN SYSTEMS

J Rodellar Benedé
Department of Applied Mathematics III
Technical University of Catalunya, Barcelona, Spain

G Leitmann
Department of Mechanical Engineering
University of California, Berkeley, USA

E P Ryan
School of Mathematical Sciences
University of Bath, Bath, UK

Abstract

Stabilization of uncertain dynamical systems on $\mathbb{R}^n$, which admit a decomposition into two coupled subsystems, of dimension n_c and n_r respectively, is studied. We refer to the n_c-dimensional subsystem as the *reduced-order system* and to the n_r-dimensional subsystem as the *residual system*: the overall n-dimensional system is termed the *full system*. Each subsystem is modelled by a differential equation with a stable linear nominal part and a nonlinear perturbation of a specified class. The output available for feedback purposes is an $\mathbb{R}^{n_c}$-valued linear combination of the state components of the full system and is subject to bounded measurement noise (with known bound). An output feedback strategy is described which guarantees the existence of a calculable global uniform compact attractor (containing the state origin) for systems of this class.

1 Introduction

Among the many approaches proposed in the literature of the past decade for the control of uncertain systems, there is one which is characterized by the following main features: (i) the systems are described (in the time domain) by linear or nonlinear differential equations; (ii) uncertainties are modelled deterministically and information on their possible "size" only is required; (iii) Lyapunov techniques are used constructively to design feedback controllers that render the systems "practically stable".

Early work using the above approach includes that of Leitmann (1978) and Corless & Leitmann (1981,1983). Since then, a variety of related problems have been addressed within the same framework (for example, see Corless & Leitmann (1990) and Leitmann (1990) for reviews and extensive bibliographies). We briefly survey some of these problems.

Concerning the question of how the uncertainties enter into the system description, much of the literature has dealt with so-called *matched uncertainties*, although some treatment have considered the *unmatched* case (Barmish and Leitmann 1982, Ryan & Corless 1984, Chen & Leitmann 1987, Goodall & Ryan 1988, amongst others).

Also, much of the literature has focussed on the design of fixed-parameter controllers, assuming that bounds on the uncertainties are known *a priori*. However, in some investigations this assumption has been relaxed, leading to the design of *adaptive controllers* (Corless & Leitmann 1983, Ryan 1991).

Problems of robustness in the presence of neglected sensor and actuator dynamics has been studied via singular perturbation analyses (Leitmann *et al.* 1986, Corless *et al.* 1990).

Finally, while many studies assume that the full state of the system is available, *output feedback* has been considered by, for example, Chen (1987a,b), Galimidi & Barmish (1986), Steinberg & Corless (1985), and Zaheb (1986).

The present paper employs the same methodology, but addresses a problem different from the above, namely, the design of practically stabilizing reduced-order linear output feedback controllers. Specifically, we focus on analysis and synthesis of *output feedback* controls for a particular class of systems which admit a decomposition of a special structure. The rationale for this study can be summarized as follows: we seek *low-order* feedback controls for *high-order* systems, the prototype problem being that of controlling structures by strategies that are based only on a certain number of modes of the system. We remark that, while we deal with large-scale systems, the approach is not one of decentralized control (Šiljak, 1978). Our stability analysis echoes, in a loose sense, that of the singular perturbation studies in Corless & Ryan (1991), Saberi & Khalil (1981), and Zinober (ed.) (1990).

2 Class of systems

The class of systems Σ to be considered are systems on $\mathbf{R}^n$ which can be decomposed into two coupled subsystems Σ_c and Σ_r (of respective dimensions n_c and n_r), henceforth referred to as the *reduced-order system* and the *residual system* respectively and described by the following controlled differential equations:

$$\Sigma_c: \qquad \dot{x}_c(t) = A_c x_c(t) + B_c u(t) + g_c(t, x(t), u(t)) \tag{1}$$

$$\Sigma_r: \qquad \dot{x}_r(t) = A_r x_r(t) + B_r u(t) + g_r(t, x(t), u(t)). \tag{2}$$

Here, $n = n_c + n_r$, $x_c(t) \in \mathbf{R}^{n_c}$, $x_r(t) \in \mathbf{R}^{n_r}$, $x(t) = (x_c(t), x_r(t)) \in \mathbf{R}^n$ (the state of the *full system*) and $u(t) \in \mathbf{R}^{N_c}$. Note that the control dimension coincides with that of the reduced-order system Σ_c. The operators A_c, B_c, A_r and B_r are assumed known and A_c, A_r satisfy the following stability assumption.

Assumption 1. (i) $\sigma(A_c) \subset \mathbf{C}^-$; (ii) $\sigma(A_r) \subset \mathbf{C}^-$.

Here, $\mathbf{C}^-$ denotes the open left-half complex plane. The subsystems are coupled through the uncertain functions g_c and g_r, which are assumed to be of the Carathéodory class (measurable in t, continuous in their other arguments) and to exhibit the following structural properties.

Assumption 2. There exist Carathéodory functions $e_c, e_r : \mathbf{R} \times \mathbf{R}^n \times \mathbf{R}^m \to \mathbf{R}^m$, $f_c : \mathbf{R} \times \mathbf{R}^n \times \mathbf{R}^m \to \mathbf{R}^{n_c}$, $f_r : \mathbf{R} \times \mathbf{R}^n \times \mathbf{R}^m \to \mathbf{R}^{n_r}$ and *known* scalars α_c^c, α_c^r, α_r^r, α_r^c, ν_c, ν_r and $\beta < 1$ such that, for almost all t, and for all (x, u) we have:
(i) $g_c(t,x,u) = B_c e_c(t,x,u) + f_c(t,x,u)$ and $g_r(t,x,u) = B_r e_r(t,x,u) + f_r(t,x,u)$;
(ii) $\|e_c(t,x,u)\| \leq \alpha_c^c\|x_c\| + \alpha_c^r\|x_r\| + \beta\|u\|$ and $\|e_r(t,x,u)\| \leq \alpha_r^r\|x_r\| + \alpha_r^c\|x_c\| + \beta\|u\|$;
(iii) $\|f_c(t,x,u)\| \leq \nu_c$ and $\|f_r(t,x,u)\| \leq \nu_r$.

Throughout, $\langle\ ,\ \rangle$ and $\|\ \|$ denote the Euclidean inner product and norm on the appropriate space: induced matrix (spectral) norms are similarly denoted.

Remarks. Assumption 2 states that the uncertain functions g_c and g_r can be decomposed into *matched* components ($B_c e_c$ and $B_r e_r$) and *unmatched* components (f_c and f_r): the unmatched components are assumed uniformly bounded, whereas the matched components are only required to be bounded affinely in $\|x_c\|$, $\|x_r\|$ and $\|u\|$.

We suppose that the available output can be expressed in the form

$$y(t) = C_c x_c(t) + C_r x_r(t) + w(t)$$

where $C_c \in \mathbf{R}^{n_c \times n_c}$ and $C_r \in \mathbf{R}^{n_c \times n_r}$ are known, and $w(\cdot) \in L^\infty(\mathbf{R})$ is an unknown disturbance (measurement noise) with known bound δ, that is,

$$\|w(t)\| \leq \delta \qquad \text{for almost all } t.$$

We assume that C_c is invertible.

Assumption 3. $C_c \in \mathbf{R}^{n_c \times n_c}$ is non-singular.

3 Problem formulation

Our objective is to assure "global uniform attractivity" of a compact set (containing the origin), in the sense of Definition 1 below, for the full system Σ by means of a linear output feedback controller with a feedback matrix based only on the known (nominal) part of the reduced-order system Σ_c and a scalar gain which may utilize information on the nominal part of the full system and on the uncertainty bounds.
We introduce some notation: for compact $\mathcal{K} \subset \mathbf{R}^n$, the (Euclidean) distance of $x \in \mathbf{R}^n$ from $\mathcal{K}$ is defined as

$$d(x, \mathcal{K}) := \min_{k \in \mathcal{K}} \|x - k\|.$$

Definition 1. A compact set $\mathcal{K} \subset \mathbf{R}^n$ is *a global uniform attractor* for the dynamical system

$$\dot{x}(t) = f(t, x(t)), \qquad x(t_0) = x^0, \tag{3}$$

if:

(i) *existence and continuation of solutions* - for every $(t_0, x^0) \in \mathbf{R} \times \mathbf{R}^n$, the initial-value problem (3) has a solution, and every solution can be extended into a solution on $[t_0, \infty)$;

(ii) *uniform boundedness of solutions* - for each $r > 0$ there exists $R > 0$ such that $\|x(t)\| < R$ for all $t \geq t_0$ on every solution $x(\cdot)$ of (3) with $\|x^0\| < r$;

(iii) *uniform stability of $\mathcal{K}$* - for each $\epsilon > 0$ there exists $\delta > 0$ such that $d(x(t), \mathcal{K}) < \epsilon$ for all $t \geq t_0$ on every solution $x(\cdot)$ of (3) with $d(x^0, \mathcal{K}) < \delta$;

(iv) *global uniform attractivity of $\mathcal{K}$* - for each $r > 0$ and $\mu > 0$, there exists $T > 0$ such that $d(x(t), \mathcal{K}) < \mu$ for all $t > t_0 + T$ on every solution $x(\cdot)$ of (3) with $\|x^0\| < r$.

4 Reduced-order output feedback control

Let $Q_c \in \mathbf{R}^{n_c \times n_c}$ and $Q_r \in \mathbf{R}^{n_r \times n_r}$ (design parameters) be symmetric and positive definite. By Assumption 1, there exists unique symmetric positive-definite solutions K_c and K_r of the Lyapunov equations

$$K_c A_c + A_c^T K_c + Q_c = 0 \qquad \text{and} \qquad K_r A_r + A_r^T K_r + Q_r = 0.$$

Now, we propose the following linear output feedback strategy:

$$u(t) = -q_\gamma(y(t)), \qquad q_\gamma(y) := \gamma B_c^T K_c C_c^{-1} y, \tag{4}$$

where γ is a scalar gain parameter (to be specified).
The overall feedback-controlled system is then described by:

$$\dot{x}(t) = F_\gamma(t, x(t)) \tag{5}$$

where

$$F_\gamma : (t, x) \mapsto \begin{bmatrix} A_c x_c - B_c q_\gamma(C_c x_c + C_r x_r + w(t)) + g_c(t, x, -q_\gamma(C_c x_c + C_r x_r + w(t)) \\ A_r x_r - B_r q_\gamma(C_c x_c + C_r x_r + w(t)) + g_r(t, x, -q_\gamma(C_c x_c + C_r x_r + w(t)) \end{bmatrix}$$

5 Stability analysis

Define the symmetric positive-definite quadratic function V as follows

$$V : x = (x_c, x_r) \mapsto V_c(x_c) + V_r(x_r)$$

where

$$V_c(x_c) = \langle x_c, K_c x_c \rangle \qquad \text{and} \qquad V_r(x_r) = \langle x_r, K_r x_r \rangle.$$

5.1 Preliminary lemma

For notational convenience, we introduce the following scalar quantities

$$a_1 := 2(1+\beta)\|B_r^T K_r\| \|B_c^T K_c C_c^{-1} C_r\|, \qquad a_2 := (1+\beta)\left[\|B_c^T K_c C_c^{-1} C_r\| + \|B_r^T K_r\|\right],$$

$$a_3 := 2(1-\beta), \quad \tilde{\alpha}_r^c := \alpha_r^c \|B_r^T K_r\|, \quad q_c := \lambda_{min}(Q_c), \quad b_1 := \lambda_{min}(Q_r) - 2\alpha_r^r \|B_r^T K_r\|,$$

where, for any symmetric positive-definite matrix, $\lambda_{min}(P)$ denotes its minimum eigenvalue:

$$\lambda_{min}(P) := \min\{\lambda | \ \lambda \in \sigma(P)\}.$$

We define the symmetric matrix $M_\gamma \in \mathbf{R}^{3\times 3}$, vector $\rho_\gamma \in \mathbf{R}^3$, and function $z : \mathbf{R}^n \to \mathbf{R}^3$ by

$$M_\gamma := \begin{bmatrix} q_c & -\tilde{\alpha}_r^c & -\alpha_c^c \\ -\tilde{\alpha}_r^c & b_1 - a_1\gamma & -(\alpha_c^r + a_2\gamma) \\ -\alpha_c^c & -(\alpha_c^r + a_2\gamma) & a_3\gamma \end{bmatrix}$$

$$\rho_\gamma := \begin{bmatrix} 2\nu_c\|K_c\| \\ 2\gamma\delta(1+\beta)\|B_r^T K_r\| \|B_c^T K_c C_c^{-1}\| + 2\nu_r\|K_r\| \\ 2\gamma\delta(1+\beta)\|B_c^T K_c C_c^{-1}\| \end{bmatrix}, \quad z : x \mapsto \begin{bmatrix} \|x_c\| \\ \|x_r\| \\ \|B_c^T K_c x_c\| \end{bmatrix}.$$

The proof of our main result (in Section 5.2 below) requires the following technical lemma.

Lemma. For almost all t, $\langle\nabla V(x), F_\gamma(t,x)\rangle \le -\langle z(x), M_\gamma z(x)\rangle + \langle\rho_\gamma, z(x)\rangle$ for all $x=(x_c,x_r)$.

Proof. The proof is essentially a matter of careful "book-keeping". First, observe that, for all $x=(x_c,x_r)$,

$$\langle K_c x_c, A_c x_c\rangle = -\tfrac{1}{2}\langle x_c, Q_c x_c\rangle \le -\tfrac{1}{2}\lambda_{min}(Q_c)\|x_c\|^2 \quad\text{and}\quad \langle K_r x_r, A_r x_r\rangle \le -\tfrac{1}{2}\lambda_{min}(Q_r)\|x_r\|^2.$$

Also, using Assumption 2 and the bound on $w(\cdot)$, we find that, for almost all t and for all $x=(x_c,x_r)$,

$$-\langle K_c x_c, B_c q_\gamma(C_c x_c + C_r x_r + w(t))\rangle \le -\gamma\|B_c^T K_c x_c\|^2 + \gamma\|B_c^T K_c x_c\|\,[\|B_c^T K_c C_c^{-1} C_r\|\|x_r\| + \delta\|B_c^T K_c C_c^{-1}\|]\,,$$

$$\begin{aligned}\langle K_c x_c, g_c(t,x,-q_\gamma(C_c x_c + C_r x_r + w(t)))\rangle \le{}& \nu_c\|K_c\|\|x_c\| + \|B_c^T K_c x_c\|\,[\alpha_c^c\|x_c\| + \alpha_c^r\|x_r\| \\ &+\beta\gamma\|B_c^T K_c x_c\| + \beta\gamma\|B_c^T K_c C_c^{-1} C_r\|\|x_r\| + \beta\gamma\delta\|B_c^T K_c C_c^{-1}\|]\,,\end{aligned}$$

$$-\langle K_r x_r, B_r q_\gamma(C_c x_c + C_r x_r + w(t))\rangle \le \gamma\|B_r^T K_r\|\|x_r\|\,[\|B_c^T K_c x_c\| + \|B_c^T K_c C_c^{-1} C_r\|\|x_r\| + \delta\|B_c^T K_c C_c^{-1}\|]\,,$$

and

$$\begin{aligned}\langle K_r x_r, g_r(t,x,-q_\gamma(C_c x_c + C_r x_r + w(t)))\rangle \le{}& \nu_r\|K_r\|\|x_r\| + \|B_r^T K_r\|\|x_r\|\,[\alpha_r^c\|x_c\| + \alpha_r^r\|x_r\| \\ &+\beta\gamma\|B_c^T K_c x_c\| + \beta\gamma\|B_c^T K_c C_c^{-1} C_r\|\|x_r\| + \beta\gamma\delta\|B_c^T K_c C_c^{-1}\|]\,.\end{aligned}$$

It now follows that, for almost all t and for all $x=(x_c,x_r)$,

$$\begin{aligned}\langle\nabla V(x), F_\gamma(t,x)\rangle \le{}& -\lambda_{min}(Q_c)\|x_c\|^2 - (\lambda_{min}(Q_r) - 2\alpha_r^r\|B_r^T K_r\|)\|x_r\|^2 - 2(1-\beta)\gamma\|B_c^T K_c x_c\|^2 \\ &+ 2\gamma(1+\beta)\|B_r^T K_r\|\|B_c^T K_c C_c^{-1} C_r\|\|x_r\|^2 + 2\alpha_r^c\|B_r^T K_r\|\|x_c\|\|x_r\| \\ &+ 2\alpha_c^c\|x_c\|\|B_c^T K_c x_c\| + 2\alpha_c^r\|x_r\|\|B_c^T K_c x_c\| \\ &+ 2\gamma(1+\beta)(\|B_c^T K_c C_c^{-1} C_r\| + \|B_r^T K_r\|)\|x_r\|\|B_c^T K_c x_c\| \\ &+ 2\nu\|K_c\|\|x_c\| + 2\nu_r\|K_r\|\|x_r\| + 2\gamma\delta(1+\beta)\|B_r^T K_r\|\|B_c^T K_c C_c^{-1}\|\|x_r\| \\ &+ 2\gamma\delta(1+\beta)\|B_c^T K_c C_c^{-1}\|\|B_c^T K_c x_c\| \\ ={}& -q_c\|x_c\|^2 - (b_1 - a_1\gamma)\|x_r\|^2 - a_3\gamma\|B_c^T K_c x_c\|^2 + 2\tilde{\alpha}_r^c\|x_c\|\|x_r\| \\ &+ 2\alpha_c^c\|x_c\|\|B_c^T K_c x_c\| + 2(\alpha_c^r + a_2\gamma)\|x_r\|\|B_c^T K_c x_c\| + \langle\rho_\gamma, z(x)\rangle \\ ={}& -\langle z(x), M_\gamma z(x)\rangle + \langle\rho_\gamma, z(x)\rangle.\end{aligned}$$

Q.E.D.

5.2 The main result

Writing

$$\psi_1 := q_c(a_1 a_3 + a_2^2),\quad \psi_2 := (\alpha_c^c)^2 a_1 + q_c b_1 a_3 - (\tilde{\alpha}_r^c)^2 a_3 - 2a_2(q_c\alpha_c^r + \alpha_c^c\tilde{\alpha}_r^c),\quad \psi_3 := (\alpha_c^c)^2 b_1 + 2\alpha_c^c\tilde{\alpha}_r^c\alpha_c^r + q_c(\alpha_c^r)^2,$$

$$\gamma_1 := \frac{\psi_2 - \sqrt{\psi_2^2 - 4\psi_1\psi_3}}{2\psi_1},\quad \gamma_2 := \frac{\psi_2 + \sqrt{\psi_2^2 - 4\psi_1\psi_3}}{2\psi_1},\quad \gamma_3 := [b_1 - q_c^{-1}(\tilde{\alpha}_r^c)^2]/a_1,$$

we arrive at the main result of the paper.

Theorem 1. If (i) $\psi_2^2 - 4\psi_1\psi_3 > 0$ and (ii) $\gamma_3 > \gamma_1$, then, for each γ with $\gamma_1 < \gamma < \min\{\gamma_2,\gamma_3\}$, the matrix M_γ is positive definite and the compact set

$$\mathcal{K}_\gamma = \{x \mid V(x) \le r_\gamma\}$$

is a global uniform attractor for the output-feedback-controlled system (5), where

$$r_\gamma := \max\{V(x) \mid x\in\mathbf{R}^n,\ \langle z(x), M_\gamma z(x)\rangle \le \langle\rho_\gamma, z(x)\rangle\}.$$

Proof. That M_γ is positive definite under the hypotheses of the theorem is straightforward to verify: this ensures that r_γ is well defined.

Define the continuously differentiable map $W:\mathbf{R}^n \to [0,\infty)$ by

$$W(x) := \begin{cases} \left[V^{\frac{1}{2}}(x) - r_\gamma^{\frac{1}{2}}\right]^2, & \text{if } x\notin\mathcal{K}_\gamma \\ 0, & \text{if } x\in\mathcal{K}_\gamma \end{cases}$$

Writing $c_1 := \min\{\lambda_{min}(K_c), \lambda_{min}(K_r)\}$ and $c_2 := \max\{\lambda_{max}(K_c), \lambda_{max}(K_r)\}$, we have that

$$c_1 d^2(x, \mathcal{K}_\gamma) \le W(x) \le c_2 d^2(x, \mathcal{K}_\gamma) \quad \text{for all } x. \tag{6}$$

Also,

$$\nabla W(x) = \begin{cases} \left[1 - (r_\gamma / V(x))^{\frac{1}{2}}\right] \nabla V(x), & \text{if } x \notin \mathcal{K}_\gamma \\ 0, & \text{if } x \in \mathcal{K}_\gamma \end{cases}$$

Note that, by definition of r_γ,

$$\left[1 - (r_\gamma / V(x))^{\frac{1}{2}}\right] > 0 \quad \text{for all } x \notin \mathcal{K}_\gamma$$

and so, by the Lemma, we have that, for almost all t,

$$\langle \nabla W(x), F_\gamma(t, x)\rangle \le -\theta(x) \quad \text{for all } x \in \mathbf{R}^n,$$

where

$$\theta(x) := \begin{cases} \left[1 - (r_\gamma / V(x))^{\frac{1}{2}}\right] \left[\langle z(x), M_\gamma z(x)\rangle - \langle \rho_\gamma, z(x)\rangle\right] > 0, & \text{if } x \notin \mathcal{K}_\gamma \\ 0, & \text{if } x \in \mathcal{K}_\gamma \end{cases}$$

Now, define a map $\phi : [0, \infty) \to [0, \infty)$ by

$$\phi(y) := \inf\{\theta(x) |\ d(x, \mathcal{K}_\gamma) \ge y\}.$$

It is not difficult to show that ϕ is continuous, monotone increasing and $\phi(y) \to \infty$ as $y \to \infty$. It follows that, for almost all t,

$$\langle \nabla W(x), F_\gamma(t, x)\rangle \le -\phi(d(x, \mathcal{K}_\gamma)) \quad \text{for all } x \in \mathbf{R}^n. \tag{7}$$

By Assumption 2 and boundedness of the measurable function $w(\cdot)$, we see that F_γ satisfies the hypotheses of the classical (Carathéodory) existence theory and so, for each $(t_0, x^0) \in \mathbf{R} \times \mathbf{R}^n$, the initial-value problem (5) has a solution and every solution can be maximally extended.

Let $(t_0, x^0) \in \mathbf{R} \times \mathbf{R}^n$ be arbitrary, and let $x : [t_0, \omega) \to \mathbf{R}^n$ be a maximal solution of (5). By (7), it follows that, for almost all $t \in [t_0, \omega)$,

$$\frac{d}{dt} W(x(t)) \le -\phi(d(x(t), \mathcal{K}_\gamma)) \le 0 \tag{8}$$

which, on integration, yields

$$d^2(x(t), \mathcal{K}_\gamma) \le c_1^{-1} W(x(t)) \le c_1^{-1} W(x^0) \le c_1^{-1} c_2 d^2(x^0, \mathcal{K}_\gamma) \quad \text{for all } t \in [t_0, \omega). \tag{9}$$

Therefore, $x(\cdot)$ is bounded and so $\omega = \infty$. This establishes property (i) of Definition 1.

Let $r > 0$ be arbitrary and let r^* be such that

$$\|x^0\| < r \quad \Longrightarrow \quad d(x^0, \mathcal{K}_\gamma) < r^*.$$

Let $R > 0$ be such that

$$d(x, \mathcal{K}_\gamma) < (c_1^{-1} c_2)^{\frac{1}{2}} r^* \quad \Longrightarrow \quad \|x\| < R.$$

Then, by (9), it follows that, for each $(t_0, x^0) \in \mathbf{R} \times \mathbf{R}^n$,

$$\|x^0\| < r \quad \Longrightarrow \quad d(x^0, \mathcal{K}_\gamma) < r^* \quad \Longrightarrow \quad d(x(t), \mathcal{K}_\gamma) \le (c_1^{-1} c_2)^{\frac{1}{2}} r^* \quad \Longrightarrow \quad \|x(t)\| < R \quad \text{for all } t \ge t_0.$$

This establishes property (ii) of Definition 1.

Now let $\epsilon > 0$ be arbitrary. Define $\delta := (c_1 c_2^{-1})^{\frac{1}{2}} \epsilon$. Then, by (9), for each $(t_0, x^0) \in \mathbf{R} \times \mathbf{R}^n$, we have

$$d(x^0, \mathcal{K}_\gamma) < \delta \quad \Longrightarrow \quad d(x(t), \mathcal{K}_\gamma) < \epsilon \quad \text{for all } t \ge t_0.$$

Therefore, property (iii) of Definition 1 holds.

It remains only to establish property (iv). Let $r, \mu > 0$ be arbitrary and define $\mu^* := \min\{r, \mu\}$. Let $\nu > 0$ be such that $\nu < (c_1 c_2^{-1})^{\frac{1}{2}} \mu^*$, and define

$$\tau := \min\{\phi(y) |\ \nu \le y \le \mu^*\}.$$

Let $t_0 \in \mathbf{R}$ be arbitrary and let $x^0 \in \mathbf{R}^n$ be such that $d(x^0, \mathcal{K}_\gamma) < r$. Define $T := \tau^{-1}c_2r^2$. Suppose that $d(x(t), \mathcal{K}_\gamma) > \nu$ for all $t \in [t_0, t_0 + T]$. Then, integrating (8), we arrive at a contradiction:

$$0 < \nu^2 < d^2(x(t_0+T), \mathcal{K}_\gamma) \leq c_1^{-1}W(x(t_0+T)) \leq c_1^{-1}[W(x^0) - \tau T] \leq c_1^{-1}[c_2 d^2(x^0, \mathcal{K}_\gamma) - c_2 r^2] < 0.$$

Therefore, there exists $t_1 \in [t_0, t_0 + T]$, such that $d(x(t_1), \mathcal{K}_\gamma) = \nu$. It immediately follows that

$$d(x(t), \mathcal{K}_\gamma) \leq (c_1^{-1}c_2)^{\frac{1}{2}} d(x(t_1), \mathcal{K}_\gamma) = (c_1^{-1}c_2)^{\frac{1}{2}}\nu < \mu^* < \mu \quad \text{for all } t \geq t_0 + T.$$

This completes the proof. Q.E.D.

Corollary. Let the hypotheses of the theorem hold. If $\rho_\gamma = 0$, then, for each γ with $\gamma_1 < \gamma < \min\{\gamma_2, \gamma_3\}$, $\{0\}$ is a globally uniformly asymptotically stable equilibrium of the output-feedback-controlled system (5).

6 References

Barmish, B.R., and G. Leitmann, 1982, "On ultimate boundedness control of uncertain systems in the absence of matching conditions", *IEEE Trans. Automatic Control*, **AC-27**, 1253.

Chen, Y.H., 1987a, "Robust output feedback controller: direct design", *Int. J. Control*, **46**, 1083; 1978b, "Robust output feedback controller: indirect design", *ibid.*, **46**, 1093.

Chen, Y.H., and G. Leitmann, 1987, "Robustness of uncertain systems in the absence of matching assumptions", *Int. J. Control*, **45**, 1527.

Corless, M., and G. Leitmann, 1981, "Continuous state feedback guaranteeing uniform ultimate boundedness for uncertain dynamic systems", *IEEE Trans. Automatic Control*, **AC-26**, 1139; 1983, "Erratum to: Continuous state feedback guaranteeing uniform ultimate boundedness for uncertain dynamic systems", *ibid.*, **AC-28**, 249; 1983, "Adaptive control of systems containing uncertain functions and unknown functions with uncertain bounds", *J. Optimiz. Theory Applic.*, **46**, 155; 1990, "Deterministic control of uncertain systems: a Lyapunov theory approach", in *Deterministic Control of Uncertain Systems* (A.S.I. Zinober, ed.), London: Peter Peregrinus.

Corless, M., G. Leitmann, and E.P. Ryan, 1990, "Control of uncertain systems with neglected dynamics", in *Deterministic Control of Uncertain Systems* (A.S.I. Zinober, ed.), London: Peter Peregrinus.

Corless, M, and E.P. Ryan, 1991, "Robust feedback control of singularly perturbed uncertain dynamical systems", *Dynamics and Stability of Systems*, to appear.

Galimidi, A.R., and B.R. Barmish, 1986, "The constrained Lyapunov problem and its application to robust output feedback stabilization", *IEEE Trans. Automatic Control*, **AC-31**, 410.

Goodall, D.P. and E.P. Ryan, 1988, "Feedback controlled differential inclusions and stabilization of uncertain dynamical systems", *SIAM J. Control & Optimiz.*, **26**, 1431.

Leitmann, G., 1978, "Guaranteed untimate boundedness for a class of uncertain linear dynamical systems", *IEEE Trans. Automatic Control*, **AC-23**, 1109; 1990, "Deterministic control of uncertain systems via a constructive use of Lyapunov stability theory", *Proc 14th IFIP Conf. on System Modelling and Optimization, Leipzig, 1989*, New York: Springer Verlag.

Leitmann, G., E.P. Ryan, and A. Steinberg, 1986, Feedback control of uncertain systems: Robustness with respect to neglected sensor and actuator dynamics", *Int. J. Control*, **43**, 1243.

Ryan, E.P., 1991, "A universal adaptive stabilizer for a class of nonlinear systems", *Systems & Control Letters*, to appear.

Ryan, E.P., and M. Corless, 1984, "Ultimate boundedness and asymptotic stability of a class of uncertain dynamical systems via continuous and discontinuous feedback control", *IMA J. Math. Control & Information*, 1, 223.

Saberi, A., and H. Khalil, "Quadratic-type Lyapunov functions for singularly perturbed systems", *IEEE Trans. Autom. Control*, **AC-26**, 505-507, 1981.

Šiljak, D.D., 1978, *Large-Scale Dynamic Systems: Stability and Structure*, New York: North Holland.

Steinberg, A., and M. Corless, 1985, "Output feedback stabilization of uncertain dynamical systems", *IEEE Trans. Automatic Control*, **AC-30**, 1025.

Zaheb, E., 1986, "A sufficient condition for output feedback stabilization of uncertain systems", *IEEE Trans. Automatic Control*, **AC-31**, 1055.

Zinober, A.S.I. (editor), 1990, *Deterministic Control of Uncertain Systems*, London: Peter Peregrinus.

CONTROLLABILITY OF A FOKKER-PLANCK EQUATION, THE SCHRODINGER SYSTEM, AND A RELATED STOCHASTIC OPTIMAL CONTROL[1].

A. Blaquière[2]

Abstract. We start from a controllability problem for a Fokker-Planck equation, termed *Problem A*. A solution (v^*, Φ^*) to that problem is constructed via a Theorem of Jamison, under proper assumptions. Theorem 2 gives a sufficiency condition concerning the given initial and terminal data for that solution to exist. Theorem 3 states that v^* is an optimal feedback control for a stochastic optimal control problem with constraint on the end-state, termed *Problem B*, and further v^* corresponds to the minimum of an entropy distance. At last Problem A is transformed into a controllability problem for a stochastic differential equation, termed *Problem C*: the solution to Problem C corresponding to the one constructed in Problem A is the Markovian process satisfying the given end conditions in a set of reciprocal processes of Jamison.

1. Introduction.

The present paper is concerned with a solution to a controllability problem for a forward parabolic equation, namely a Fokker-Planck equation, met in the course of a joint-research reported in [1], [2]. The aim is to find a control $v : [0, T] \otimes E^n \rightarrow E^n$ such that the equation

$$\partial\Phi/\partial t = - \operatorname{div}(\Phi v) + \frac{1}{2}\Delta\Phi,$$

has a solution satisfying both given initial and terminal data

$$\Phi(0,.) = \Phi_0, \quad \text{and} \quad \Phi(T,.) = \Phi_1 .$$

---

[1]Presented at the FOURTH WORKSHOP ON CONTROL MECHANICS, January 22-23, 1991, University of Southern California, Los Angeles, California (USA).

[2]University Paris 7, Laboratoire d'Automatique Théorique, Tour 14-24, 2 Place Jussieu, 75251 PARIS CEDEX 05 (FRANCE).

This problem deserves interest not only in the area of the control of parabolic partial differential equations, but also for the control of a most important hyperbolic one, namely the equation of Schrödinger. This is so because the equation of Schrödinger can be written in the form of a system of two equations, one of which is the continuity equation of hydrodynamics and the other one has similarities with the Hamilton-Jacobi equation of classical mechanics. These are the equations (C) and (J) of Louis de Broglie (C for "continuity", J for "Jacobi")[3]. The equation (C) itself can be put in the form of a Fokker-Planck equation. It follows that the very important and difficult problem of controlling a Schrödinger's equation, in quantum mechanical engineering, can be viewed as a pair of sub-problems, namely

1. the control of a Fokker-Planck equation; and
2. the fitting of the phase of the wave-function to initial and terminal data.

Of course, there is an interplay of parabolic equations and stochastic differential equations, so that, if one focusses one's attention on the first sub-problem, one may be concerned with the control of a stochastic differential equation. An original feature of our approach is that it started with a pair of stochastic optimal control problems, one "forward" in time, one "backward" in time, so as to account for a type of time-reversibility similar to the one met in quantum mechanics.

We belatedly discovered its connection with a programme initiated by Schrödinger in 1931[4] : the aim of Schrödinger was to construct some unconventional diffusion processes associated with the classical heat equation, in such a way that their properties are as close as possible to the ones of the probabilist concepts involved in quantum mechanics. This programme has challanged mathematicians. It has been extensively developed and still originates deep questions. On the way from the 1931-Schrödinger's paper to modern approaches are the ones of Bernstein[5], Burling [6], Jamison [7] and others : the central idea in them is the one of *reciprocal processes*. Recently, papers of Wakolbinger [8], and Dawson-Gorostiza-Wakolbinger [9], relying on the *method of large deviations* along Föllmer's views [10] make a connection between Schrödinger's problem and a problem of stochastic calculus of variations. This new bridge permits a deeper insight into our approach, which is however different, and, in particular, exhibits its links with a problem of *minimum entropy distance* .

The present paper starts from the above mentioned controllability problem for a Fokker-Planck equation, termed *Problem A*. A solution (v^*, Φ^*) to that problem is constructed via a Theorem of Jamison, under proper assumptions on the initial and the terminal data. Theorem 2 gives a sufficiency condition concerning the initial and the terminal data for that solution to exist. We then have the following question : *As the solution we have constructed is not unique, what does characterize that solution among the set of all solutions to Problem A* ? An answer to that question is provided by Theorem 3, which states that v^* is an optimal feedback control for a stochastic optimal control problem with constraint on the end-state, termed *Problem B*, and further v^* corresponds to the minimum of an entropy distance. At last Problem A is transformed into a controllability problem for a stochastic differential equation, termed *Problem C*: the solution to Problem C corresponding to the one we have constructed in Problem A appears to be the Markovian process satisfying the given end conditions in a set of reciprocal processes of Jamison.

2. Notation and Assumptions

E^n denotes n-dimensional Euclidean space, $\mathcal{B}(E^n)$ its σ-field of Borel sets, [0, T] a compact interval in E^1. Definitions of *stochastic process, Brownian motion, Wiener process* will be taken from [15] as will other elements of our framework which will be noted below.

We assume a given underlying probability space (Ω, F, P) endowed with a nondecreasing family $\{F_t: 0 \le t \le T\}$ of sub-σ-algebras of F. For a stochastic process β defined on some subinterval I of [0, T], we denote by F^β the smallest σ-algebra with respect to which β is measurable, and by $\{F_t^\beta : t \in I\}$ the corresponding nondecreasing family of sub-σ-algebras of F^β.

Let $T < \Theta$, and let V be a nonnegative function defined on the closed strip $S = [0, \Theta] \otimes E^n$. We shall need assumptions (A1) and (A2) of Ref. [13], namely

(A1) $V : S \to R_+$ (R_+ *denoting nonnegative real numbers) is bounded and continuous on the strip* S *and satisfies a Hölder condition with respect to* x *on* E^n.

(A2) V *is uniformly Hölder continuous in* (s,x) *in compact subsets of* S.

Under assumption (A1), the inhomogeneous equation of heat transfer $\frac{\partial \phi}{\partial t} = \frac{1}{2}\Delta\phi - V\phi$ and its adjoint have a fundamental solution which we shall denote by $p(s,x;t,y)$ ($0 \le s < t \le T$, $x,y \in E^n$).

3. Controllability Problem for a Forward Parabolic Equation

Problem A

Let μ_0 and μ_1 be given probability measures on $\mathcal{B}(E^n)$ having densities Φ_0 and Φ_1 , respectively, with respect to the n-dimensional Lebesgue's measure λ. Find $v : [0, T] \otimes E^n \to E^n$ such that the equation

$$\partial\Phi/\partial t = -\operatorname{div}(\Phi v) + \frac{1}{2}\Delta\Phi, \qquad (1)$$

has a solution satisfying

$$\Phi(0, .) = \Phi_0 , \qquad (2)$$

$$\Phi(T, .) = \Phi_1 . \qquad (3)$$

A solution to that controllability problem, considered in [2], relies on a theorem of Jamison [7], namely

<u>Theorem 1.(Jamison)</u> Suppose M is a σ-compact metric space, that μ_0 and μ_1 are probability measures on its σ-field $\mathcal{B}$ of Borel sets, and that q is an everywhere continuous, strictly positive function on $M\otimes M$. Then there is a unique pair μ, π of measures on $\mathcal{B}\otimes\mathcal{B}$ for which

(a) μ is a probability measure and π is a σ-finite product measure.

(b) $\mu(B\otimes M) = \mu_0(B)$, $\qquad \mu(M\otimes B) = \mu_1(B)$, $\qquad B \in \mathcal{B}$.

(c) $d\mu/d\pi = q$.

Let (A1) hold. Let $M = E^n$, $\mathcal{B} = B(E^n)$ and $q(x,y) = p(0,x;T,y)$, $(x,y) \in E^n\otimes E^n$. By Theorem 1, there exist measures v_0 and v_1 on $B(E^n)$ such that

$$\mu_0(B_0) = \int_{B_0} dv_0(x) \int_{E^n} p(0,x;T,y)\, dv_1(y) \;, \qquad B_0 \in B(E^n)\,,$$

$$\mu_1(B_1) = \int_{B_1} dv_1(y) \int_{E^n} p(0,x;T,y)\, dv_0(x) \;, \qquad B_1 \in B(E^n).$$

Obviously, the pair (v_0, v_1) is not unique : for any pair (k_0, k_1) of real numbers with $k_0k_1 = 1$, the factor measures $k_0 v_0$, k_1v_1, will produce the same π. *In the case* π *is finite*, one can eliminate this inconvenience by normalizing the v_0, v_1, by the condition

$$\int dv_0(x) = \int dv_1(y) = [\int d\pi(x,y)]^{1/2}.$$

When μ_0 and μ_1 have densities Φ_0 and Φ_1, respectively, with respect to λ, then, as shown in [7], v_0 and v_1 have nonnegative densities ρ_0 and ρ_1, respectively, with respect to λ, satisfying

$$\Phi_o = \rho_o(x) \int_{E^n} p(0,x;T,y)\, \rho_1(y)\, dy \ , \qquad (4)$$

$$\Phi_1 = \rho_1(x) \int_{E^n} p(0,y;T,x) \rho_o(y)\, dy \ . \qquad (5)$$

In the case v_o *and* v_1 *are finite* , the densities ρ_o and ρ_1 are integrable so that uniqueness of the solution (ρ_o, ρ_1) to the system (4), (5) can be ensured again by the normalization condition

$$\int_{E^n} \rho_o(x)\, dx = \int_{E^n} \rho_1(y)\, dy \ .$$

Now, let us recall the construction of a solution to Problem A following Ref. [2] :

If (A1) holds and ρ_o *and* ρ_1 *are continuous and bounded* then (see [11]) the functions ρ and ρ_* given by

$$\rho(s,x) = \int_{E^n} p(s,x;T,y)\, \rho_1(y)\, dy \qquad \text{in } [0, T)\otimes E^n, \qquad (6)$$

$$\rho_*(t,y) = \int_{E^n} p(0,x;t,y)\, \rho_o(x)\, dx \qquad \text{in } (0, T]\otimes E^n, \qquad (7)$$

with $\rho(T,y) = \rho_1(y)$, $\rho_*(0,x) = \rho_o(x)$, are solutions to the Cauchy problems

$$\frac{\partial \phi}{\partial s} = -\frac{1}{2}\Delta\phi + V\phi \qquad \text{in } [0, T)\otimes E^n , \qquad (8)$$

$$\phi(T,x) = \rho_1(x) \qquad \text{in } E^n \qquad (9)$$

and

$$\frac{\partial \phi_*}{\partial t} = \frac{1}{2}\Delta\phi_* - V\phi_* , \qquad \text{in } (0, T]\otimes E^n , \qquad (10)$$

$$\phi_*(0,y) = \rho_o(y) \qquad \text{in } E^n , \qquad (11)$$

respectively.

Further, the functions ρ and ρ_* satisfy $\rho(s,x) > 0$ in $[0, T)\otimes E^n$, and $\rho_*(t,y) > 0$ in $(0, T]\otimes E^n$.

One can verify by direct computation that (1) is satisfied for $v = v^*$ and $\Phi = \Phi^*$ given by

$$v^*(t,x) = (1/\rho(t,x)) \operatorname{grad} \rho(t,x) , \qquad \text{in } [0, T)\otimes E^n , \qquad (12)$$

$$\Phi^* (t,x) = \rho_*(t,x)\, \rho(t,x) , \qquad \text{in } [0, T]\otimes E^n , \qquad (13)$$

and that Φ^* satisfies the end conditions (2) and (3). Therefore v^* is a solution to Problem A.

At this point we address the following questions :

1. *Under which conditions on* Φ_0 *and* Φ_1 *can we assert that* ρ_0 *and* ρ_1 *are continuous and bounded*? and

2. *Since, as pointed out in* [2], *the solution we have constructed is not unique, what does characterize that solution among the set of all solutions to Problem* A?

An answer to question 1 is provided by

Theorem 2. If (A1) holds and Φ_0 and Φ_1 are continuous with compact support, then ρ_0 and ρ_1 are continuous with compact support.

The proof of Theorem 2 is given in Appendix 1.

Question 2 will be studied in the subsequent paragraphs. We will see that v^* is the solution to a stochastic optimal control problem, and that this problem is equivalent to a problem of minimum entropy distance.

4. Stochastic Optimal Control Problem

We shall assume

(A3) $\Phi_0 = d\mu_0/d\lambda$ *and* $\Phi_1 = d\mu_1/d\lambda$ *are continuous with compact support.*

Let there be given a measurable space (Ω, F) endowed with an increasing family $\{F_t, 0 \le t \le T\}$ of sub-σ-algebras of F. Let (A1) and (A3) hold. Then, by Theorem 2, the functions ρ_0 and ρ_1 satisfying the system of integral equations (4), (5), are nonnegative and continuous with compact support. Now let us consider

Problem B

Let

$$J(s,x;T,v) = E_{sx} \{ \int_s^T L(t,\xi(t),v(t,\xi(t)))dt + W_T(\xi(T)) \} . \qquad (14)$$

where ξ is a n-dimensional diffusion process in the weak sense with drift v and diffusion coefficient $D \equiv 1/2$. Let L be given by $L(t,x,v) = (1/2)v^2+V(t,x)$ and let W_T be a real valued function defined on the bounded subset Σ of E^n, namely $\Sigma = \{y: \rho_1(y)>0\}$, by $W_T(y) = -\log \rho_1(y)$. The problem is to choose v in a set $\mathcal{V}^*$ of admissible feedback controls with range in E^n such that the end conditions $\xi(s) = x$, $\xi(T) \in \Sigma$ hold P_{sx} - a.s. and the expectation $J(s,x;T,v)$ is minimized, for all $(s,x) \in [0, T)\otimes E^n$.

Since Σ is bounded, the usual theorems of existence of an optimal feedback control [12], [13], do not apply. A problem with end constraints on the state of this type has been studied in [14], here we will use another method.

4.1.Definition of the class $\mathcal{V}^*$

Let us denote by $\mathcal{V}$ the class of all functions v such that

(a) v is a Borel measurable function from $[0, T)\otimes E^n$ into E^n;

(b) For each s in [0, T) there exists a σ-algebra F^s contained in F, and for each (s,x) in $[0, T)\otimes E^n$ there exists a probability P_{sx} on (Ω, F^s) and an F_t n-dimensional standard Wiener process w for P_{sx} - say w_{sx} - such that

$$d\xi(t) = v(t,\xi(t))dt + \sqrt{2D}\, dw_{sx}(t) , \qquad s\le t\le T, \qquad (15)$$

$$\xi(s) = x , \qquad (16)$$

has a solution, unique in probability law;

(c) $E_{sx} |W_T(\xi(T))| < \infty$;

(d) $$E_{sx} \int_s^T |v(t,\xi(t))|^2 \, dt < \infty .$$

Except for minor changes, these are usual conditions in the definition of an admissible feedback control when the stochastic optimal control problem has no constraints on the end state [12]. In the present case where $\xi(T) \in \Sigma$ P_{sx} - a.s., with W_T defined on Σ, we shall restrict the class $\mathcal{V}$ as follows.

For any given F_t- n-dimensional standard Wiener process $w°$ with respect to some probability $P°$ on (Ω, F), consider the system of stochastic differential equations

$$d\eta(t) = dw°(t) , \qquad s \le t , \qquad (17)$$

$$\eta(s) = x , \qquad (18)$$

in $(\Omega, F, P°)$.

The solution η_{sx} to that system depends on $w°$.

We denote by $\mathcal{V}^*$ the class of all functions $v \in \mathcal{V}$ such that, for any $w°$ and for each (s,x) in $[0, T)\otimes E^n$, there is a continuous $P°$-martingale

$$z_{sx} = (z_{sx}(t), F_t^{w°-w°(s)}), \qquad s \le t \le T , \qquad \text{such that}$$

$$z_{sx}(t) = \exp \left(\int_s^t v(r, \eta_{sx}(r)) \, dw°(r) - (1/2) \int_s^t | v(r, \eta_{sx}(r)) |^2 dr \right), \quad s \le t < T, \quad (19)$$

where $\eta_{sx} = x + w° - w°(s)$.

<u>Lemma 1.</u> Assume (A1)-(A3) hold. Then, v^* given by (12) or equivalently by

$$v^*(t,x) = - \operatorname{grad} W(t,x) \qquad \text{in } [0, T)\otimes E^n ,$$

with

$$W(t,x) = - \log \rho(t,x)$$

belongs to $\mathcal{V}^*$, and the corresponding martingale family - say $\{z^*_{sx}\}$ - is given by

$$z^*_{sx}(t) = \frac{\rho(t, \eta_{sx}(t))}{\rho(s,x)} \exp \left(- \int_s^t V(r, \eta_{sx}(r)) \, dr \right), \qquad s \le t \le T. \qquad (20)$$

See Appendix 2.

<u>4.2.A Property of the class $\mathcal{V}^*$</u>

Let $v \in \mathcal{V}^*$. Then, according to Theorem 5.7 of Lipster-Shiryayev [15], for any (s,x) in $[0, T)\otimes E^n$, there is an $F^{w°-w°(s)}$ - adapted process

$(\gamma_{sx}(t,\omega), F_t^{w°-w°(s)})$, $s \le t \le T$, such that

$$P^\circ\left(\int_s^T \gamma_{sx}^2(r,\omega)\, dr < \infty \right) = 1 \qquad (21)$$

and such that, for all $s \le t \le T$

$$z_{sx}(t) = 1 + \int_s^t \gamma_{sx}(r,\omega)\, dw^\circ(r) . \qquad (22)$$

The représentation given by (22) is unique.

One can easily prove that (19) and the uniqueness of the representation given by (22) entail

$$\gamma_{sx}(t) = z_{sx}(t)\; v(t, \eta_{sx}(t)) , \qquad s \le t < T , \quad P^\circ\text{- a.s.} \qquad (23)$$

Since z_{sx} is a P°-martingale on $[\, s, T \,]$ and $z_{sx}(s) = 1$, it satisfies $E^\circ z_{sx}(T) = 1$, then on the measurable space (Ω , F^s) there is defined a probability measure P_{sx} with

$$d\, P_{sx} = z_{sx}(T, \omega)\, dP^\circ, \qquad (24)$$

and, according to the generalized theorem of Girsanov (Theorem 6.2 of Lipster-Shiryayev [15]), on the probability space (Ω, F^s , P_{sx}) the random process $w_{sx} = (w_{sx}(t), F)$, $s \le t \le T$, with

$$w_{sx}(t) = w^\circ(t) - \int_s^t \overset{+}{z}_{sx}(r)\; \gamma_{sx}(r)\, dr , \qquad (25)$$

where $\overset{+}{z}_{sx}(r) = z_{sx}^{-1}(r)$ if $z_{sx}(r) > 0$, and $\overset{+}{z}_{sx}(r) = 0$ with $z_{sx}(r) = 0$, is a Wiener process (with respect to the measure P_{sx}).

In view of (19), (23) and (24), for $s \le t < T$, the system (17), (18) rewrites

$$d\eta(t) = v(t, \eta(t))\, dt + dw_{sx}(t) \qquad (26)$$

$$\eta(s) = x, \qquad (27)$$

that is

$$\eta_{sx}(t) = x + \int_s^t v(r, \eta_{sx}(r))\, dr + w_{sx}(t), \qquad s \le t < T, \quad P_{sx}\text{- a.s.} \qquad (28)$$

Since η_{sx} and w_{sx} are continuous on $[\, s, T]$,

$$\int_s^t v(r, \eta_{sx}(r))\, dr \to \int_s^T v(r, \eta_{sx}(r))\, dr , \qquad P_{sx}\text{- a.s.} , \qquad \text{as } t \uparrow T,$$

that is

$$d\eta(t) = v(t, \eta(t))\, dt + dw_{sx}(t)\,, \qquad s \le t \le T, \tag{29}$$

$$\eta(s) = x, \tag{30}$$

in (Ω, F, P_{sx}).

Therefore, since uniqueness in probability law in condition (b) above implies that the expected system performance J in (14) depends only on v and on the initial data (s,x), for given (s,x) and $v \in \mathcal{V}^*$ the value of J is

$$J(s,x,T,v) = E_{sx} \left[\int_s^T L(t,\eta_{sx}(t),v(t,\eta_{sx}(t)))dt + W_T(\eta_{sx}(T)) \right]. \tag{31}$$

with P_{sx} given by (24) and η_{sx} by (29), (30).

5.Existence of an Optimal Feedback Control to Problem B

By condition (d) in the definition of an admissible feedback control v

$$E_{sx} \int_s^T |v(t,\eta_{sx}(t))|^2\, dt < \infty \quad . \tag{32}$$

Therefore

$$E_{sx} \int_s^T v(t,\eta_{sx}(t))\; dw_{sx}(t) = 0\,. \tag{33}$$

Now, by (23) and (25),

$$(1/2)\int_s^t |v(r,\eta_{sx}(r))|^2\, dr + \int_s^t v(r,\eta_{sx}(r))\; dw_{sx}(r) =$$

$$= (1/2)\int_s^t |v(r,\eta_{sx}(r))|^2\, dr + \int_s^t v(r,\eta_{sx}(r))\; dw^\circ(r) - \int_s^t |v(r,\eta_{sx}(r))|^2\, dr =$$

$$= \int_s^t v(r,\eta_{sx}(r))\; dw^\circ(r) - (1/2)\int_s^t |v(r,\eta_{sx}(r))|^2\, dr\,, \quad s \le t < T\,. \tag{34}$$

By (32), $\int_s^t v(r,\eta_{sx}(r))\; dw_{sx}(r)$ and $\int_s^t |v(r,\eta_{sx}(r))|^2\, dr$ are continuous

functions in $t \in [\, s, T\,]$, P_{sx} - a.s. Therefore

$$\int_s^t v(r,\eta_{sx}(r))\; dw^\circ(r) - (1/2)\int_s^t |v(r,\eta_{sx}(r))|^2\, dr$$

is a continuous function in $t \in [\, s, T\,]$, P_{sx} - a.s. , and tends to

$$\int_s^T v(r,\eta_{sx}(r))\; dw^\circ(r) - (1/2)\int_s^T |v(r,\eta_{sx}(r))|^2\, dr, \qquad P_{sx} \text{ - a.s. , as } t \uparrow T\,.$$

Then, by the continuity of the martingale z_{sx}, and by (19)

$$z_{sx}(t) \to z_{sx}(T) = \exp\left\{ \int_s^T v(r,\eta_{sx}(r))\; dw^\circ(r) - (1/2)\int_s^T |v(r,\eta_{sx}(r))|^2\, dr \right\} > 0\,,$$

P_{sx} - a.s. , as $t \uparrow T$.

Accordingly, from (24),

$$\exp \{ \int_s^T v(r,\eta_{sx}(r)) \, dw^\circ(r) - (1/2)\int_s^T |v(r,\eta_{sx}(r))|^2 \, dr \} = \frac{dP_{sx}}{dP^\circ}, \quad P_{sx} \text{ - a.s.} \tag{35}$$

At last we have

$$(1/2)\int_s^T |v(r,\eta_{sx}(r))|^2 \, dr + \int_s^T v(r,\eta_{sx}(r)) \, dw_{sx}(r) =$$

$$\int_s^T v(r,\eta_{sx}(r)) \, dw^\circ(r) - (1/2)\int_s^T |v(r,\eta_{sx}(r))|^2 \, dr = \log \frac{dP_{sx}}{dP^\circ}, \quad P_{sx} \text{ - a.s.} \tag{36}$$

By taking expectations in both sides of (36), on account of (33), we get

$$E_{sx}\{ (1/2)\int_s^T |v(r,\eta_{sx}(r))|^2 \, dr \} = E_{sx} \log \frac{dP_{sx}}{dP^\circ}. \tag{37}$$

By substitution in (31) we obtain

$$J(s,x;T,v) = E_{sx} \log \{ \frac{dP_{sx} \, \rho(s,x)}{dP^\circ \exp[- \int_s^T V(r,\eta_{sx}(r))dr] \, \rho(T,\eta_{sx}(T))} \} - \log \rho(s,x). \tag{38}$$

Now, by Lemma 1, under assumptions (A1)-(A3), for $v = v^*$ we obtain

$$dP_{sx} = dP^*_{sx} = \frac{\rho(T,\eta_{sx}(T))}{\rho(s,x)} \exp[- \int_s^T V(r,\eta_{sx}(r))dr] \, dP^\circ , \tag{39}$$

so that (38) becomes

$$J(s,x;T,v) = E_{sx} \log \frac{dP_{sx}}{dP^*_{sx}} - \log \rho(s,x), \tag{40}$$

and, for $v = v^*$

$$J(s,x;T,v^*) = - \log \rho(s,x) = W(s,x). \tag{41}$$

By Jensen's inequality it turns out that

$$E_{sx} \log \frac{dP_{sx}}{dP^*_{sx}} \geq 0 ,$$

so that (40) and (41) imply that

$$J(s,x;T,v^*) \le J(s,x;T,v) \qquad \text{for all } v \in \mathcal{V}^*, \tag{42}$$

and (42) holds for all (s,x) in $[0, T)\otimes E^n$.

Summarizing, we have

Theorem 3. Let (A1)-(A3) hold. Then v* is an optimal feedback control for the stochastic optimal control problem B. Further for v = v*, the *entropy distance* $H(P^*_{sx} | P_{sx}) = E_{sx} \log [dP^*_{sx}/dP_{sx}]$ is minimum.

6. Stochastic Representation for Problem A

Assume (A1)-(A3) hold. It is easy to prove that, if η is the solution to

$$d\eta(t) = dw^\circ(t), \qquad s \le t,$$
$$\eta(0) = \eta^\circ, \qquad \text{a.s.,} \qquad \text{in } (\Omega, F, P^\circ),$$

for given n-dimensional random vector η° with distribution measure μ_0,

then

1. η satisfies

$$d\eta(t) = v^*(t, \eta(t))\, dt + dw^*(t), \qquad 0 \le t \le T,$$
$$\eta(0) = \eta^\circ, \qquad \text{a.s.,} \qquad \text{in } (\Omega, F, P^*),$$

where P* and w* are given by

$$dP^* = \frac{\rho(T, \eta(T))}{\rho(0,\eta(0))} \exp [- \int_0^T V(r, \eta(r))\, dr]\, dP^\circ,$$

$$w^*(t) = w^\circ(t) - \int_0^t v^*(r, \eta(r))\, dr, \qquad 0 \le t \le T.$$

2. The stochastic process η in (Ω, F, P^*) is a Markov process in the sense of Fleming-Rishel [12].

3. Its transition function, computed as in Ref. [2] (Appendix I of that Ref.) has the density q* given by

$$q^*(s,x;t,y) = \frac{\rho(t,y)}{\rho(s,x)} p(s,x;t,y), \qquad 0 \le s < t \le T.$$

It follows that the stochastic process η in (Ω, F, P^*) is equivalent to a *reciprocal process* of Jamison [7] with the *(joint) endpoint distribution density* $[\, \rho_0\, p(0,x;T,y)\, \rho_1]$ (that is, it has the same transition function).

4. The distribution measure of $\eta(T)$ is μ_1.

Consequently, if (A1)-(A3) hold, then v^* is also a solution to Problem C, namely

Problem C

Find $v : [0, T] \otimes E^n \to E^n$ such that the stochastic differential equation

$$d\xi(t) = v(t, \xi(t))\, dt + dw(t), \qquad 0 \le t \le T,$$

has a solution in the weak sense - say (P, ξ, w) - satisfying

$P\{\xi(0) \in B\} = \mu_0(B)$,

$P\{\xi(T) \in B\} = \mu_1(B)$,

for all $B \in \mathcal{B}(E^n)$.

Problem C has been considered in Ref. [2] under stronger assumptions.

Appendix 1: Proof of Theorem 2.

Let μ_0 and μ_1 be probability measures on $\mathcal{B}(E^n)$ with $d\mu_0/d\lambda = \Phi_0$, $d\mu_1/d\lambda = \Phi_1$, with λ the n-dimensional Lebesgue's measure. Suppose Φ_0 and Φ_1 are continuous with compact supports.

Denote by A_0 and A_1 the compact supports of Φ_0 and Φ_1, respectively, let A be a compact subset of E^n containing $A_0 \cup A_1$, let $K = A \otimes A$, and let $\mathbf{B} = \{ B \cap A : B \in \mathcal{B}(E^n) \}$. $\mathbf{B}$ is the class of Borel subsets of A, and $\mathbf{B} \otimes \mathbf{B}$ is the class of Borel subsets of K. On K, $p(0,.;T,.)$ is bounded above and away from zero below; that is

$$0 < a \le p(0,x;T,y) \le b < \infty \qquad (1.1)$$

for some a, b and for all $(x,y) \in K$.

Since Φ_0 and Φ_1 have compact supports in A, μ_0 and μ_1 are probability measures on $\mathbf{B}$. By Theorem 1 with M = A, there exists a σ-finite product measure π° on $\mathbf{B}\otimes\mathbf{B}$ and a probability measure μ° on $\mathbf{B}\otimes\mathbf{B}$ such that

(i) $$\mu^\circ(B\otimes A) = \mu_0(B)$$

$$\mu^\circ(A\otimes B) = \mu_1(B) , \qquad B \in \mathbf{B},$$

(ii) $$\frac{d\mu^\circ}{d\pi^\circ} = p(0,.;T,.) \qquad \text{on } K.$$

From condition (ii) and (1.1) we deduce

$$a\ \pi^\circ(K) \le \int_K p(0,x;T,y)\ d\pi^\circ(x,y) = \mu^\circ(K) = 1 \le b\ \pi^\circ(K)\ .$$

Hence π° satisfies

$$\frac{1}{b} \le \pi^\circ(K) \le \frac{1}{a}$$

and, accordingly, its normalized factor measures ν°_0, ν°_1 satisfy

$$\frac{1}{\sqrt{b}} \le \nu^\circ_i(A) \le \frac{1}{\sqrt{a}}, \qquad i = 0, 1. \qquad (1.2)$$

We extend μ° and π° to all of $\mathcal{B}(E^n)\otimes\mathcal{B}(E^n)$ by setting them equal to zero on sets $\Gamma \in \mathcal{B}(E^n)\otimes\mathcal{B}(E^n)$ disjoint from K. Let us denote by μ^*, π^*, these extensions, and by $\overset{*}{\nu}_0$, $\overset{*}{\nu}_1$ the extensions of ν°_0, ν°_1, respectively. π^*, like π°, is a *finite* product measure, satisfying

$$\frac{1}{b} \le \pi^*(E^n\otimes E^n) \le \frac{1}{a} \qquad (1.3)$$

and $\overset{*}{\nu}_0$, $\overset{*}{\nu}_1$ satisfy

$$\frac{1}{\sqrt{b}} \le \overset{*}{\nu}_i(E^n) \le \frac{1}{\sqrt{a}}, \qquad i = 0, 1. \qquad (1.4)$$

μ^* is a probability measure on $\mathcal{B}(E^n)\otimes\mathcal{B}(E^n)$, and (i), (ii) are replaced by

(i)' $$\mu^*(B\otimes E^n) = \mu_0(B)$$

$$\mu^*(E^n\otimes B) = \mu_1(B) , \qquad B \in \mathcal{B}(E^n),$$

(ii)' $$\frac{d\mu^*}{d\pi^*} = p(0,.;T,.) \qquad \text{on } E^n\otimes E^n, \quad \pi^* \text{ - a.s.}$$

By Theorem 1, measures μ^* and π^* on $\mathcal{B}(E^n)\otimes\mathcal{B}(E^n)$ satisfying these conditions are unique and, accordingly, the normalized factor measures ν_0^*, ν_1^* are unique. In other words, ν_0^* and ν_1^* are the unique normalized solutions of the system

$$\mu_0(B_0) = \int_{B_0} d\nu_0^*(x) \int_{E^n} p(0,x;T,y)\, d\nu_1^*(y) , \qquad B_0 \in \mathcal{B}(E^n) , \tag{1.5}$$

$$\mu_1(B_1) = \int_{B_1} d\nu_1^*(y) \int_{E^n} p(0,x;T,y)\, d\nu_0^*(x) , \qquad B_1 \in \mathcal{B}(E^n). \tag{1.6}$$

It follows from (1.5), (1.6) that

$$\frac{d\mu_i}{d\nu_i^*} = h_i , \qquad i = 0, 1,$$

where

$$h_0(x) = \int_{E^n} p(0,x;T,y)\, d\nu_1^*(y) , \tag{1.7}$$

$$h_1(y) = \int_{E^n} p(0,x;T,y)\, d\nu_0^*(x) , \quad x,y \in E^n. \tag{1.8}$$

Since the ν_i^*, $i = 0, 1$, are finite measures and $p(0,.;T,.)$ is strictly positive, jointly continuous and bounded on $E^n\otimes E^n$, *functions* h_i, $i = 0, 1$, *are strictly positive and continuous on* E^n. Let us prove that $\nu_i^* << \mu_i$, $i = 0, 1$. From conditions (ii)', (1.1) and the definition of π^*, we deduce

$$a\, \pi^*(\Gamma) \le \mu^*(\Gamma) = \int_\Gamma p(0,x;T,y)\, d\pi^*(x,y) = \int_{\Gamma\cap K} p(0,x;T,y)\, d\pi^*(x,y) \le b\, \pi^*(\Gamma) ,$$

$$\text{for } \Gamma \in \mathcal{B}(E^n)\otimes\mathcal{B}(E^n) .$$

Therefore

$$\mu^*(\Gamma) = 0 \Leftrightarrow \pi^*(\Gamma) = 0 , \tag{1.9}$$

that is, measures μ^* and π^* are equivalent.

It follows from (1.9) that

$$\mu_0(B) = \mu^*(B\otimes E^n) = 0 \Rightarrow \pi^*(B\otimes E^n) = 0 ,$$

and

$$\pi^*(B\otimes E^n) = \overset{*}{\nu}_0(B)\ \overset{*}{\nu}_1(E^n).$$

Further, from the definition of $\overset{*}{\nu}_1$ we have

$$\frac{1}{\sqrt{b}} \le \overset{*}{\nu}_1(E^n) = \nu_1^\circ(A) \le \frac{1}{\sqrt{a}}, \qquad i=0, 1.$$

Therefore

$$\mu_0(B) = 0 \Rightarrow \overset{*}{\nu}_0(B) = 0,$$

so that $\overset{*}{\nu}_0 << \mu_0$. By similar argument $\overset{*}{\nu}_1 << \mu_1$. Since $\mu_i << \overset{*}{\nu}_i$, and $\overset{*}{\nu}_i << \mu_i$,

μ_i and $\overset{*}{\nu}_i$ are equivalent, i = 0, 1. Therefore

$$\frac{d\overset{*}{\nu}_i}{d\mu_i} = \left(\frac{d\mu_i}{d\overset{*}{\nu}_i}\right)^{-1} = h_i^{-1}, \qquad i=0, 1.$$

Since $\mu_i << \lambda$ and $\overset{*}{\nu}_i << \mu_i$, we have $\overset{*}{\nu}_i << \lambda$, i = 0, 1. Let

$$\rho_i = \frac{d\overset{*}{\nu}_i}{d\lambda} \qquad i = 0, 1.$$

The system (1.5), (1.6) is equivalent to

$$\Phi_0(x) = \rho_0(x) \int_{E^n} p(0,x;T,y)\ \rho_1(y)\ dy\ , \tag{1.10}$$

$$\Phi_1(y) = \rho_1(y) \int_{E^n} p(0,x;T,y)\ \rho_0(x)\ dx\ , \qquad x,y \in E^n, \tag{1.11}$$

with

$$\rho_i = \frac{d\overset{*}{\nu}_i}{d\mu_i}\ \frac{d\mu_i}{d\lambda} = h_i^{-1}\ \Phi_i\ , \qquad i=0, 1. \tag{1.12}$$

Since the functions Φ_i, i = 0, 1, are nonnegative and continuous on E^n with compact supports in A, and the functions h_i, i = 0, 1, are strictly positive and continuous on E^n, it follows from (1.12) that the functions ρ_i, i = 0, 1, solutions to system (1.10), (1.11) are *nonnegative and continuous with compact supports in* A, which ends the proof of Theorem 2.

Bounds of the functions h_i^{-1}, $i = 0, 1$, can be computed as follows :

Consider for instance the function h_o. By (1.1), (1.7) and the definition of v_1^*, we have

$$a \quad v_1^\circ(A) \leq h_0(x) = \int_{E^n} p(0,x;T,y) \; dv_1^*(y) = \int_A p(0,x;T,y) \; dv_1^\circ(y) \leq b \; v_1^\circ(A) \; ,$$

for $x \in A$.

Then, using (1.2), we have

$$\frac{a}{\sqrt{b}} \leq h_o(x) \leq \frac{b}{\sqrt{a}} \qquad \text{for } x \in A. \qquad (1.13)$$

Since a and b are strictly positive constants, we deduce from (1.13)

$$0 < \frac{\sqrt{a}}{b} \leq h_o^{-1} \leq \frac{\sqrt{b}}{a} \qquad \text{for } x \in A. \qquad (1.14)$$

A similar argument holds concerning h_1^{-1}.

The proof of Theorem 2 relies on condition (1.1). If the support of one of the functions Φ_o, Φ_1 is not compact, this condition is dropped. Then the product measure π is not necessarily finite and the functions h_o and h_1 (as well as the functions ρ and ρ_* given by (6) and (7)) will in general neither be bounded nor continuous.

Appendix 2

Assume (A1)-(A3) hold. Let

$$z^*_{sx}(t) = \frac{\rho(t, \eta_{sx}(t))}{\rho(s,x)} \; \exp\left(- \int_s^t V(r, \eta_{sx}(r)) \, dr \right) .$$

Let us prove first that $z^*_{sx}(t)$, $s \le t \le T$, *is a* $(P^\circ, \mathcal{F}_t^{w^\circ - w^\circ(s)})$-*martingale.*

Under (A1) and (A3), ρ is continuous and bounded on $[\ 0,\ T\ ]\otimes E^n$ and, for any $\alpha>0$ sufficiently small, is of class $C^{1,2}$ on $[\ 0,\ T-\alpha\ ]\otimes E^n$. By an application of Itô's formula to ρ and of Theorem 1.2.8 of Stroock and Varadhan [16], one can prove easily that z^*_{sx} is a $(P^\circ, \mathcal{F}_t^s)$-martingale on $[\ s,\ T-\alpha]$.

Accordingly, for any given $t\in(s,\ T)$, $A \in \mathcal{F}_t^s$ and $t \le t' < T$,

$$\int_A z^*_{sx}(t')\ dP^\circ = \int_A z^*_{sx}(t)\ dP^\circ = \text{const.} \tag{2.1}$$

By the continuity of the stochastic process $z^*_{sx}(t)$ on $[\ s,\ T\ ]$, for $\omega \in \Omega$,

$$z^*_{sx}(t') \to z^*_{sx}(T)\ ,\quad \text{as } t' \uparrow T\ ,\quad P^\circ\text{ - a.s.},$$

and since, further, z^*_{sx} is uniformly bounded, by the Lebesgue's theorem of dominated convergence,

$$\int_A z^*_{sx}(t')\ dP^\circ \to \int_A z^*_{sx}(T)\ dP^\circ. \tag{2.2}$$

By (2.1) and (2.2)

$$\int_A z^*_{sx}(t')\ dP^\circ = \int_A z^*_{sx}(T)\ dP^\circ, \tag{2.3}$$

and since (2.3) holds for all $A \in \mathcal{F}_t^s$,

$$E^\circ\ [\ z^*_{sx}(T) \mid \mathcal{F}_t^s\ ] = z^*_{sx}(t)\ . \tag{2.4}$$

We have thus proved that z^*_{sx} is a $(P^\circ, \mathcal{F}_t^s)$-martingale on $[\ s,\ T\ ]$. Then, since z^*_{sx}, like η_{sx}, is $\mathcal{F}^{w^\circ - w^\circ(s)}$-measurable and since $\mathcal{F}^{w^\circ - w^\circ(s)}$ is contained in $\mathcal{F}_t^s$, $z^*_{sx}(t)$, $s \le t \le T$, *is a* $(P^\circ, \mathcal{F}_t^{w^\circ - w^\circ(s)})$-*martingale* .

Then we define P^*_{sx} by

$$d\ P^*_{sx} = z^*_{sx}(T)\ dP^\circ\ .$$

The proof that

$$z^*_{sx}(t) = \exp\ (\ \int_s^t v^*(r,\ \eta_{sx}(r))\ dw^\circ(r)\ -\ (1/2)\ \int_s^t |\ v^*(r,\ \eta_{sx}(r))\ |^2 dr\ ),\quad s \le t < T\ ,$$

is the same as in [2].

Under (A1)-(A3), by arguments paralleling the ones of Ref. [13], one can see easily that, for any $\alpha>0$ sufficiently small,

$$E^*_{sx} \int_s^{T-\alpha} | v^*(t, \eta_{sx}(t)) |^2 dt < \infty , \qquad (2.5)$$

and that

$$W(s,x) = J(s,x;T-\alpha,v^*) = E^*_{sx} \{ \int_s^{T-\alpha} [(1/2)| v^*(t, \eta_{sx}(t)) |^2 + V(t, \eta_{sx}(t))] dt +$$

$$+W(T-\alpha, \eta_{sx}(T-\alpha)) \} . \qquad (2.6)$$

Now we will consider the three terms in the right-hand side of (2.6) and let $\alpha\downarrow 0$.

Concerning the third term, we have

$$E^*_{sx} W(T-\alpha,\eta_{sx}(T-\alpha)) = \int_\Omega [- \log \rho(T-\alpha, \eta_{sx}(T-\alpha))] dP^*_{sx}(\omega),$$

and

$$P^*_{sx} \{\eta_{sx}(T-\alpha) \in B\} = \int_{\eta_{sx}(T-\alpha)\in B} [\frac{\rho(T-\alpha, \eta_{sx}(T-\alpha))}{\rho(s,x)} \exp (-\int_s^{T-\alpha} V(r,\eta_{sx}(r))dr] dP^\circ =$$

$$= \int_B \frac{\rho(T-\alpha, y)}{\rho(s,x)} E^\circ[\exp (-\int_s^{T-\alpha} V(r,\eta_{sx}(r))dr \mid \eta_{sx}(T-\alpha)=y] \; k(s,x;T-\alpha,y) \; dy =$$

$$= \int_B \frac{\rho(T-\alpha, y)}{\rho(s,x)} \; p(s,x;T-\alpha,y) \; dy \; .$$

Therefore

$$E^*_{sx} W(T-\alpha,\eta_{sx}(T-\alpha)) = - \int_{E^n} [\log \rho(T-\alpha, y) \frac{\rho(T-\alpha, y)}{\rho(s,x)}] \; p(s,x;T-\alpha,y) \; dy \; .$$

Let $\alpha\downarrow 0$. Since $\log \rho(T-\alpha, y) \frac{\rho(T-\alpha, y)}{\rho(s,x)}$ is bounded and

$$0 < p(s,x;T-\alpha,y) \leq K \exp -a[|y-x|^2/ T-s] ,$$

where K and a are positive constants, for α sufficiently small, by the Lebesgue's theorem of dominated convergence we have

$$\lim E^*_{sx} W(T-\alpha,\eta_{sx}(T-\alpha)) = - \int_{E^n} [\log \rho(T, y) \frac{\rho(T, y)}{\rho(s,x)}] \; p(s,x;T,y) \; dy =$$

$$= E^*_{sx} W(T, \eta_{sx}(T)) < \infty .$$

Concerning the second term in the right-hand side of (2.6), since V is continuous and bounded on $[0, T]\otimes E^n$, by the theorem of bounded convergence we have

$$\lim E^*_{sx} \int_s^{T-\alpha} V(t, \eta_{sx}(t))\, dt = E^*_{sx} \int_s^T V(t, \eta_{sx}(t))\, dt < \infty .$$

Concerning the first term in the right-hand side of (2.6), we have $E^*_{sx} \int_s^{T-\alpha} (1/2)| v^*(t, \eta_{sx}(t)) |^2 dt =$

$$= W(s,x) - E^*_{sx} W(T-\alpha, \eta_{sx}(T-\alpha)) - E^*_{sx} \int_s^{T-\alpha} V(t, \eta_{sx}(t))\, dt .$$

Since ρ is nonnegative and bounded on $[\ 0,\ T\ ] \otimes E^n$, $W = -\log \rho$ is bounded below and, accordingly, $-\ E^*_{sx} W(T-\alpha, \eta_{sx}(T-\alpha))$ is bounded above by a constant not depending on α. Likewise, since V is bounded,

$- E^*_{sx} \int_s^{T-\alpha} V(t, \eta_{sx}(t))\, dt$ is bounded above by a constant not depending on α.

Therefore,

$E^*_{sx} \int_s^{T-\alpha} (1/2)| v^*(t, \eta_{sx}(t)) |^2 dt$ is bounded above by a constant not depending on α. It then follows from the monotone convergence theorem that

$$\lim E^*_{sx} \int_s^{T-\alpha} (1/2)| v^*(t, \eta_{sx}(t)) |^2 dt = E^*_{sx} \int_s^T (1/2)| v^*(t, \eta_{sx}(t)) |^2 dt < \infty .$$

What remains to be proved is that condition (b) in the definition of $\mathcal{V}^*$ is satisfied for $v = v^*$.

Since, as we have seen above, $z^*_{sx}(t)$, $s \le t \le T$, is a $(P^\circ, \mathcal{F}_t^{w^\circ - w^\circ(s)})$-martingale, it follows from Theorem 5.7 of Lipster-Shiryayev together with the generalized theorem of Girsanov [15], by the same argument as in paragraph 4.2, that η_{sx} is a solution to system (15), (16) with $P_{sx} = P^*_{sx}$ and $w_{sx} = w^*_{sx}$ given by

$$w^*_{sx}(t) = w^\circ(t) - \int_s^t v^*(r, \eta_{sx}(r))\, dr , \qquad s \le t \le T , \qquad P^*_{sx} \text{ - a.s.}$$

Uniqueness of that solution on $[\ s,\ T-\alpha\ ]$ for any $\alpha>0$ is proved by the same argument as Lemma 6.B of Ref. [13], then uniqueness of that solution on

[s, T] is a direct consequence of the continuity (and therefore of the separability) of the stochastic process η_{sx} on [s, T] . Uniqueness in probability law follows.

Indeed, denote by C^s the space of all continuous functions x from [s, T] into E^n and by B its Borel σ-algebra. Define a continuous process X(t) = X(t, x) on [s, T] by X(t, x) = x(t), t ∈ [σ, T], and let

$$P^\circ_{sx}\{\Gamma\} = P^\circ\{\ \omega : \ \eta_{sx}(\,.\,,\omega) \in \Gamma\ \}\,.$$

By the definition of η_{sx}, P°_{sx} is the Wiener measure starting at (s,x).

By Lemma 1.6 of Dynkin [17]

$$\int_{\eta_{sx}(.,\omega)\in\Gamma} \frac{\rho(T,\ \eta_{sx}(T))}{\rho(s,x)} \ \exp\left(-\int_s^T V(r,\ \eta_{sx}(r))\,dr\right) dP^\circ = P^*_{sx}\ \{\eta_{sx}(\,.\,,\omega) \in \Gamma\} =$$

$$= \int_\Gamma \frac{\rho(T,\ x(T))}{\rho(s,x)} \ \exp\left(-\int_s^T V(r,\ x(r))\,dr\right) dP^\circ_{sx}\ . \qquad (2.7)$$

The right-hand side of (2.7) is independent of (w°,P°). It rewrites $\Pi^*_{sx}(\Gamma)$, where Π^*_{sx} is the distribution measure of η_{sx} with respect to P^*_{sx} .

Acknowledgment The author is indebted to M. Sigal-Pauchard and A. Benchettah for their valuable comments.

References

1. K. Kime and A. Blaquière, From two Stochastic Optimal Control Problems to the Schrödinger Equation, in *Modeling and Control of Systems in Engineering, Quantum Mechanics, Economics and Biosciences.* Lecture Notes in Control and Information Sciences 121. Springer-Verlag Berlin Heidelberg 1989.

2. A. Blaquière, Girsanov Transformation and two Stochastic Optimal Control Problems. The Schrödinger System and Related Controllability Results, in *Modeling and Control of Systems in Engineering, Quantum Mechanics, Economics and Biosciences.* Lecture Notes in Control and Information Sciences 121. Springer-Verlag Berlin Heidelberg 1989.

3. L. de Broglie, Une tentative d'interprétation causale et nonlineaire de la mécanique ondulatoire, Gauthier-Villars, Paris, 1956.

4. E. Schrödinger, *Sitzungsbericht der Preußischen Akademie, Phys. Math.* Classe, 144 (1931); "Une analogie entre la mécanique ondulatoire et quelques problèmes de probabilités en physique classique", *Annales de l'Institut Henri Poincaré*, 11, 300 (1932).

5. S. Bernstein, "Sur les liaisons entre les grandeurs aléatoires", *in Verh. des intern. Mathematikerkongt.* Zürich, Band 1 (1932).

6. A. Beurling, "An Automorphism of Product Measures", *Annals of Mathematics*, Vol. 72, No. 1, July 1960.

7. B. Jamison, "Reciprocal Processes", *Z. Wahrscheinlichkeitstheorie ver. Gebiete* 30, 65 (1974).

8. A. Wakolbinger, A Simplified Variational Characterisation of Schrödinger Processes, *J. Math. Phys.*, 30, 2943, 1989.

9. D. Dawson, L. Gorostiza, and A. Wakolbinger, Shrödingér processes and large deviations, *J. Math. Phys.* 31 (10), October 1990.

10. H. Föllmer, "Random fields and diffusion processes", Ecole d'été de Saint Flour 1986, Lecture Notes in Mathematics, Springer-Verlag Berlin.

11. A.M. Il'in, A.S. Kalashnikov, and O.A. Oleinik, Linear Equations of the Second Order of Parabolic Type, in *Russian Mathematical Surveys*, edited by K.A. Hirsch, Vol. XVII, Macmillan and Co. Ltd, London, 1962.

12. W. H. Fleming and R. Rishel, Deterministic and Stochastic Optimal Control, Springer-Verlag, Berlin, 1975.

13. A. Blaquière, Sufficiency Conditions for Existence of an Optimal Feedback Control in Stochastic Mechanics, *Dynamics and Control* 1, 7-24, 1991.

14. W. H. Fleming and Sheunn-Jyi Sheu, Stochastic Variational Formula for Fundamental Solutions of Parabolic PDE, *Appl. Math. Optim.* 13 : 193-204, 1985.

15. R. S. Lipster and A. N. Shiryayev, Statistics of Random Processes I, General Theory, Springer-Verlag, Berlin Heidelberg, 1977.

16. D. W. Stroock and S. R. S. Varadhan, Multidimensional Diffusion Processes, Springer-Verlag, Berlin Heidelberg, 1979.

17. E. B. Dynkin, Théorie des processus markoviens, Dunod, Paris, 1963.

18. M. Nagasawa, "Transformations of Diffusion and Schrödinger Processes", Probab. Th. Rel. Fields 82, 109-136 (1989).

STABILITY ROBUSTNESS OF LINEAR FEEDBACK CONTROLLED FLEXIBLE MECHANICAL SYSTEMS†

M. Corless, D. Da
School of Aeronautics & Astronautics
Purdue University
West Lafayette, Indiana 47907
USA

ABSTRACT

We consider linear mechanical systems containing flexible components and subject to linear memoryless feedback controllers. In general, if a stabilizing controller is designed based on a rigidified model, i.e., a model in which some of the flexible components are assumed rigid, it is not true that this controller also stabilizes the 'real' flexible system. We present a condition which guarantees that a stabilizing controller whose design is based on a rigidified model also stabilizes the flexible system, provided that the components, whose flexibilities are neglected in the rigidified model, are sufficiently stiff. This condition involves the location of the rate sensors and is independent of the location of the displacement sensors. For a natural class of systems and controllers, the condition is automatically satisfied when the rate sensors are collocated with the actuators.

† Based on research supported by the U.S. National Science Foundation under grants MSM-87-06927 and MSS-90-57079.

1. INTRODUCTION

In designing a stabilizing controller for a flexible mechanical system one usually considers a rigidified model in which some of the flexible components are considered rigid. The natural question that arises is whether a controller, whose design is based on the rigidified model, also stabilizes the original system. A singular perturbation approach is one approach to answer this question for flexible mechanical systems whose 'neglected' elements are sufficiently stiff. By 'neglected' elements we mean those flexible elements which are considered rigid in the rigidified model. We use the term rigidified rather than rigid since this model may contain other flexible elements. Ficola *et al* [5] appear to be the first to use a singular perturbation approach to treat the control problem for elastic robots. After that, many papers have been published on utilizing singular perturbation theory to deal with flexible mechanical control systems; some of them are Kokotovic [7-9], Khorasani [6], Marino [10], Spong [12, 13], McClamroch [11] and Corless [2, 3].

A singularly perturbed system is one whose description depends on a scalar parameter $\mu > 0$; this dependency is such that setting $\mu = 0$ results in a system of lower order than that for $\mu > 0$. For small $\mu > 0$, it is customary to determine the qualitative behavior of a singularly perturbed system by studying two associated lower order subsystems, namely, the reduced order system ($\mu = 0$) and the boundary layer system. The latter subsystem is obtained by letting $\mu = 0$ in the 'fast time' scale. One method of controller design for such systems is to base the design on the reduced order model. To guarantee stability robustness with respect to the dynamics which were not taken into account in the controller design, it is customarily required that the boundary layer system be asymptotically stable.

In treating a flexible mechanical system as a singularly perturbed system, one lets $\mu = \kappa^{-1/2}$ where $\kappa > 0$ is a stiffness parameter associated with the 'neglected' elements; as κ increases the 'neglected' elements become stiffer. The reduced order system corresponds to the rigidified model of the system, i.e., the system model obtained by considering the 'neglected' elements rigid. To obtain a boundary layer system which is asymptotically stable one usually assumes the 'neglected' elements have damping coefficients which are proportional to $\kappa^{1/2}$. In this paper, we consider linear flexible mechanical systems in which the damping coefficients of the 'neglected' elements remain constant as κ increases. In this case, the boundary layer system is just marginally stable; i.e., stable, but not asymptotically stable.

We present a condition which guarantees that a stabilizing controller, whose design is based on a rigidified model also stabilizes the flexible system, provided that the 'neglected' components are sufficiently stiff. This condition involves the location of the rate sensors and is independent of the location of the displacement sensors. For a natural class of systems and controllers, the condition is automatically satisfied when the rate sensors are collocated with the actuators.

2. SYSTEM DESCRIPTION

For the class of flexible mechanical systems under consideration, we let $q(t) \in \mathbb{R}^N$ denote a vector of generalized coordinates which describe the configuration of the system at time t. We assume that, when some of the flexible elements (hereafter called the 'neglected' elements) of the system are modelled as rigid elements, this can be represented by linear constraints on the generalized coordinates, i.e., there exists a matrix $S \in \mathbb{R}^{L\times N}$ with

$$\operatorname{rank}(S) = L < N \qquad (2.1)$$

such that

$$Sq = 0 \,. \qquad (2.2)$$

Also we suppose that there are no other possible kinematical constraints on the system.

Then the systems under consideration are described by

$$M\ddot{q} + C\dot{q} + Kq = Wu + S^T\lambda \qquad (2.3a)$$

$$y_1 = D_1 q \qquad (2.3b)$$

$$y_2 = D_2 \dot{q} \qquad (2.3b)$$

where $u \in \mathbb{R}^m$ is a vector of control inputs; $y_1 \in \mathbb{R}^{P_1}$, $y_2 \in \mathbb{R}^{P_2}$ represent measurements which are based on displacement and rate sensors, respectively; $M \in \mathbb{R}^{N\times N}$ is the system mass matrix, hence it is symmetric and positive definite, i.e.,

$$M^T = M > 0 \,; \qquad (2.4)$$

the matrices $C, K \in \mathbb{R}^{N\times N}$ are such that $-C\dot{q} - Kq$ represents all the generalized forces acting on the system except the generalized forces due to the control inputs and the 'neglected' elements mentioned above; $W \in \mathbb{R}^{N\times m}$ is the influence matrix; the vector $S^T\lambda$ represents the generalized forces exerted by the 'neglected' elements; and $D_1 \in \mathbb{R}^{P_1\times N}$, $D_2 \in \mathbb{R}^{P_2\times N}$ are determined by the type and location of the displacement and rate sensors, respectively.

2.1. Rigidified Model

Consider first the situation in which the 'neglected' elements are assumed rigid. Then (2.2) holds. Condition (2.1) guarantees that there exists a matrix $U \in \mathbb{R}^{N\times(N-L)}$ such that

$$SU = 0 \tag{2.5}$$

and

$$\text{rank}\,(U) = N - L\,. \tag{2.6}$$

Define

$$V \triangleq M^{-1}S^{T}(SM^{-1}S^{T})^{-1/2}\,. \tag{2.7}$$

Then $V \in \mathbb{R}^{N\times L}$ and

$$\begin{bmatrix} U^T \\ V^T \end{bmatrix} M\,[U \;\; V] = \begin{bmatrix} \overline{M} & 0 \\ 0 & I \end{bmatrix}, \quad \overline{M} \triangleq U^T M U\,; \tag{2.8}$$

hence the square matrix [U V] is invertible and we can introduce a coordinate transformation defined by

$$q = U\phi + V\theta = [U \;\; V]\begin{bmatrix} \phi \\ \theta \end{bmatrix}\,. \tag{2.9}$$

Under the new coordinates system (2.2),(2.3) can be described by

$$\overline{M}\ddot{\phi} + \overline{C}\dot{\phi} + \overline{K}\phi = \overline{W}u \tag{2.10a}$$

$$\theta = 0 \tag{2.10b}$$

$$y_1 = \overline{D}_1\phi \tag{2.10c}$$

$$y_2 = \overline{D}_2\dot{\phi} \tag{2.10d}$$

where

$$\overline{M} \triangleq U^T M U \;\; ; \;\; \overline{C} \triangleq U^T C U \;\; ; \;\; \overline{K} \triangleq U^T K U\,;$$

$$\overline{W} \triangleq U^T W \;\; ; \;\; \overline{D}_1 \triangleq D_1 U \;\; ; \;\; \overline{D}_2 \triangleq D_2 U\,.$$

We call this the rigidified model.

2.2. Flexible Model

We suppose that the effect of the flexibilities of the 'neglected' elements can be represented by letting

$$\lambda = -\kappa K_0(Sq) - C_0(S\dot{q}) \tag{2.11}$$

where $K_0 \in \mathbb{R}^{L\times L}$ is symmetric and positive definite and $C_0 \in \mathbb{R}^{L\times L}$, i.e.,

$$K_0^T = K_0 > 0\,; \tag{2.12}$$

$\kappa \in (0, \infty)$ is a parameter characterizing the stiffness of the elements; see [1]. Clearly, the larger κ is, the stiffer these elements are. Thus the flexible model is given by

$$M\ddot{q} + (C + S^T C_0 S)\dot{q} + (K + \kappa S^T K_0 S)q = Wu \tag{2.13a}$$

$$y_1 = D_1 q \tag{2.13b}$$

$$y_2 = D_2 \dot{q} \tag{2.13c}$$

3. MAIN RESULTS

3.1. The Main Theorem

Suppose one designs a stabilizing controller

$$u = -F_1 y_1 - F_2 y_2 \tag{3.1}$$

for the rigidified model (2.10), where $F_1 \in \mathbb{R}^{m\times P_1}$ and $F_2 \in \mathbb{R}^{m\times P_2}$. Then the feedback controlled rigidified model is

$$\overline{M}\ddot{\phi} + (\overline{C} + \overline{W}F_2\overline{D}_2)\dot{\phi} + (\overline{K} + \overline{W}F_1\overline{D}_1)\phi = 0\,. \tag{3.2}$$

If one uses controller (3.1) on the flexible model (2.13), then the feedback controlled flexible model is given by

$$M\ddot{q} + (C + S^T C_0 S + WF_2 D_2)\dot{q} + (K + \kappa S^T K_0 S + WF_1 D_1)q = 0\,. \tag{3.3}$$

In the following we present an assumption which guarantees that any stabilizing controller (3.1) for the rigidified model (2.10) also stabilizes the flexible model (2.13), provided the stiffness parameter κ is sufficiently large.

Assumption 1. *The matrix* F_2, *given in (3.1), is such that the following inequality is satisfied:*

$$C_0 + P^T CP + P^T WF_2 D_2 P > 0 \tag{3.4}$$

where $P \in \mathbb{R}^{N \times L}$ *is given by*

$$P \triangleq M^{-1} S^T (SM^{-1} S^T)^{-1} . \tag{3.5}$$

Our main result is given by the following theorem.

Theorem. *Suppose the controller (3.1) satisfies Assumption 1 and the feedback controlled rigidified model (3.2) is asymptotically stable. Then there exists* $\kappa^* > 0$ *such that for all* $\kappa > \kappa^*$ *the feedback controlled flexible model (3.3) is asymptotically stable.*

Proof. We provide a proof in the next section.

3.2. Collocation of Rate Sensors and Actuators

If the rate sensors of the system are collocated with the actuators, we have

$$D_2 = W^T \tag{3.6}$$

and hence $F_2 \in \mathbb{R}^{m \times m}$ is a square matrix. We have the following corollary for this special case.

Corollary. *Suppose (3.6) holds, the damping terms satisfy* $C_0 + P^T CP > 0$, *and* $F_2 \geq 0$. *Then any stabilizing controller (3.1) for the rigidified model (2.10) also stabilizes the flexible model (2.13) provided* κ *is sufficiently large.*

4. PROOF OF THE THEOREM

A singular perturbation approach is used in the proof of the theorem. Before presenting this proof, we introduce some preliminary results on singularly perturbed systems.

4.1. Singularly Perturbed Systems

We consider singularly perturbed systems described by

$$\dot{x} = A_{11}(\mu)x + A_{12}(\mu)z \tag{4.1a}$$

$$\mu\dot{z} = A_{21}(\mu)x + A_{22}(\mu)z \tag{4.1b}$$

where $t \in \mathbb{R}$ is the time variable; $x(t) \in \mathbb{C}^n$ and $z(t) \in \mathbb{C}^m$ describe the state of the system; $A_{11}(\cdot)$, $A_{12}(\cdot)$, $A_{21}(\cdot)$, and $A_{22}(\cdot)$ are matrix functions which are differentiable at zero; and $\mu \in (0,\infty)$ is the singular perturbation parameter.

A reduced order system associated with (4.1) is obtained by letting $\mu = 0$ in (4.1). Hence a reduced order system can be described by

$$\dot{x} = A_{11}(0)x + A_{12}(0)z \tag{4.2a}$$

$$0 = A_{21}(0)x + A_{22}(0)z\,. \tag{4.2b}$$

In order to obtain a unique reduced order system for (4.1), we make the following assumption.

Assumption S1. $A_{22}(0)$ *is nonsingular.*

The unique reduced order system guaranteed by the above assumption is given by

$$\dot{x} = \bar{A}x \tag{4.3}$$

$$z = Hx \tag{4.4}$$

where

$$\bar{A} \triangleq A_{11}(0) - A_{12}(0)A_{22}(0)^{-1}A_{21}(0) \tag{4.5}$$

$$H \triangleq -A_{22}(0)^{-1}A_{21}(0). \tag{4.6}$$

We make the following assumption on the reduced order system (4.3).

Assumption S2. *The reduced order system (4.3) is asymptotically stable.*

In the following we present a condition on a matrix pair (M, N). Consider any matrix pair (M, N) where $M, N \in \mathbb{C}^{m\times m}$ and M is stable.† Then there exists a nonsingular matrix $T \in \mathbb{C}^{m\times m}$ such that

$$T^{-1}MT = \text{blockdiag}\{M_0, j\omega_1 I_1, \cdots, j\omega_l I_l\} \tag{4.7}$$

where $j \triangleq \sqrt{-1}$; $M_0 \in \mathbb{C}^{m_0 \times m_0}$ is asymptotically stable‡; $I_i \in \mathbb{R}^{m_i \times m_i}$, $i = 1, 2, \ldots, l$, are identity matrices; $j\omega_i$, $i = 1,2,\ldots,l$, are the imaginary eigenvalues of M with $\omega_i \neq \omega_j$ for $i \neq j$; and $m_0 \geq 0$, $m_i > 0$, $i = 1, 2, \ldots, l$, are integers with $\sum_{i=0}^{l} m_i = m$. Let

$$\begin{bmatrix} * & * & * & * \\ * & N_1 & * & * \\ * & * & \ddots & * \\ * & * & * & N_l \end{bmatrix} = T^{-1}NT \tag{4.8}$$

where $N_i \in \mathbb{C}^{m_i \times m_i}$, $i = 1, 2, \ldots, l$. The following definition describes a condition on the matrix pair (M, N).

Definition 1. The matrix pair (M, N) satisfies *Condition C1* iff

(i) M is stable;

(ii) each N_i, $i = 1, 2, \ldots, l$, given by (4.7) and (4.8) is asymptotically stable.

We define

$$\bar{A}_{22} \triangleq A'_{22}(0) + A_{22}(0)^{-1}A_{21}(0)A_{12}(0) , \tag{4.9}$$

where $A'_{22} \triangleq \dfrac{dA_{22}}{d\mu}$ and make the following assumption.

---

† We define a square matrix M to be stable if the system described by $\dot{x} = Mx$ is stable. This is equivalent to the requirement that all the eigenvalues of M be in the closed left half of the complex plane and those eigenvalues which are on the imaginary axis have the same algebraic and geometric multiplicities.

‡ We define a square matrix M to be asymptotically stable if the system described by $\dot{x} = Mx$ is asymptotically stable. This is equivalent to the requirement that all the eigenvalues of M be in the open left half of the complex plane.

Assumption S3. *Condition C1 is satisfied by the matrix pair* $(A_{22}(0), \bar{A}_{22})$.

The following lemma yields a stability result for system (4.1).

Lemma 1. (Corollary 4.1 in [4]) *Consider a singularly perturbed system described by (4.1). Suppose Assumptions S1-S3 are satisfied. Then there exists* $\mu^* > 0$ *such that system (4.3) is asymptotically stable for all* $\mu \in (0, \mu^*)$.

If a matrix pair (M, N) has the following structure

$$M = \begin{bmatrix} 0 & I \\ -K & 0 \end{bmatrix}; \qquad N = \begin{bmatrix} 0 & 0 \\ 0 & C \end{bmatrix}, \tag{4.10}$$

where $K, C \in \mathbb{R}^{n \times n}$ and $I \in \mathbb{R}^{n \times n}$ is an identity matrix, then the following lemma provides a convenient way of checking the satisfaction of Condition C1 by (M, N).

Lemma 2. *Condition C1 is satisfied by (M, N) given in (4.10) iff*

(i) K is diagonalizable with positive real eigenvalues and

(ii) Condition C1 is satisfied by the matrix pair (jK, C).

Proof. The Appendix contains a proof.

4.2. Proof of the Theorem.

The feedback controlled flexible mechanical system (3.3) can be represented by a singularly perturbed system by first introducing the coordinate transformation (2.9) which yields

$$\overline{M}\ddot{\phi} = G_{11}\dot{\phi} + G_{12}\dot{\theta} + H_{11}\phi + H_{12}\theta \tag{4.11a}$$

$$\ddot{\theta} = G_{21}\dot{\phi} + G_{22}\dot{\theta} + H_{21}\phi + H_{22}\theta - \kappa\hat{K}\theta \tag{4.11b}$$

where

$$\begin{bmatrix} G_{11} & G_{12} \\ G_{21} & G_{22} \end{bmatrix} \triangleq - \begin{bmatrix} U^T \\ V^T \end{bmatrix} (C + S^T C_0 S + WF_2D_2)[U \;\; V] \tag{4.11c}$$

$$\begin{bmatrix} H_{11} & H_{12} \\ H_{21} & H_{22} \end{bmatrix} \triangleq - \begin{bmatrix} U^T \\ V^T \end{bmatrix} (K + WF_1D_1)[U \;\; V] \tag{4.11d}$$

$$\hat{K} \triangleq V^T S^T K_0 S V = (SM^{-1}S^T)^{½} K_0 (SM^{-1}S^T)^{½T} . \tag{4.11e}$$

Now let $\mu \triangleq \kappa^{-½}$, and introduce the states

$$x \triangleq \begin{bmatrix} \phi \\ \dot{\phi} \end{bmatrix} , \quad z \triangleq \begin{bmatrix} \mu^{-2}\theta \\ \mu^{-1}\dot{\theta} \end{bmatrix} . \tag{4.12}$$

Then, system (3.3) can be represented by the singularly perturbed system (4.1) with

$$A_{11}(\mu) = \begin{bmatrix} 0 & I \\ \overline{M}^{-1}H_{11} & \overline{M}^{-1}G_{11} \end{bmatrix}, \quad A_{12}(\mu) = \begin{bmatrix} 0 & 0 \\ \mu^2\overline{M}^{-1}H_{12} & \mu\overline{M}^{-1}G_{12} \end{bmatrix} \tag{4.13a}$$

$$A_{21}(\mu) = \begin{bmatrix} 0 & 0 \\ H_{21} & G_{21} \end{bmatrix}, \qquad A_{22}(\mu) = \begin{bmatrix} 0 & I \\ \mu^2H_{22} - \hat{K} & \mu G_{22} \end{bmatrix} . \tag{4.13b}$$

Hence the theorem is proven if the system (4.1), (4.13) is asymptotically stable for all μ sufficiently small; this is demonstrated by using Lemmas 1, 2 as follows.

Since

$$A_{22}(0) = \begin{bmatrix} 0 & I \\ -\hat{K} & 0 \end{bmatrix} \tag{4.14}$$

with $\hat{K}$ positive definite, Assumption S1 is satisfied.

Since $A_{12}(0) = 0$, the unique reduced order system for (4.1), (4.13) is given by

$$\dot{x} = A_{11}(0)x \tag{4.15}$$

which is equivalent to the rigidified model (3.2). Hence the asymptotic stability of (3.2) guarantees that Assumption S2 is satisfied.

By considering Lemma 1, system (4.1), (4.13) is asymptotically stable for all μ sufficiently small if Condition C1 is satisfied by the matrix pair $(A_{22}(0), \bar{A}_{22})$. Since $A_{22}(0)$ has the structure (4.14) with $\hat{K}$ positive definite and $\bar{A}_{22}$ is given by

$$\bar{A}_{22} \triangleq A'_{22}(0) + A_{22}(0)^{-1}A_{21}(0)A_{12}(0) = A'_{22}(0) = \begin{bmatrix} 0 & 0 \\ 0 & G_{22} \end{bmatrix}, \tag{4.16}$$

it follows from Lemma 2 that Condition C1 is satisfied by $(A_{22}(0), \bar{A}_{22})$ iff it is satisfied by $(j\hat{K}, G_{22})$. We now prove that $(j\hat{K}, G_{22})$ satisfies Condition C1.

Since $\hat{K}$ is positive definite and symmetric, there exists a unitary matrix $T \in \mathbb{R}^{n \times n}$ such that

$$T^T\hat{K}T = \text{diag}\{k_1, k_2, \ldots, k_n\} \tag{4.17}$$

where k_i, i=1,...,n are the positive real eigenvalues of $\hat{K}$. Since any diagonal block sub-matrix of a positive definite matrix is positive definite and all the eigenvalues of a positive definite matrix have positive real parts, we conclude that Condition C1 is satisfied by $(j\hat{K}, G_{22})$ if

$$-T^TG_{22}T > 0 \tag{4.18a}$$

i.e.,

$$T^TV^T(C + WF_2D_2 + S^TC_0S)VT > 0 \ . \tag{4.18b}$$

Considering (2.7), (4.18b) is equivalent to

$$T^T(SM^{-1}S^T)^{1/2}[C_0 + P^T(C + WF_2D_2)P](SM^{-1}S^T)^{1/2T}T > 0 \tag{4.19}$$

where P is given by (3.5). Thus (3.4) guarantees that Condition C1 is satisfied by $(j\hat{K}, G_{22})$.

5. APPENDIX: PROOF OF LEMMA 2.

Necessity. Suppose Condition C1 is satisfied by a matrix pair (M, N) which has the structure given in (4.10). Then by the definition of Condition C1, the system

$$\dot{x} = Mx \tag{A.1}$$

is marginally stable. This is equivalent to marginal stability of the system

$$\ddot{\xi} + K\xi = 0 \ . \tag{A.2}$$

Let $T \in \mathbb{C}^{n\times n}$ be a nonsingular matrix such that

$$T^{-1}KT = J \tag{A.3}$$

where J is in Jordan form and its diagonal elements are the eigenvalues of K. Introducing the state transformation

$$y = T^{-1}\xi \tag{A.4}$$

to system (A.2) yields

$$\ddot{y} + Jy = 0 \ . \tag{A.5}$$

The stability of (A.2) implies the stability of (A.5); this implies that J is a diagonal matrix with positive real diagonal terms. Hence (i) is proven. Also jK is stable.

We let

$$J = \Lambda^2 \tag{A.6}$$

where

$$\Lambda \triangleq \text{blockdiag}\,\{\omega_1 I_1,\ \omega_2 I_2,\ \ldots,\ \omega_l I_l\} \tag{A.7}$$

$I_i \in \mathbb{R}^{n_i \times n_i}$, $i = 1,\ldots,l$ are identity matrices, and

$$\hat{T} \triangleq \begin{bmatrix} T & T \\ jT\Lambda & -jT\Lambda \end{bmatrix}. \tag{A.8}$$

Then it can be readily seen that

$$\hat{T}^{-1} M \hat{T} = \text{blockdiag}\{\ j\Lambda,\ -j\Lambda\} \tag{A.9}$$

and

$$\hat{T}^{-1} N \hat{T} = ½ \begin{bmatrix} \Lambda^{-1}(T^{-1}CT)\Lambda & * \\ * & \Lambda^{-1}(T^{-1}CT)\Lambda \end{bmatrix}. \tag{A.10}$$

Suppose

$$\begin{bmatrix} C_1 & * & * & * \\ * & C_2 & * & * \\ * & * & \ddots & * \\ * & * & * & C_l \end{bmatrix} = T^{-1}CT \tag{A.11}$$

where C_i, $i = 1, \ldots, l \in \mathbb{C}^{n_i \times n_i}$. Then we also have

$$\Lambda^{-1}(T^{-1}CT)\Lambda = \begin{bmatrix} C_1 & * & * & * \\ * & C_2 & * & * \\ * & * & \ddots & * \\ * & * & * & C_l \end{bmatrix}. \tag{A.12}$$

Hence satisfaction of Condition C1 by (M, N) implies that C_i, $i = 1, \ldots, l$, are asymptotically stable; this guarantees that Condition C1 is also satisfied by (jK, C).

Sufficiency. Suppose K is diagonalizable with positive real eigenvalues and Condition C1 is satisfied by (jK, C). Then we can define T and $\hat{T}$ as in (A.3), (A.6)-(A.8); hence (A.9) and (A.10) hold. (A.9) implies M is stable. (A.10)-(A.12) then imply that (M, N) satisfies Condition C1.

REFERENCES

1. M. Corless, Modelling 'flexible constraints' in mechanical systems, *Twentieth Midwestern Mechanics Conference"*, 1987.

2. M. Corless, Stability robustness of linear feedback controlled mechanical systems in the presence of unmodelled flexibilities, *Proceedings of the Conference on Decision and Control*, 1988.

3. M. Corless, Controllers which guarantee robustness with respect to unmodelled flexibilities for a class of uncertain mechanical systems, *International Journal of Adaptive Control and Signal Processing* Vol. **4** pp. 565-579, 1990.

4. D. Da and M. Corless, Lyapunov functions for linear singularly perturbed systems with marginally stable boundary layer systems, *Proceedings of the American Control Conference*, 1991.

5. A. Ficola, R. Marino, and S. Nicosia, A singular perturbation approach to the dynamic control of elastic robots, *Proceedings of the 21st Annual Allerton Conference on Communication, Control, and Computing*, 1983.

6. K. Khorasani, A slow manifold approach to linear equivalents of nonlinear singularly perturbed systems, *Automatica* Vol. **25** pp. 301-306, 1989.

7. P. V. Kokotovic, Applications of Singular Perturbation Techniques to Control Problems, *SIAM Review* Vol. **26** pp. 396-416 , 1984.

8. P.V. Kokotovic and H.K. Khalil, *Singular Perturbations in Systems and Control,* IEEE Press, New York , 1986.

9. P.V. Kokotovic, H.K. Khalil, and J. O'Reilly, *Singular Perturbation Methods in Control: Analysis and Design,* Academic Press , 1986.

10. R. Marino and S. Nicosia, On the feedback control of industrial robots with elastic joints: a singular perturbation approach, *University of Rome* Vol. **R-84.01** 1984.

11. N. H. McClamroch, A singular perturbation approach to modeling and control of manipulators constrained by a stiff environment, *IEEE Conference on Decision and Control,* 1989.

12. M. W. Spong, Modeling and control of elastic joint robots, *Journal of Dynamic Systems, Measurement and Control* Vol. **109** pp. 310-319, 1987.

13. M. W. Spong, K. Khorasani, and P. V. Kokotovic, An integral manifold approach to the feedback control of flexible joint robots, *IEEE Journal of Robotics and Automation* Vol. **RA-3** pp. 291-300, 1987.

PREDICTIVE CONTROLLER AND ESTIMATOR FOR NASA DEEP SPACE NETWORK ANTENNAS

W. Gawronski
Jet Propulsion Laboratory, California Institute of Technology, Pasadena, California, U.S.A.

1. Introduction

The recent pointing requirements for the X-band (8.4-GHz) frequency Deep Space Network antennas, as well as the expectation of the future K-band (32-GHz) capability, dictate a need for high-performance controllers for the antenna azimuth and elevation drives. This paper presents a new design procedure for a predictive controller that significantly improves antenna tracking performance. The predictive controller uses future values of the stored input command to generate the control signal. For antennas tracking stars or spacecraft, these values are known in advance, hence the predictive control scheme is easily implemented in this case.

The tracking control problem is a non-trivial extension of the regulator problem, widely investigated in the control literature. Different solutions of the tracking problem have been presented, see Kwakernaak and Sivan [1], Anderson and Moore [2], Astrom [3], Goodwin and Sin [4], Lewis [5], Albertos and Ortega [6], Clarke and Mohtadi [7], Clarke *et al.* [8], and Ortega and Galindo [9]. Predictive controllers are analyzed in many papers, e.g. Goodwin and Sin [4], Albertos and Ortega [6], Clarke and Mohtadi [7], Mohtadi *et al.* [8], Ortega and Galindo [9], Reid *et al.* [10], Xi [11], where CARIMA (Controlled Auto-Regressive and Integrated Moving-Average) models are used for system description. Although the CARIMA approach is widely used for predictive controller design, the state space description serves as a standard tool for system analysis and design. Papers of Albertos and Ortega [6], Clarke and

Mohtadi [7] and Xi [11] introduce interpretations of CARIMA modeling in state space. This paper presents a state-space predictive controller design. This tool standardization provides a new insight into system properties, improves system design, and simplifies analysis.

Also presented below are new input reference schemes and weighting procedures. Various input reference horizons are used to determine the control signal increment, leaving the choice up to the designer, thus improving prediction accuracy. Introduction of a weighting matrix configuration which includes a forgetting factor significantly improves system performance.

The predictive control law is applicable to plants with all state variables available for measurement. This is not a typical situation, although for an observable system the linear quadratic (LQ) estimator estimates states from the plant output. The LQ estimator is too slow, however, for predictive control purposes, thus a predictive estimator is developed in this paper, which speeds up the estimation process, enhancing system performance.

The predictive controller is designed for tracking control of the NASA/JPL 70-m aperture antenna, known as the DSS-14 antenna. On-axis- (or servo-) tracking is considered, where the output is taken on the encoder, or tachometer. Simulation results show significant improvement in tracking performance over the LQ controller and estimator presently in use, while the performance robustness to the parameter variations and the disturbance suppression properties are found to be fairly good.

2.Output Prediction for a Linear System

A plant model with nu inputs, ny outputs, and n states is considered.

$$x(i+1)=Ax(i)+Bu(i), \qquad y(i)=Cx(i)+Du(i), \quad x(0)=0, \tag{1}$$

where $x \in R^{n}$, $u \in R^{nu}$, $y \in R^{ny}$. The task is to predict the output for NY steps ahead, given the input for NU steps ahead. The integer NY is the length of the output horizon, and NU is the length of the input horizon, and for casual systems $NU \leq NY$. In this case, input and output sequences (customarily termed horizons) rather then input and output values are used in the predictive control analysis and design. In this paper, three types of input sequences are defined. First, the input horizon $U(i)$, consisting of inputs up to $NU-1$ steps ahead

$$U^{T}(i)=[u_{i}^{T}(0),\ u_{i}^{T}(1),\dots,\ u_{i}^{T}(NU-1)], \tag{2a}$$

where $u_i(k)$ is the predicted input at instant i with k steps ahead. The preceding input horizon $U(i-1)$ is a horizon predicted at a previous time instant. Note that it is not a delayed prediction at instant i, i.e. $u_{i-1}(k)\neq u_i(k-1)$. The predicted input horizon is not known in advance, although for a closed-loop system it is not an arbitrary one. Hence the predicted input horizon can be decomposed into a sum of a known input $U_r(i)$, called the reference horizon, and the increment $\Delta U(i)$

$$U(i)=U_r(i)+\Delta U(i) \tag{2b}$$

In this paper, the reference input horizon $U_r(i)$ is chosen to be identical to the preceding input horizon $U(i-1)$ for the first NR time instants, and constant for the remaining $NU-NR$ instants

$$u_r(k)=\begin{cases} u_{i-1}(k) & \text{for } k=1,\dots,NR \\ u_{i-1}(NR) & \text{for } k=NR+1,\dots,NU \end{cases} \tag{3a}$$

where the integer $NR\leq NU$ is the length of the reference horizon. Thus

$$U_r(i)=EU(i-1), \qquad E=\begin{bmatrix} I_{N1} & 0 & 0 \\ 0 & J & 0 \end{bmatrix}, \tag{3b}$$

where $J=[I_{nu}\ \ I_{nu}\ \dots\ I_{nu}]^{T}$ (I_{nu} repeated $NU-NR+1$ times), I_{N1} is the identity matrix of dimension $N1$, and $N1=(NR-1)\times nu$.

The input increment horizon $\Delta U(i)$ is defined with respect to the reference horizon $U_r(i)$, c.f. (2b), hence

$$\Delta U(i)=U(i)-U_r(i)=U(i)-EU(i-1) \tag{4}$$

The sequences $U(i)$, $U(i-1)$, $U_r(i)$, and $\Delta U(i)$ are shown in Fig.1.

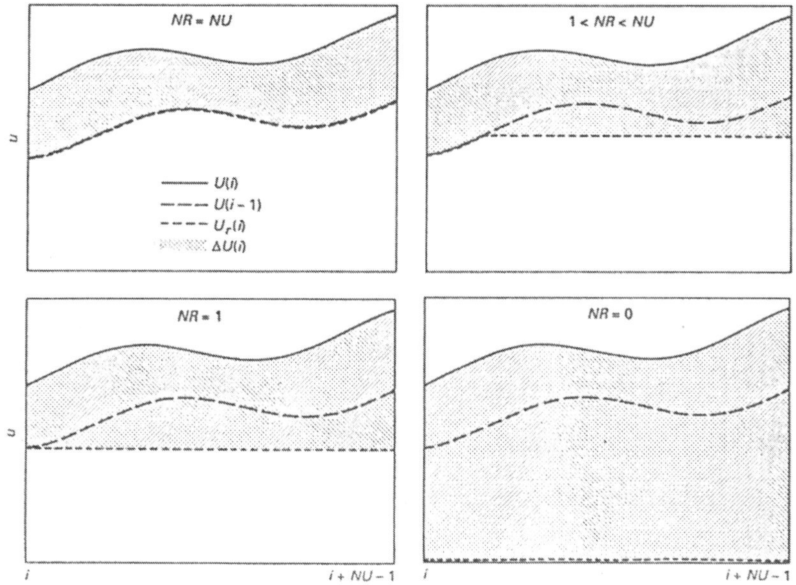

Fig.1. Input horizons.

Two output sequences are introduced: the output horizon Y

$$Y(i)^T=[y_i^T(1),\ y_i^T(2),\ldots,\ y_i^T(NY)], \tag{5a}$$

and the predicted output horizon $\bar{Y}$

$$\bar{Y}^T(i)=[\bar{y}_i^T(1),\ \bar{y}_i^T(2),\ldots,\ \bar{y}_i^T(NY)], \tag{5b}$$

The latter is an output of the system with the reference horizon U_r as an input. The components $y_i(k)$, $\bar{y}_i(k)$ are the output and predicted output, respectively, at instant i with k steps ahead. Note that although the system output at instant i with k steps ahead is equal to the output at instant $i+l$ with k-l steps ahead: $y_i(k)=y_{i+1}(k-l)=y(i+l)$, $l<k$, the same is not true for the predicted output.

The output horizon is obtained from the plant model (1), for $k=1,\dots,NY$

$$y(i+k)=CA^{k}x(i)+CA^{k-1}Bu(i)+\dots+CBu(i+k-1) \tag{6a}$$

The predicted output $\bar{y}_i(k)$ is defined as a system response to the reference horizon input $U_r(i)$. Thus for $k=1,\dots,NY$

$$\bar{y}_i(k)=CA^{k}x(i)+CA^{k-1}Bu_r(1)+\dots+CBu_r(k), \tag{7}$$

and (6a) is now

$$y(i+k)=\bar{y}(k)+CA^{k-1}B\Delta u_i(0)+\dots+(CA^{k-NR-1}B+\dots+CB)\Delta u_i(NR-1) \tag{6b}$$

Denoting the Markovian matrix

$$G=\begin{bmatrix} g_1 & 0 & \dots 0 \\ g_2 & g_1 & \dots 0 \\ \vdots & \vdots & \vdots \\ g_{NU} & g_{NU-1} & \dots g_1 \\ \vdots & \vdots & \vdots \\ g_{NY} & g_{NY-1} & \dots g_{NY-NU+1} \end{bmatrix} \tag{8}$$

where $g_i=CA^{i-1}B$ is the *i-th* Markov parameter, the output horizon from (6b) is

$$Y(i)=\bar{Y}(i)+G\Delta U(i) \tag{9}$$

The predicted output horizon is determined from (7)

$$\bar{Y}(i)=Hx(i)+GU_r(i)=Hx(i)+GEU(i-1)=Hx(i)+FU(i-1) \tag{10a}$$

where

$$F=GE, \qquad H=\mathcal{O}_{NY-1}A, \tag{10b}$$

E given by (3c), and $\mathcal{O}_{NY-1}=[C^T\ (CA)^T\ \dots\ (CA^{NY-1})^T]^T$.

The input increment horizon depends on the length of the reference horizon. In particular, for $NR=1$ one obtains $\Delta u_i(k)=u_i(k)-u_{i-1}(1)$, $k=1,\dots,NU$, and $F^T=[g_1^T, g_1^T+g_2^T, \ldots, g_1^T+..+g_{NY}^T]$, the case of the generalized predictive control of Clarke and Mohtadi (1989), where the control increments are defined with respect to the last input command. For $NR=NU$ one obtains $E=I$, $F=G$, and $\Delta U(i)=U(i)-U(i-1)$, where the control increment is defined with respect to the preceding control over the whole length of the input horizon. For the input increment determined with respect to the zero reference input, the input increment horizon is equal to the input horizon, $\Delta U(i)=U(i)$, one obtains $F=0$, and this case is denoted with $NR=0$. In this case the output is predicted from the system state only, otherwise predicted from the state and the system input.

3. Predictive Control

The basic task for the predictive controller is to assure that the future output Y will closely follow the input command Y_o

$$Y_o^T(i)=[y_{oi}^T(1),\ y_{oi}^T(2),\ldots,\ y_{oi}^T(NY)]. \tag{11a}$$

where $y_{oi}(k)$ is the command signal at instant i with k steps ahead. The tracking error is measured with respect to the input command, thus

$$\varepsilon(i)=Y_o(i)-Y(i) \tag{11b}$$

where $\varepsilon^T(i)=[\varepsilon_i^T(1),\ \varepsilon_i^T(2),\ldots,\varepsilon_i^T(NY)]$, and $\varepsilon_i(k)$ is the error at instant i with k steps ahead. The task is to minimize the tracking error while the input remains bounded, i.e. to minimize the performance index J

$$J=tr[\varepsilon^T(i)Q\varepsilon(i)+\Delta U^T(i)R\Delta U(i)] \tag{12}$$

where $tr(.)$ denotes trace of a square matrix, Q, R are symmetric positive definite matrices; Q is the tracking error weighting matrix, while R is the control effort weighting matrix. The necessary condition for the extreme point to exist is that $\partial J/\partial\Delta U=0$. From (12) and (9) it yields that

$$\Delta U(i)=K(Y_o(i)-\bar{Y}(i))=K\bar{\varepsilon}(i) \tag{13a}$$

where

$$K=(G^TQG+R)^{-1}G^TQ \tag{13b}$$

and $\bar{\varepsilon}(i)=Y_o(i)-\bar{Y}(i)$ is the predicted output error.

The resulting control increment $\Delta U(i)$ covers the whole input horizon; for the control purposes, however, only the first component (the current control increment) is used. Let k denote the first nu rows of K,

$$k= eK, \qquad e=[I_{nu} \; 0 \; 0 \; \dots \; 0] \tag{14}$$

then the control increment at instant i is

$$\Delta u(i)=k(Y_o(i)-\bar{Y}(i))=k\bar{\varepsilon}(i) \tag{15}$$

and the control input, from (4), is $u(i)=u(i-1)+\Delta u(i)$. Combining it with (15), and (10) one obtains the control command at moment i

$$u(i)=u(i-1)+k(Y_o(i)-\bar{Y}(i))=u(i-1)+kY_o(i)-kHx(i)-kFU(i-1)$$

and with $u(i-1)=eU(i-1)$, the above equation yields

$$u(i)=kY_o(i)-kHx(i)+(e-kF)U(i-1) \tag{16}$$

Thus, the command $u(i)$ depends on the preceding input horizon $U(i-1)$, on the actual state $x(i)$, and on the control command $Y_o(i)$ up to NY steps ahead.

The closed-loop system equations are obtained by combining the plant equation (1) with the controller equation (16). Introducing a new state variable $U_o(i)$, such that $U_o(i+1)=U(i)$, one obtains

$$x(i+1)=(A-BkH)x(i)+B(e-kF)U_o(i)+BkY_o(i)$$

$$U_o(i+1)=-KHx(i)+(I_N-KF)U_o(i)+KY_o(i) \tag{17}$$

$$y(i)=Cx(i)$$

and $N=NU\times nu$. With the new state variable $z^T=[x^T,\ U_o^T]$, the closed-loop equations are as follows

$$z(i+1)=A_c z(i)+B_c Y_o(i), \qquad y(i)=C_c z(i) \tag{18a}$$

where

$$A_c=\begin{bmatrix} A\text{-}BkH & B(e\text{-}kF) \\ \text{-}KH & I_N\text{-}KF \end{bmatrix}=\begin{bmatrix} (I_n\text{-}Bk\,\mathcal{O}_{NY\text{-}1})A & B(e\text{-}kF) \\ \text{-}KH & I_N\text{-}KF \end{bmatrix}, \tag{18b}$$

$$B_c^T=[Bk\quad K]^T, \quad C_c=[C\quad 0] \tag{18c}$$

One can see that the control command $u(i)$ is now

$$u(i)=kY_o(i)+[\text{-}kH\quad e\text{-}kF]z(i), \tag{19}$$

and is fully recovered from the input command and the current state of the system.

The block diagram of the closed-loop system (18) is presented in Fig.2. The system consists of the plant, the predictor (PRD), the controller (CO), and the command horizon generator (CHG). The predictor structure is shown in Fig.3a. The command horizon generator generates the input horizon $Y_o(i)$ from the input command $y_o(i)$ at each instant i. Its structure is shown in Fig.3b, and the related state-space representation is as follows

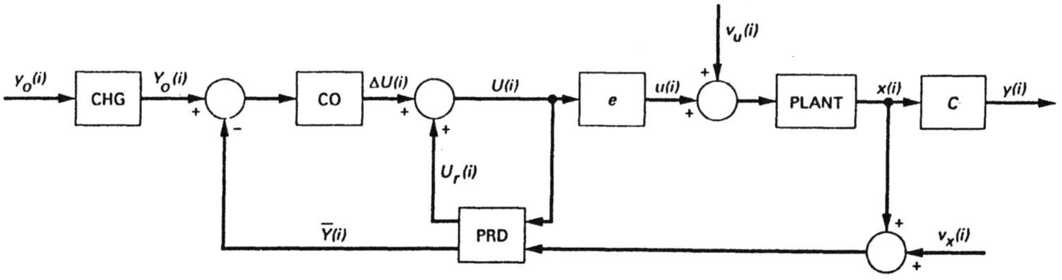

Fig.2. Predictive control system.

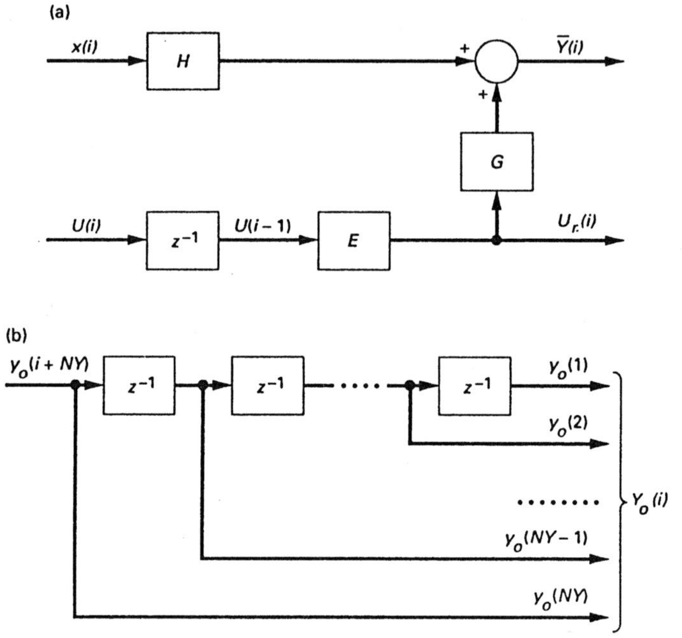

Fig.3. Predictor (a), and command horizon generator (b).

$$A_{rh}=\begin{bmatrix}0 & I & 0 \dots 0\\ 0 & 0 & I \dots 0\\ \vdots & \vdots & \vdots\\ 0 & 0 & 0 \dots I\\ 0 & 0 & 0 \dots 0\end{bmatrix},\quad B_{rh}=\begin{bmatrix}0\\0\\ \vdots\\0\\I\end{bmatrix},\quad C_{rh}=\begin{bmatrix}I & 0 \dots 0\\ 0 & I \dots 0\\ \vdots & \vdots\\ 0 & 0 \dots I\\ 0 & 0 \dots 0\end{bmatrix},\quad D_{rh}=\begin{bmatrix}0\\0\\ \vdots\\0\\I\end{bmatrix} \tag{20}$$

where I is an identity matrix of order ny, and A_{rh}, B_{rh} have NY-1 rows, while C_{rh}, D_{rh} have NY columns.

The weighting matrices R, Q are tuning parameters in the optimal design, see for example Anderson and Moore [2], or Maciejowski [12]. They are adjusted until satisfactory results are obtained. Although they are not "active" in the search for the optimal solution, their choice significantly influences the performance and stability of the system. Although a general procedure for a reasonable choice of the weighting matrices is not yet known, a simplified procedure presented here significantly improves system performance. A diagonal matrix $R=\rho I$ has been chosen as an input weighting matrix, where $\rho>0$ is a positive scalar, while the output error weighting matrix Q has the following structure

$$Q=diag(q,\ \alpha q,\ \alpha^2 q,\dots,\ \alpha^{NY-1}q). \tag{21}$$

Recall that $\varepsilon(i)=[\varepsilon_i^T(1),\ \varepsilon_i^T(2),\dots,\ \varepsilon_i^T(NY)]^T$, hence the diagonal component $\alpha^{k-1}q$ is the weight of $\varepsilon_i(k)$, the kth component of the error $\varepsilon(i)$. The scalar α is a forgetting factor, and α^{k-1} is a weight of the output error at the $i+k$th time instant. The most recent output is given a unit weight, and the future outputs are penalized (in fact - awarded, as it is shown later) exponentially. With this arrangement, the choice of R and Q reduces to the choice of parameters ρ, α, and q.

A random simulation approach is used for evaluation of performance robustness of the closed-loop system. A statistical approach to system robustness testing was used by Ray and Stengel [13]. In our approach the perturbed plant matrix A_{dp} is obtained by adding a random increment ΔA_d to the nominal matrix A_d, thus $A_{dp}=A_d+\Delta A_d$. The perturbation is characterized by a normal random variable $\theta=\|A_{dp}\|_2$, with the mean value $\theta_o=\|A_d\|_2$, and standard deviation $\sigma_\theta=(E(\theta-\theta_o)^2)^{1/2}$. The system performance is characterized by the normalized tracking error $\varepsilon=\|y-y_o\|_2/\|y_o\|_2$, a ratio of norm of the tracking error to the norm of the tracking command. The tracking error is simulated for N perturbation samples. From simulations, the histogram, and the probability distribution function $F(\varepsilon_r)=prob(\varepsilon<\varepsilon_r)$ are obtained. The distribution of the tracking error is a statistical measure of performance robustness to modeling errors, and it is a base for determination of the probability confidence for robustness requirements.

Next the system disturbances are discussed. Two sources of disturbances are considered: measurement noise $v_y(i)$ and input disturbances $v_u(i)$ (see Fig.2). The input disturbances are represented in the closed-loop system by the triple (A_c,B_y,C_c), and the measurement noise by the triple (A_c,B_u,C_c), where

$$B_y=\begin{bmatrix} A\text{-}BkH \\ \text{-}KH \end{bmatrix},\quad B_u=\begin{bmatrix} B(e-kF)e_v \\ (I\text{-}KF)e_v \end{bmatrix},$$

and $e_v^T=[I_{nu}\ \ 0]$. For high-frequency disturbances, the disturbance rejection properties of the system significantly improve when a lowpass filter, as in Fig.4a, is applied. In this case, the plant states obtained from the plant model (PM) are subtracted from the measured states. The resulting signal passes through a lowpass filter (LPF), and is added to the states previously removed. The block diagram of the filter is shown in Fig.4b.

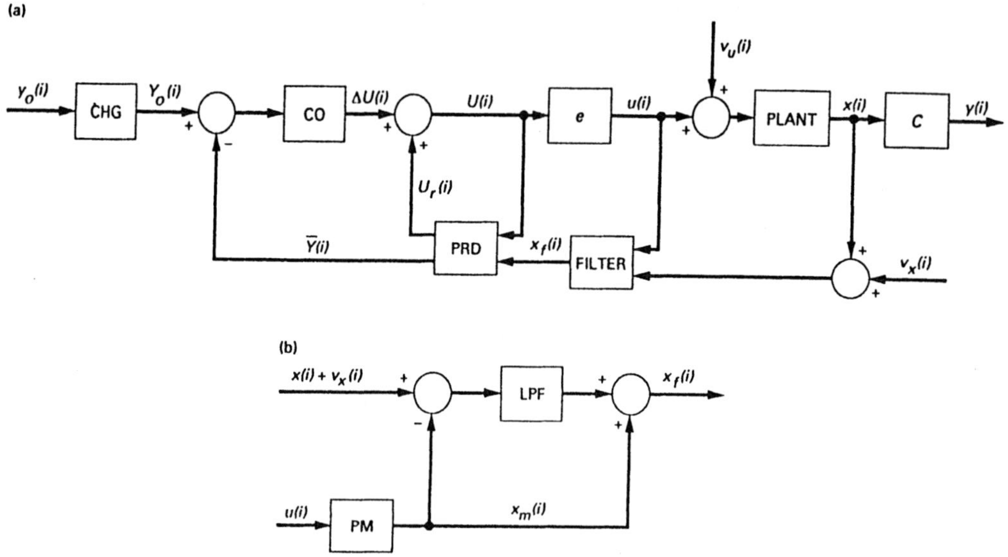

Fig.4. Predictive control system with a measurement noise filter (a), and the filter (b).

4. Predictive Estimation

The implementation of the predictive controller depends on the availability of the plant states for measurement. In most cases they are not available, and a state estimator has to be implemented. Linear quadratic estimators are too slow for predictive control systems, and a predictive scheme is utilized in the design of the estimator. Thus, a new estimator with dynamics comparable to the predictive controller is developed.

Denote $y_f(i)$ as the free motion output of the plant (its output for zero input), and $Y_f(i)$ as the free output horizon

$$Y_f(i)=[y_f(i)\ y_f(i+1),\dots,\ y_f(i+NY)]$$

and

$$\hat{Y}_f(i)=[\hat{y}_f(i)\ \hat{y}_f(i+1),\dots,\ \hat{y}_f(i+NY)],$$

the estimated free motion horizon. The estimate $\hat{x}(i)$ of the plant state $x(i)$ is determined from the input and output horizons. From the state- space model (1) one obtains a system of equations

$$CA^k x(i)=y(i+k)-\sum_{j=1}^{k} CA^{j-1}Bu(i+k-j), \qquad \text{for } k=0,1,2,\dots,NY,$$

which can be written as follows

$$\mathcal{O}_{NY}x(i)=Y_f(i)=Y_e(i)-G_eU(i) \tag{22a}$$

where

$$\mathcal{O}_{NY}=\begin{bmatrix}C\\H\end{bmatrix},\qquad Y_e(i)=\begin{bmatrix}y(i)\\Y(i)\end{bmatrix},\qquad G_e=\begin{bmatrix}0\\G\end{bmatrix} \tag{22b}$$

and H, G, Y are given in (10a), (8), and (5a), respectively. The variable Y_e is the augmented output horizon, composed of the current output $y(i)$ and the output horizon $Y(i)$.

The plant state is estimated from its input and output. The same input is applied to the plant and the estimator. Consequently, the free motion estimation error is minimized, such that for a symmetric positive weighting matrix Q_e the estimation index

$$J_e=\|Y_f(i)-\hat{Y}_f(i)\|^2_{Q_e}=\|\mathcal{O}_{NY}(x(i)-\hat{x}(i))\|^2_{Q_e} \tag{23}$$

is minimal, From the necessary condition for minimum one obtains

$$\hat{x}(i)=\mathcal{O}^{+}_{NY}(Y_e(i)-G_eU(i)) \tag{24}$$

where $\mathcal{O}^{+}_{NY}=(\mathcal{O}^{T}_{NY}Q_{o}\mathcal{O}_{NY})^{-1}\mathcal{O}^{T}_{NY}Q_{o}$. The above equation indicates that the state estimate is determined from input and output horizons. The input horizon is available right after the controller output (see Fig.2). The output horizon *Y(i)*, not available directly, can be obtained from the plant model

$$X(i+1)=AX(i)+BU(i), \qquad Y(i)=CX(i)+DU(i) \tag{25}$$

The block diagram of the estimator is shown in Fig.5a. The scheme, similar to the LQ estimation scheme, uses the available input and output signals and the plant model to generate the estimate. The block diagram of the predictive control system with the predictive estimator (EST) is shown in Fig.6.

Unlike the LQ estimator, the predictive estimator does not have filtering properties, since its output $\hat{x}(i)$ is proportional to a noisy signal $y(i)$, see Eq. (24). This shortcoming is removed through filtering the error signal $\varepsilon_{ny}(i)=y(i)-y_{n}(i)$ by a proper filter, obtaining the filtered error $\varepsilon_{yf}(i)$, where $y_{n}(i)$ is the output from the plant model. In most cases, for the output error being a high-frequency noise, a lowpass filter is required. In the filtered output, obtained by adding the nominal output to the filtered error, $y_{filt}(i)=y_{n}(i)+\varepsilon_{yf}(i)$ most of the noise power is removed. The estimator with a filter is shown in Fig.5b.

5. Applications

The performance of the predictive controller and estimator is tested through tracking simulations of the DSS-14 NASA/JPL 70-meter Deep Space Network antenna (see Fig.7). The existing control scheme for the 70-meter antennas (see Alvarez and Nickerson [14]), is based on the LQ controller design and the internal model principle [15]. This principle leads to the model augmented with integrators as presented by Athans [16], Fukata *et al.* [17], Johnson [18], Porter [19], Yahagi [20], and Young and Willems [21]. The augmentation ensures zero mean value of the constant rate tracking error. As a result, the LQ controller is designed for a constant tracking command, and a significant tracking error can be expected for relatively fast commands or varying rate commands. In this section the performance of the predictive controller is compared with that of the LQ controller in the tracking environment.

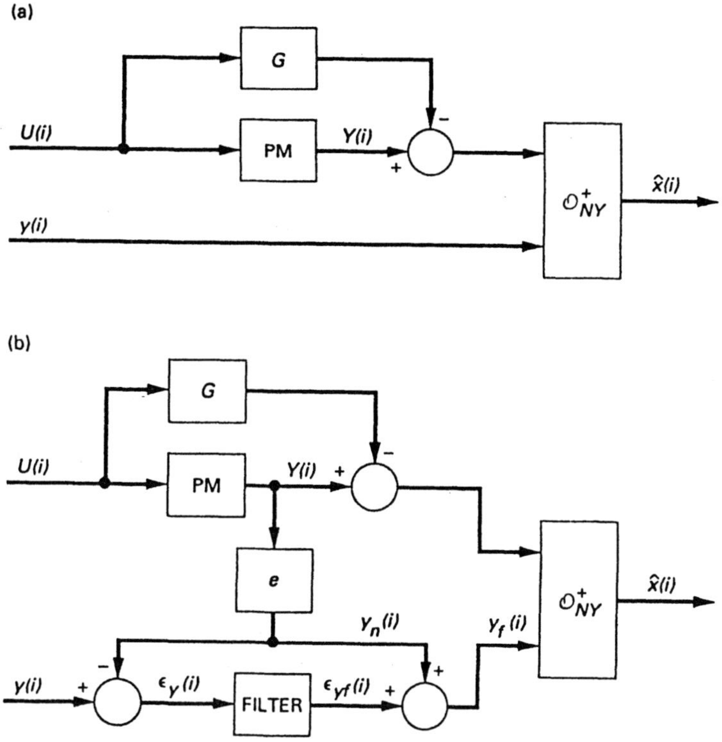

Fig.5. Predictive estimator without measurement noise filter (a), and with measurement noise filter (b).

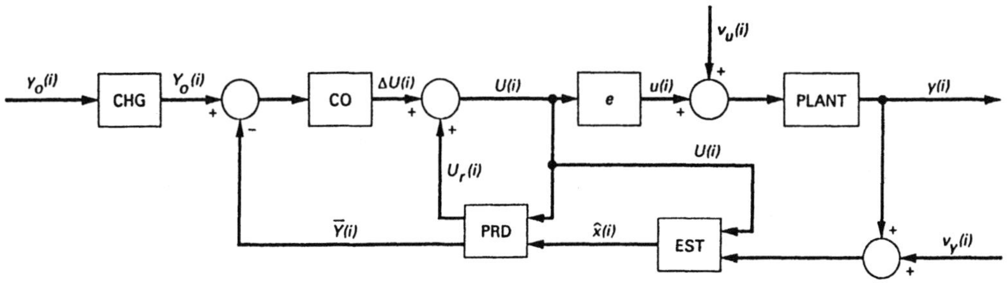

Fig.6. Predictive control and estimation system.

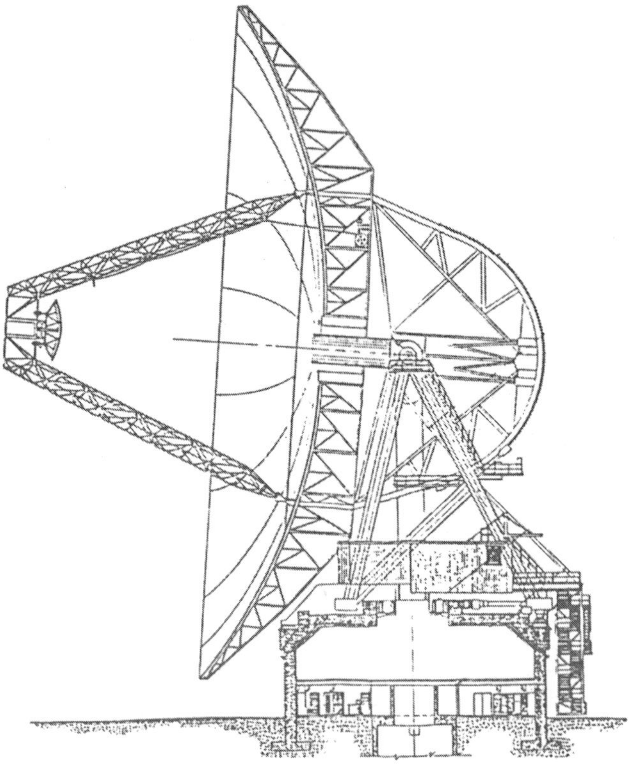

Fig.7 NASA/JPL DSS-14 antenna.

5.1 Plant Model

The state-space model of the DSS-14 antenna, by Alvarez and Nickerson [14], is a four-state model, with the position rate u as an input and the position rate y as an output. Its discrete-time representation (A_d, B_d, C_d), with the sampling period $\Delta t = 0.05$ sec, is obtained from the continuous-time representation in Ref. [14]

$$A_d = \begin{bmatrix} 0.0468 & 0 & 0 & 0 \\ 0 & 0.5443 & 0.3474 & 0 \\ 0 & -0.3474 & 0.5443 & 0 \\ 0 & 0 & 0 & 0.8872 \end{bmatrix}, \qquad B_d = \begin{bmatrix} 0.0113 \\ 0.0025 \\ 0.0399 \\ 0.0538 \end{bmatrix},$$

$$C_d = [0.7239 \;\; 9.2260 \;\; 0 \;\; 1.1421].$$

The model (A_d, B_d, C_d) is augmented with integrators, so that its output consists of the angular position, the integral of the position, and the position rate. Denoting x_d the state of the system (A_d, B_d, C_d), and x_{po}, x_{ipo}, the position and the integral of the position, respectively, one obtains

$$x_{ipo}(i+1) = x_{ipo}(i) + \Delta t x_{po}(i)$$

$$x_{po}(i+1) = x_{po}(i) + \Delta t C_d x_d(i)$$

$$x_d(i+1) = A_d x_d(i) + B_d u(i)$$

For the state variable $x^T = [x_{ipo}, x_{po}, x_d^T]$ the triple (A,B,C)

$$A = \begin{bmatrix} 1 & \Delta t & 0 \\ 0 & 1 & \Delta t C_d \\ 0 & 0 & A_d \end{bmatrix}, \quad B = \begin{bmatrix} 0 \\ 0 \\ B_d \end{bmatrix}, \quad C = \begin{bmatrix} 1 & 0 & 0 \\ 0 & 1 & 0 \\ 0 & 0 & C_d \end{bmatrix}$$

forms the resulting state-space representation of the plant.

5.2 Weighting Matrices, Input and Output Horizons

The weighting matrices R and Q have been chosen from series of simulations designed to obtain small output error, while the control effort is maintained within reasonable limits. For simulation purposes a piecewise-linear profile of the position command is chosen: linear increase followed by linear decrease and the final constant value (see Fig.8, solid line). The command rate 4 mdeg/sec is a typical sidereal tracking rate. For a diagonal control weighting matrix $R=\rho I$, the parameter $\rho=0.01$ is chosen. The tracking error weighting matrix Q is considered as in (21). The component q, in the form $q=diag(q_i, q_p, q_r)$, represents the weight of the integral, position, and rate components of the output. From a series of simulations for tracking the command as in Fig.8, the following choices of weight are recommended: $q_i=10$, $q_p=1$, and $q_r=0.1$, and the value of the parameter α is determined. The plot of the Euclidean norm of the tracking error $\|y-y_o\|_2$ vs α is shown in Fig.9 for different lengths of output horizon, and for lengths of input and reference horizon equal to the lengths of the output horizon. The plot shows the minimal tracking error obtained for $\alpha=6.2$, for $NY=NU=NR=6$.

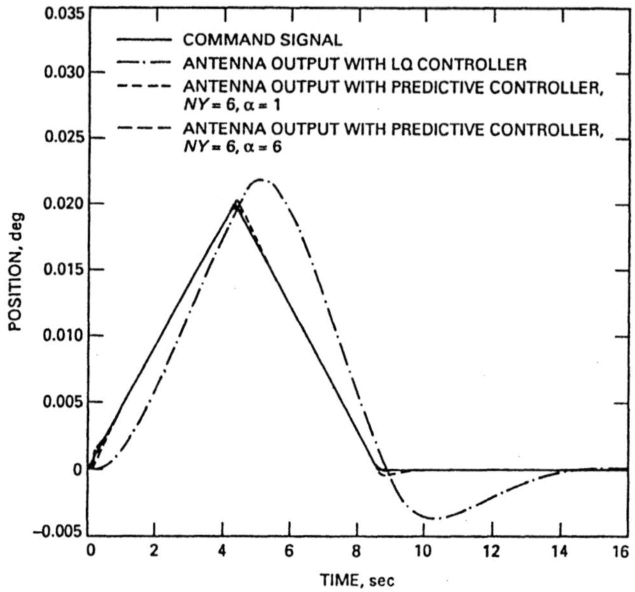

Fig.8. The command signal and antenna outputs.

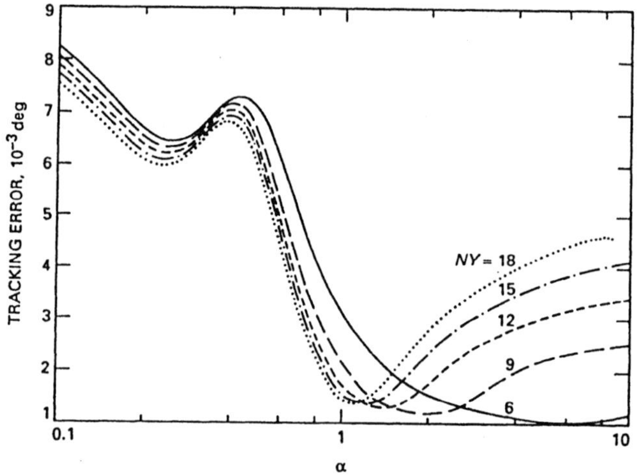

Fig.9. Tracking error vs forgetting factor α, for for different lengths of the output horizon NY.

Next, for $NR=NU=NY$, the values of α for which the tracking error is minimal are plotted in Fig.10. The figure features the forgetting factor close to 1 for long enough horizons. In this case, time weighting does not improve the tracking error. However, for short horizons, the proper choice of α is a critical factor that reduces the error dramatically. The impact of the length of the input reference horizon *NR*, and of the output horizon *NY*, on the tracking error is plotted in Figs.11a, b, where for $NR \geq n/2$ and $NY \geq 2n$ (where $n=6$ is the number of plant state variables) the performance error is close to the minimal one. Note that the forgetting factor in prediction is greater than one, in contrast to the factor in estimation, which is smaller than one (c.f. Ref. [4]).

5.3 Antenna Performance

The performance of the DSS-14 antenna with the tracking command as in Fig.8 is evaluated. The following parameters of the predictive controller have been chosen: $NR=NU=NY=n=6$, and weighting matrices with $\rho=0.01$, $q=diag(10, 1, 0.1)$, and $\alpha=6.2$. The command and the position of the antenna with the predictive controller for $\alpha=6.2$, and for $\alpha=1$, and for the antenna with the LQ controller are shown in Fig.8. The prediction errors for the above three cases ($\alpha=6.2$, $\alpha=1$, and LQ controller) are shown in Fig.12. The figures show better performance of the predictive controllers than the LQ controller with comparable control effort in all three cases. Also, the predictive controller performance with the weighted output error ($\alpha>1$) is better than the performance of a predictive controller without weighting ($\alpha=1$).

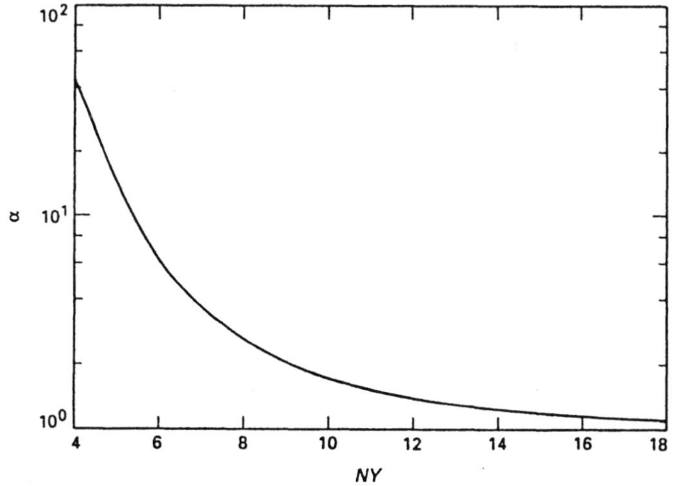

Fig.10. Forgetting factor α, for which the minimal tracking error is achieved vs length of the output horizon *NY*.

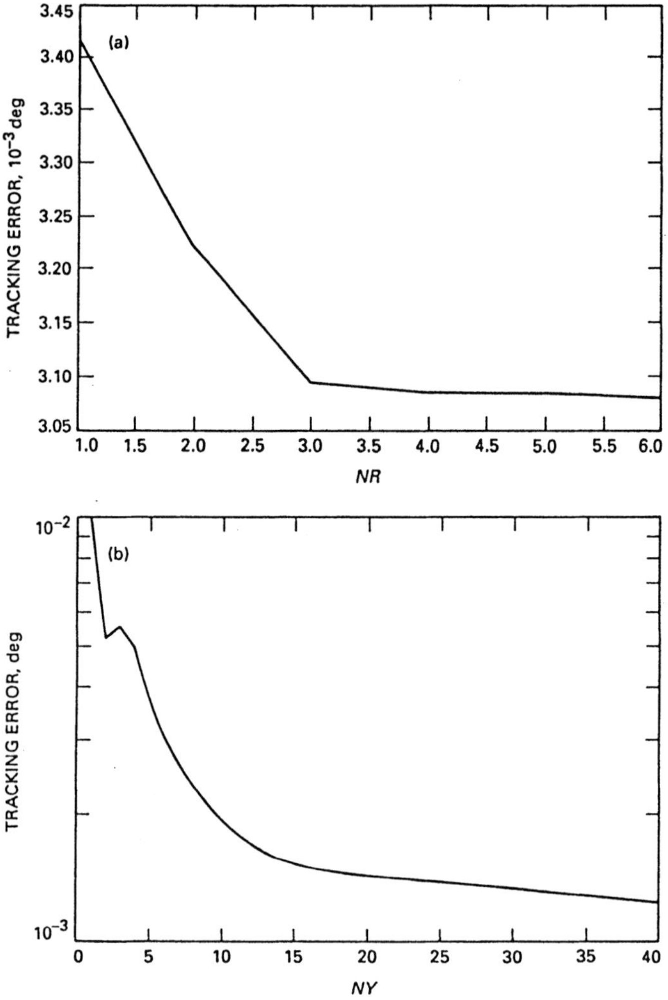

Fig.11. Tracking error vs length of the reference horizon *NR* (a), and vs length of the output horizon *NY* (b).

For the output horizon *NY*=4, the tracking error has its minimum for α=6.4. These two parameters are used in further simulations, since it is reasonable to have the length of the output horizon as small as possible: the dimension of the controller as well as the complexity of the system depend on *NY*. The step responses and frequency response plots of the closed-loop system with predictive and the LQ controller are compared in Figs.13 and 14. Fig.13

shows the settling time and the overshoot for the system with the predictive controller significantly reduced in comparison with the LQ controller (in fact, there is no overshoot in the predictive case). Similarly, Fig.14 shows improved tracking performance: the magnitude of the closed-loop transfer function equal to 1 over wider bandwidth, and roll-off rate improved for the system with predictive controller (with $NY=NU=NR=4$, and $\alpha=6.4$, as well as with $NY=NU=NR=6$, and $\alpha=5$) when compared to the system with the LQ controller.

The piecewise constant rate command, as well as the unit step command, are rather dramatic scenarios for the DSS-14 antenna, although the step command is often used in testing as an offset from the nominal trajectory. The raised cosine command, as in Fig.15 solid line, is close to the real elevation or azimuth trajectory of the antenna (conscan-like tracking). The antenna response to this command is shown in Fig.15. The output of the predictive control system overlaps the command, while the output of the LQ control system is plotted by a dashed line. The tracking error, the difference between the output and the command, is shown in Fig.16a for the LQ control system, and in Fig.16b for the predictive control system. The LQ tracking error is of order 10^{-4}, while the predictive tracking error is of order 10^{-7}. The control effort, in both cases, is almost the same, as in Fig.17.

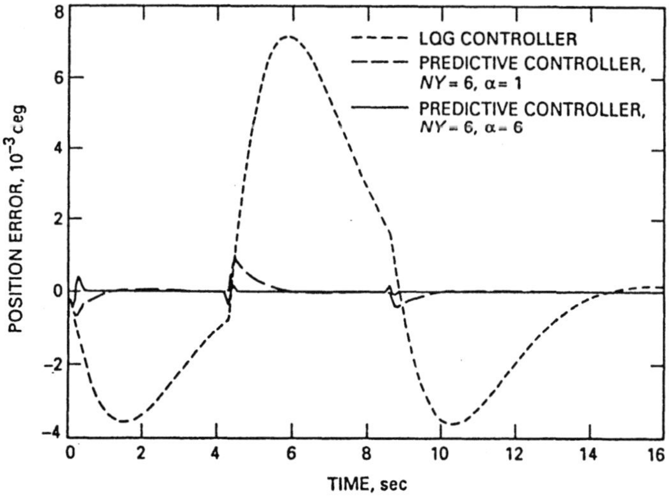

Fig.12. Tracking error.

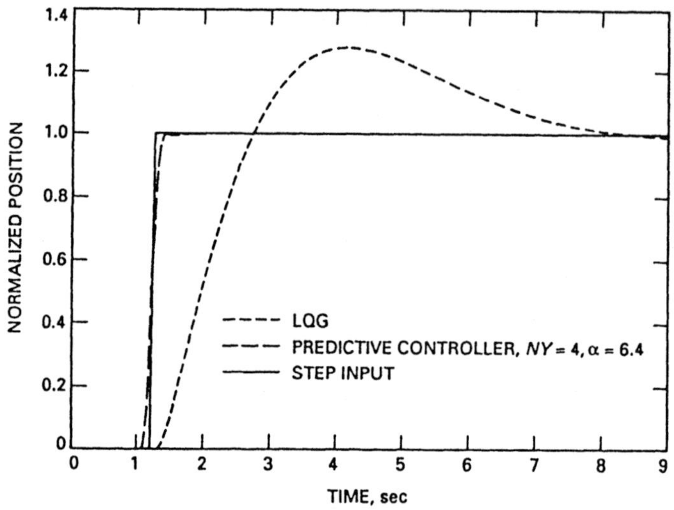

Fig.13. Closed-loop system step response.

5.4 Robustness and Disturbance Suppression

A random simulation approach is used for evaluation of performance robustness of the closed-loop system. The perturbed plant matrix $A_{dp}=A_d+\Delta A_d$ is characterized by normal distribution, with mean value $\theta_o=\|A_d\|_2$, and standard deviation $\sigma_\theta=0.2\theta_o$ (20% of the norm of the nominal value). The tracking error due to step command is simulated for $N=6000$ samples. The histogram of the normalized tracking error ε is shown in Fig.18a, and the samples are located in Fig. 18b. The nominal tracking error $\varepsilon_n=\varepsilon(\theta_o)=\|y(\theta_o)-y_o\|_2/\|y_o\|_2=0.0405$ is obtained for the nominal plant matrix. From the histogram, the probability distribution function $F(\varepsilon_r)=prob(\varepsilon<\varepsilon_r)$ is obtained, as shown in Fig 18c. From the plot, the probability that the output error exceeds 20%, 40%, or 100% of the nominal error is less that 5%, 1% or 0.03%, respectively. The mean value of the normalized error is $\varepsilon_o=0.04258$, and the standard deviation $\sigma_\varepsilon=0.003513$. The non-symmetric distribution function is approximated with the Gumbel distribution [22] $F_g(\varepsilon_r)=\exp(-\exp(-\sigma(\varepsilon_r-\mu)))$, with the scale $\sigma=480$, and location $\mu=0.0415$, as in Fig.18c. A set of step responses of the closed-loop systems with plant modeling errors is shown in Fig.19 for the 20% standard deviation of the error from the nominal value ($\sigma_\theta = 0.2\theta_o$). The plot shows small settling time and small overshoot.

Two sources of antenna disturbances are studied: the input disturbances v_u, and output disturbances v_u (measurement noise), as shown in Fig.2. The

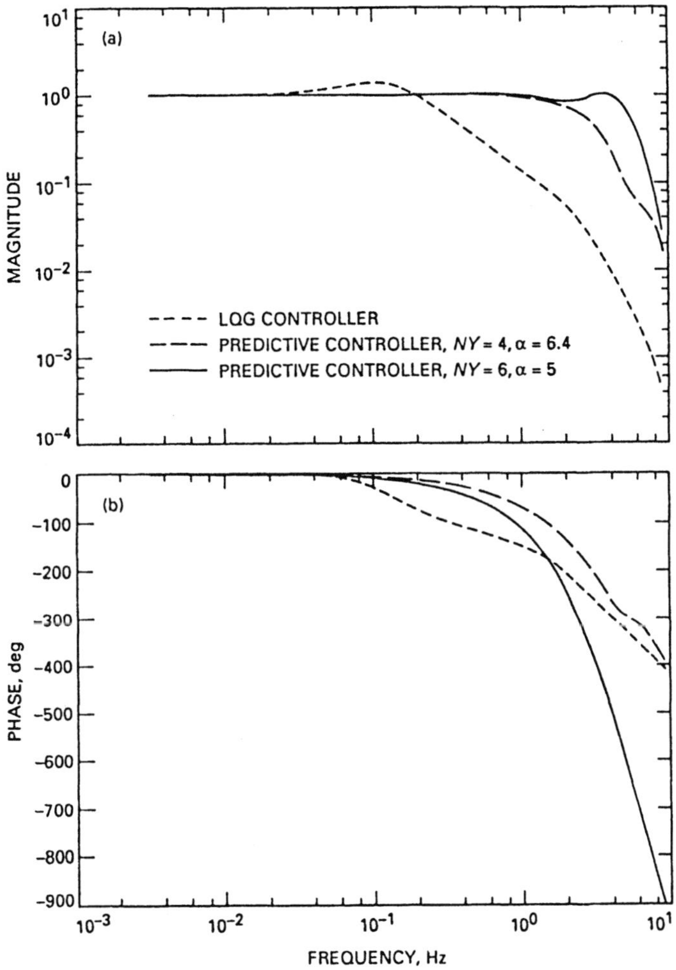

Fig.14. Closed-loop system frequency response.

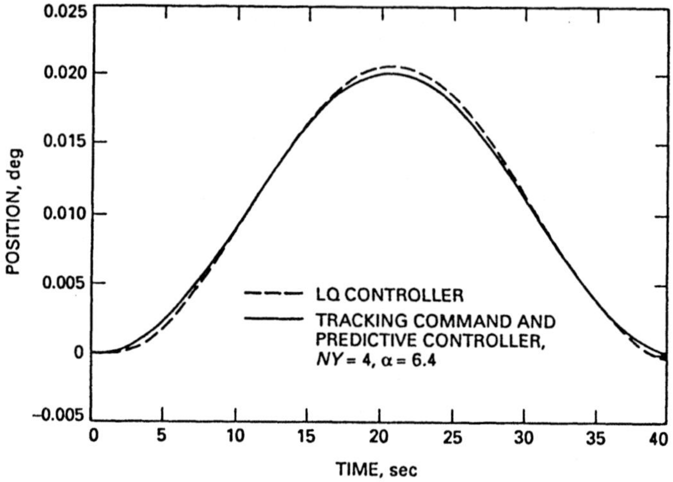

Fig.15. Closed-loop system tracking performance.

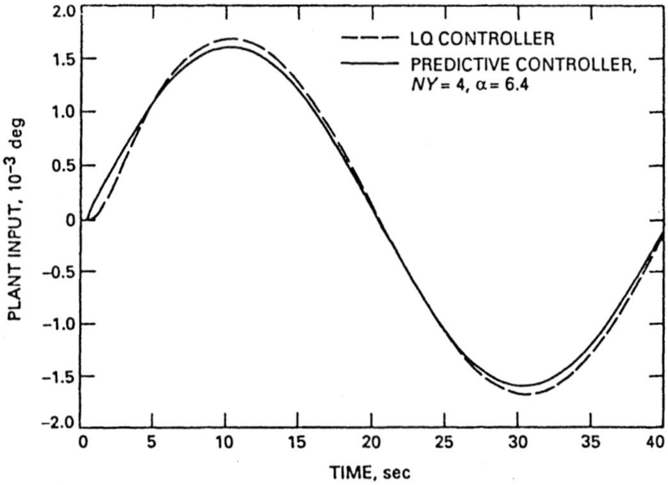

Fig.16. Plant input.

disturbance transfer functions, from v_u to y, and from v_x to y, are shown in Fig.20, the latter one for the position and rate measurement noise. It follows from the transfer function plots that the input disturbances and rate measurement noise are significantly suppressed, while the position measurement noise is amplified over certain frequency ranges.

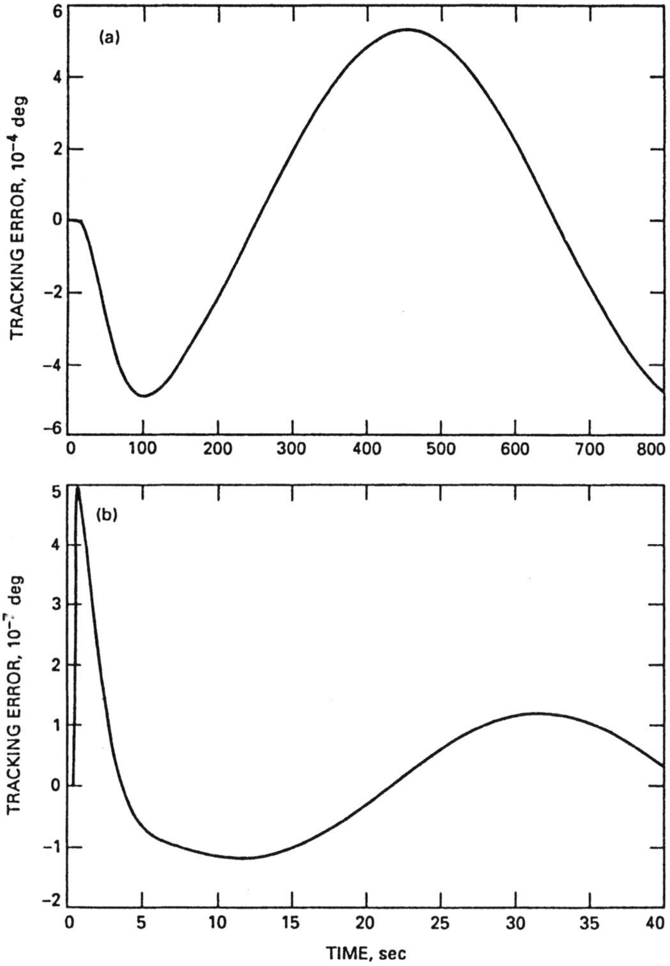

Fig.17. Tracking error for the LQ system (a), and the predictive system with *NY=4*, $\alpha=6.4$ (b).

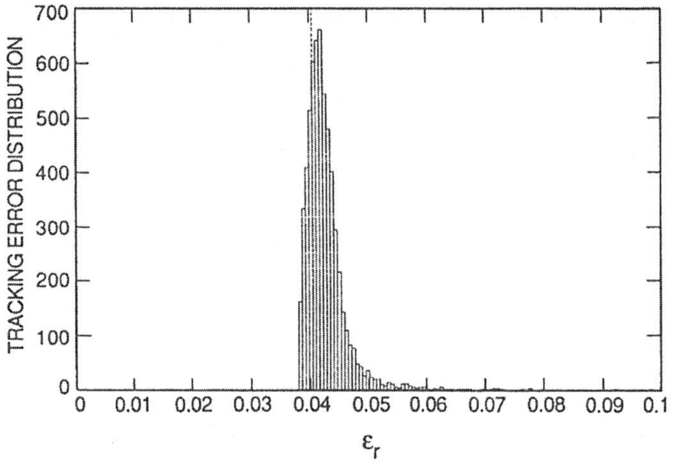
TRACKING ERROR DISTRIBUTION
700
600
500
400
300
200
100
0
0
0.01
0.02
0.03
0.04
0.05
0.06
0.07
0.08
0.09
0.1
ε_r

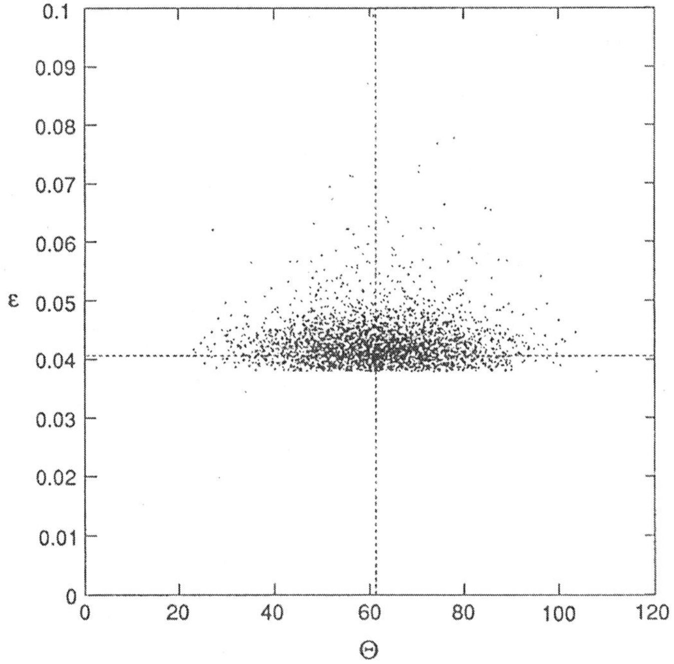
0.1
0.09
0.08
0.07
0.06
ε
0.05
0.04
0.03
0.02
0.01
0
0
20
40
60
80
100
120
Θ

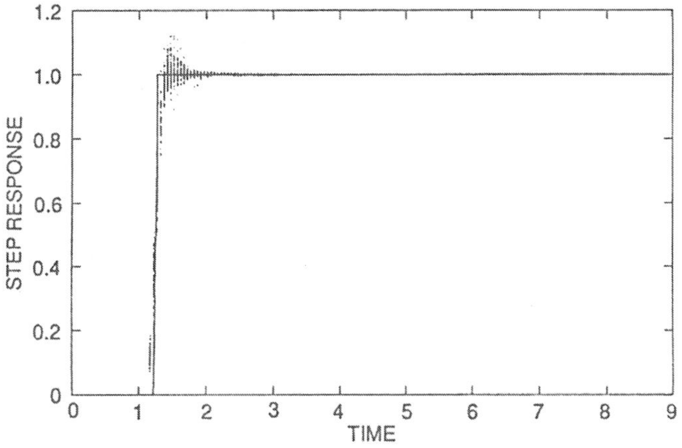

Fig.18. Histogram of tracking error (a), samples of model error and tracking performance (b), distribution functions of the tracking error and its Gumbel approximation (c).

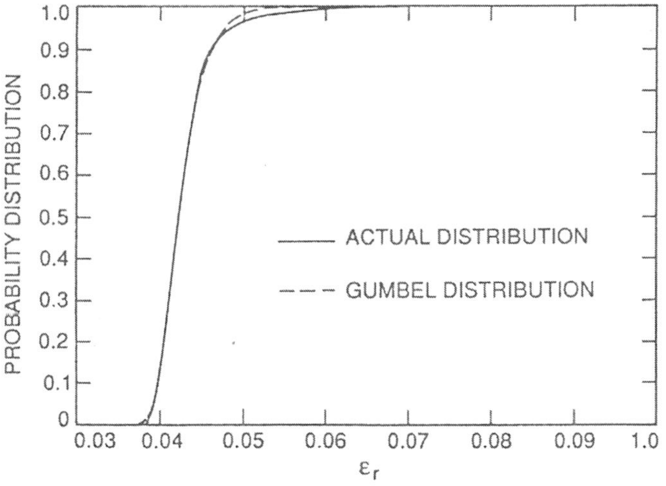

Fig.19. Step responses of deviated models.

The nature of the antenna disturbances is not satisfactorily known, thus their general properties are outlined here. Input disturbances, such as wind or thermal forces, are low-frequency signals. Measurement noise, on the other hand, is a high-frequency signal (high when compared to the antenna fundamental frequency, which is less than 1 Hz). Simulations are designed so that the above properties of the disturbances are preserved. Namely, input disturbances have properties of white noise (white noise contains low- and high-frequency components; the latter one does not affect the system performance anyway). High-frequency noise, with frequency components over 3 Hz, is applied as output disturbance. The system response to disturbances for different signal-to-noise ratios is simulated, and results compared in Fig.21. The tracking error subject to the input noise is much smaller than that subject to the measurement noise. This property is explained with the filtering property of the plant: the low-frequency components of the input

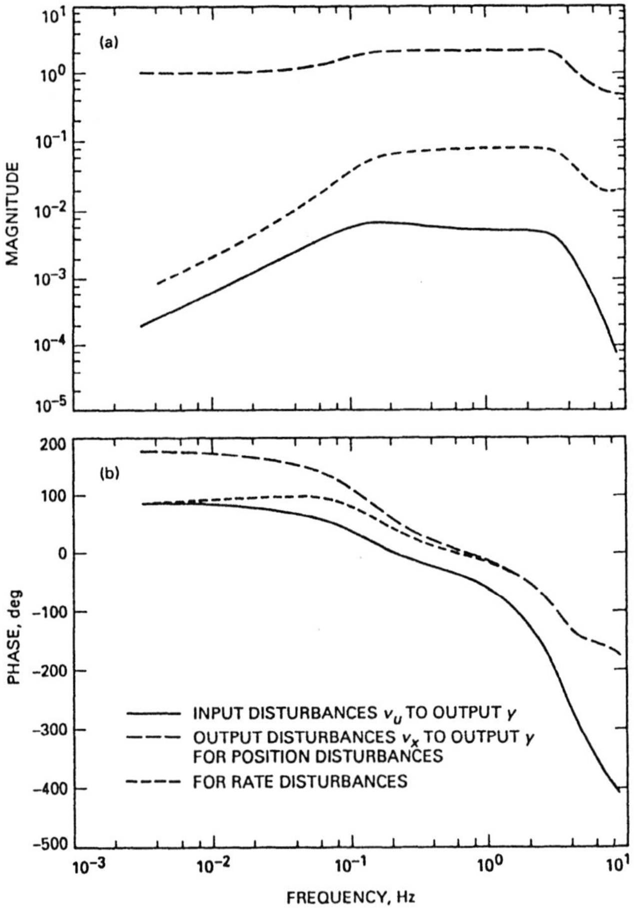

Fig.20. Transfer functions: input disturbances v_u to output y, output disturbances v_x to output y for position and rate disturbances.

noise are filtered out before entering the predictor. The tracking errors for the input noise with N/S ratio=0.1, and for the measurement noise with N/S ratio=0.01, are shown in Figs.22 and 23. The effect of the measurement noise is reduced by applying a filter, as in Fig.4. The tracking error due to the measurement noise is reduced significantly, as in Fig.24.

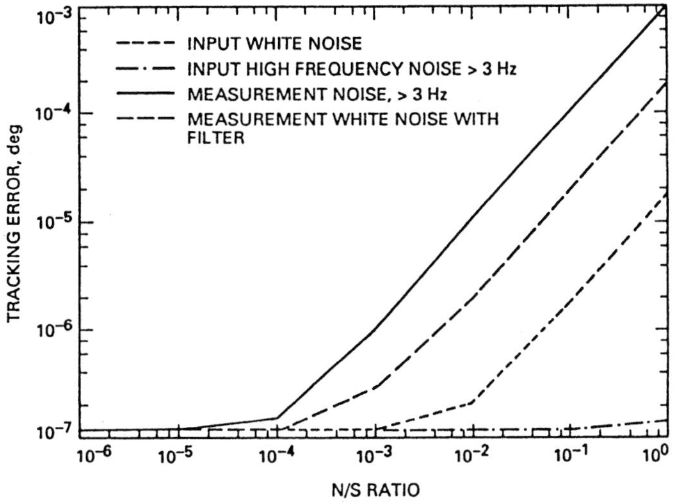

Fig.21. Tracking errors due to disturbances.

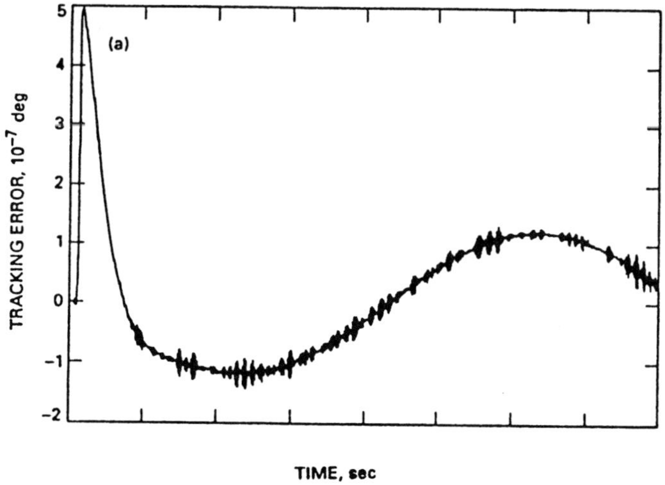

Fig.22. Tracking error for the white noise input disturbances with noise-to-signal ratio=0.1.

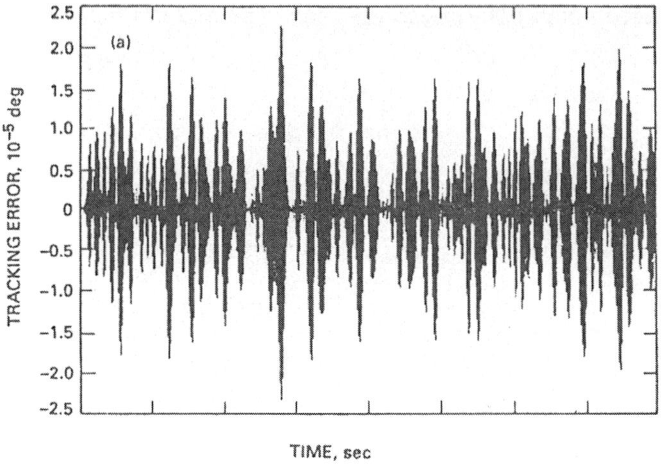

Fig.23. Tracking error for the measurement noise with noise-to-signal ratio=0.01.

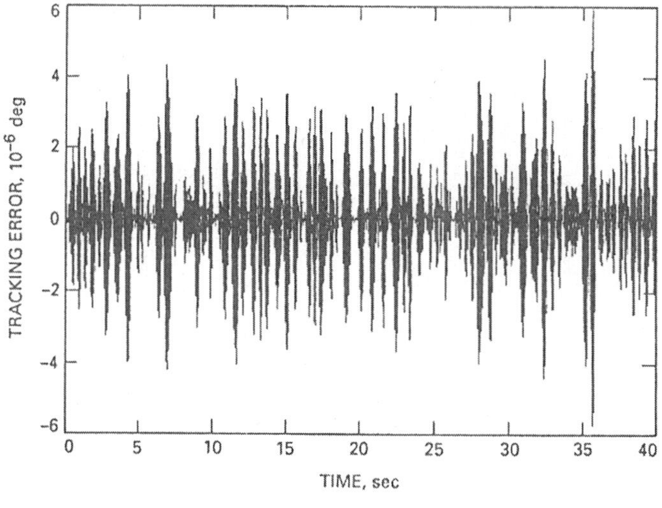

Fig.24. Tracking error for the white noise output disturbances and the system with a filter, noise-to-signal ratio=0.01.

5.5 Predictive Estimator

In order to evaluate the performance of the predictive estimator, it is compared to the well known performance of the LQ estimator. For the plant model (A_d, B_d, C_d) the output is simulated for the unit-step input and zero initial conditions. For estimation purposes the initial conditions have been shifted to $[0.1\ \ 0.1\ \ 0.1\ \ 0.1]^T$, and the identity weighting matrix $(Q_e = I)$ has been assumed. The estimation results are shown in Figs.25a,b. The LQ estimator needs approximately 2 sec to reach an acceptable estimation error, while the predictive estimator determines the states in virtually no time. In the case of noisy output, with noise-to-signal ratio 0.01, one obtains the estimation error for the LQ estimator as in Fig.26a, while the norm of the estimation error for the predictive estimator is of order 0.01, which is too large to consider it suitable for prediction purposes. The estimation error is reduced by applying a filter, and the error in this case is shown in Fig.26b. As a result, the maximum error of the predictive estimator with a filter is much smaller than the residual error of the LQ estimator, even after 4 seconds of the LQ estimator in action.

6. Conclusions

In this paper a modified state-space predictive controller is introduced, and a predictive estimator presented to complement the design of a predictive control law. This approach has been used for the design of tracking controllers for the NASA/JPL 70-m antennas. Several tracking scenarios have been tested (step input, constant rate rise and fall, raised cosine trajectory), and significant improvement of performance has been observed. A wider bandwidth and improved roll-off rate is obtained for the predictive closed-loop system in comparison with the LQ system. The predictive control system is robust to the plant parameter variations. Shifts of plant poles of 20% of their nominal values has a tracking error of the same order as for a nominal plant. Its disturbance suppression properties have also been simulated and found good for input disturbances and for the measurement noise, if the measurement noise spectrum is higher than the plant fundamental frequency. The system disturbance suppression properties can be enhanced if a disturbance filter is applied. For the predictive estimator, the estimation error and time to reach stationary error are much smaller than for the LQ estimator.

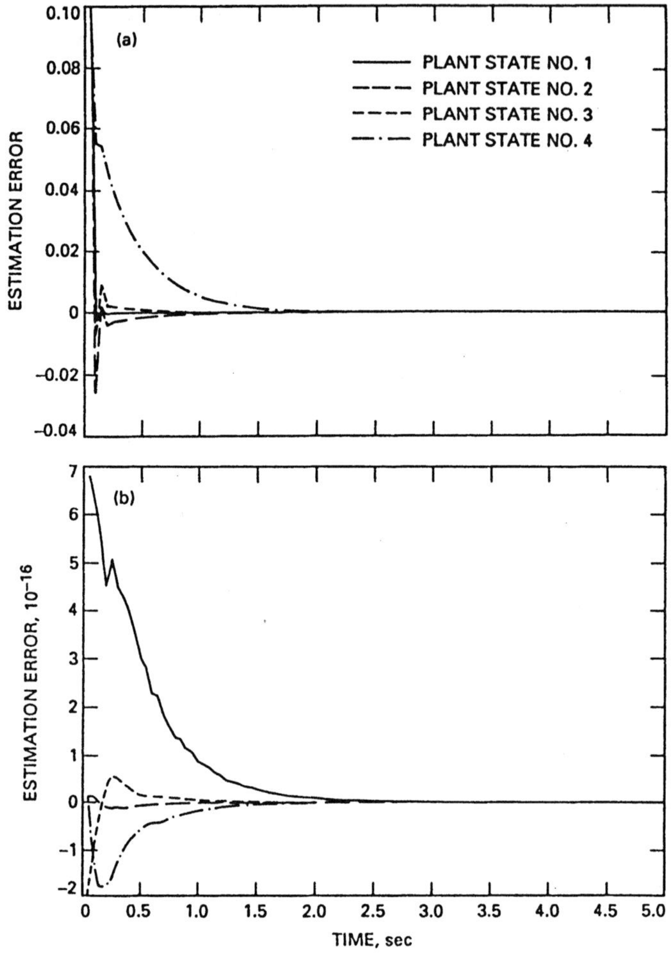

Fig.25. Estimation error for the LQ estimator (a), and for the predictive estimator (b).

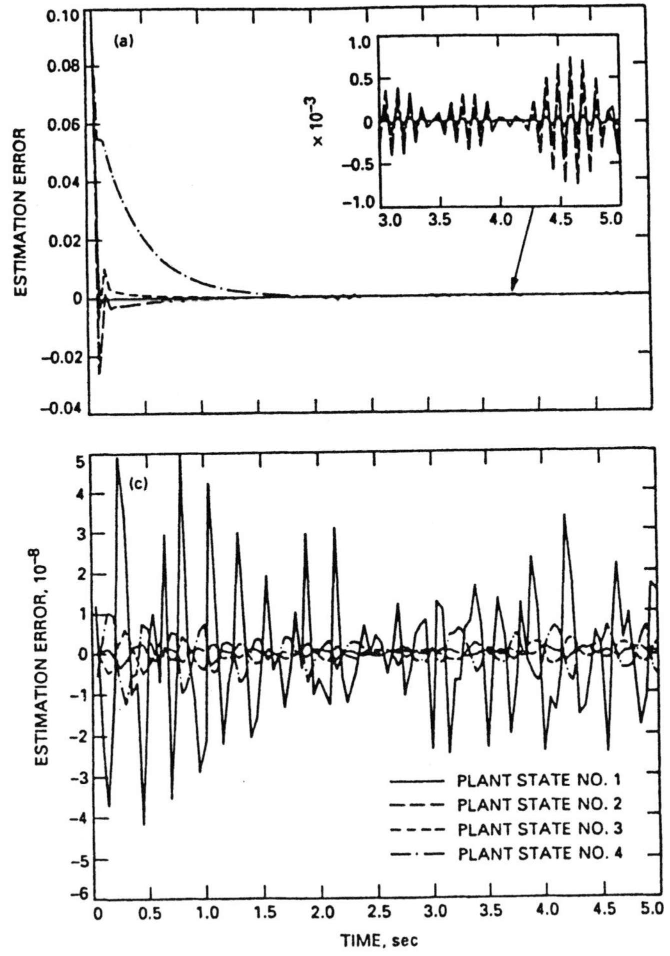

Fig.26. Estimation error in presence of measurement noise: for the LQ estimator (a), and the predictive estimator with filter (b).

Acknowledgments

The author thanks Ben Parvin for recognizing the importance of predictive control in Deep Space Network antenna design, and introducing technical aspects of antenna control, and Leon Alvarez for discussions about the DSS-14 antenna model. This research was performed at the Jet Propulsion Laboratory, California Institute of Technology, under a contract with the National Aeronautics and Space Administration.

References

[1] Kwakernaak, H., and Sivan, R.: *Linear Optimal Control Systems.* Wiley, New York, 1972.

[2] Anderson, B.D.O., and Moore, J.B.: *Optimal Control,* Prentice Hall, Englewood Cliffs, 1990.

[3] Astrom, K.J., and Wittenmark, B.: *Adaptive Control,* Addison-Wesley, Reading, 1989.

[4] Goodwin, G.C., and Sin, K.S.: *Adaptive Filtering Prediction and Control.* Prentice-Hall, Englewood Cliffs, 1984.

[5] Lewis, F.L.: *Optimal Control,* Wiley-Interscience, New York, 1986.

[6] Albertos, P., and Ortega, R.: On Generalized Predictive Control: Two Alternative Formulations. *Automatica,* vol.25, no.5, 1989, pp.753-755.

[7] Clarke, D.W. and Mohtadi, C.: Properties of Generalized Predictive Control. *Automatica,* vol.25, no.6, 1989, pp.859-875.

[8] Clarke, D.W., Mohtadi, C., and Tuffs, P.S.: Generalized Predictive Control - Part I and II. *Automatica,* vol.23, no.2, 1987, pp.137-160;

[9] Ortega, R., and Galindo, G.S.: Globally Convergent Multistep Receding Horizon Adaptive Controller. *Int. J. Control,* vol.49, no.5, 1989, pp.1655-1664.

[10] Reid, J.G., Chaffin, D.E., and Silverthorn, J.T.: Output Predictive Algorithmic Control: Precision Tracking with Application to Terrain Following. *J. Guidance, Control, and Dynamics,* vol.4, no.5, 1981, pp.502-509.

[11] Xi, Y.: New Design Method for Discrete-Time Multi-Variable Predictive Controllers. *Int. J. Control,* vol.49, No.1, 1989, pp.45-56.

[12] Maciejowski, J.M.: *Multivariable Feedback Design.* Addison - Wesley, Wokingham, 1989.

[13] Ray, L.R., and Stengel, R.F.: Stochastic Performance Robustness of Aircraft Control Systems. *Proc. AIAA Guidance, Navigation and Control Conf.,* pp.863-873, Portland, OR, 1990.

[14] Alvarez, L.S., and Nickerson, J.: Application of Optimal Control Theory to the Design of the NASA/JPL 70-Meter Antenna Axis Servos. *TDA Progress Report,* vol.42-97, 1989.

[15] Francis, B.A., and Wonham, W.M.: The Internal Model Principle of Control Theory. *Automatica,* vol.12, pp.457-465, 1976.

[16] Athans, M.: On the Design of PID Controllers Using Optimal Linear Regulator Theory. *Automatica,* vol.7, pp.643-647, 1971.

[17] Fukata, S., Mohri, A., and Takata, M.: On the Determination of the Optimal Feedback Gains for Multivariable Linear Systems Incorporating Integral Action. *Int. J. Control*, vol.31, no.6, pp.1027-1040, 1980.

[18] Johnson, C.D.: Optimal Control of the Linear Regulator with Constant Disturbances, *IEEE Trans. Automatic Control*, vol.13, pp.416-421, 1968.

[19] Porter, B.: Optimal Control of Multivariable Linear Systems Incorporating Integral Feedback. *Electronics Letters*, vol.7, no.8, 1971.

[20] Yahagi, T.: Optimal Output Feedback Control in the Presence of Step Disturbances. *Int. J. Control*, vol.26, no.5, pp.753-762, 1977.

[21] Young, P.C., and Willems, J.C.: An Approach to the Linear Multivariable Servomechanism Problem. *Int. J. Control*, vol.15, no.5, 1972, pp.961-979.

[22] Gumbel, E.J.: *Statistics of Extremes*, Columbia University Press, New York, 1958.

ROBUST CONTROL SYNTHESIS FOR UNCERTAIN/NONLINEAR DYNAMICAL SYSTEMS*

Stefen Hui
Department of Mathematical Sciences
San Diego State University
San Diego, CA 92182

Stanislaw H. Żak
School of Electrical Engineering
Purdue University
West Lafayette, IN 47907

Abstract

This paper addresses the problem of robust output-feedback controller design for uncertain and nonlinear dynamic systems. First a robust state-feedback nonlinear control law is synthesized. This control strategy practically stabilizes the closed-loop system. Then a state estimator for the nonlinear/uncertain plant is designed and its performance analyzed. Finally the control law and the estimator are combined and the practical stability region is estimated. The results are illustrated by their application to a benchmark problem for robust control design proposed by Wie and Bernstein (1990).

1. Introduction

In recent years, different approaches to the problem of robust control design for uncertain dynamical systems have been proposed. Among these different approaches is the method of deterministic control and included in this broad category are

a) deterministic control using Lyapunov functions (see e.g., Corless and Leitmann (1981), Ryan and Corless (1984), and Lindorff (1967)) and

b) variable structure control (DeCarlo et al (1988), El-Ghezawi et al (1983), Utkin (1977), Utkin (1978)),

among other methods.

The approach we propose in this paper can be viewed as a combination of the above mentioned techniques. In particular our technique is based on the combined state estimator-controller synthesis, and which involves three basic steps. The first step is the design of the controller assuming the availability of the states of the plant. The second step is the design of a state estimator, also referred to as a state observer (Luenberger (1971)). The estimator should only use the input and output of the plant. The final step is to combine the controller and the estimator obtained in the first two steps. In this paper the controller synthesis is based on Variable Structures System (VSS) Theory. In the state estimator design analysis, we employ Lyapunov functions. See Bondarev et al (1985) for a discussion of the advantages in using an asymptotic state estimator with a variable structure controller.

The results are then applied to obtain a solution to a benchmark problem for robust control design proposed by Wie and Bernstein (1990). For the convenience of the reader, we now briefly describe this benchmark problem. This will also serve to illustrate the class of systems that we consider in this paper. As we proceed with the design, we shall refer to this problem to illustrate the results obtained.

*This work was supported by the School of Electrical Engineering of Purdue University, West Lafayette, IN 47907.

The benchmark problem is a two mass-spring combination as depicted in Fig. 1. This system can

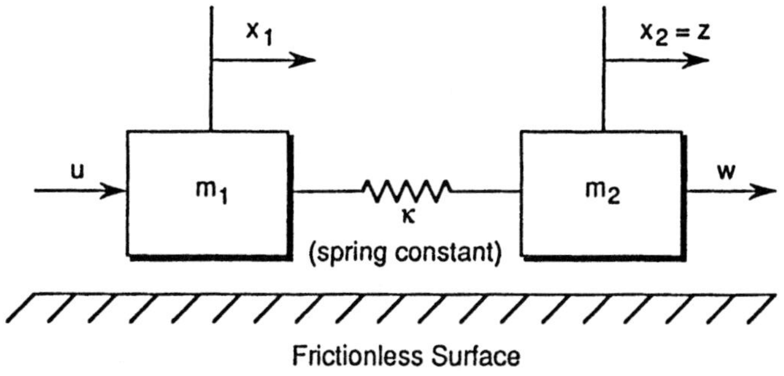

Fig. 1. System of the benchmark problem.

be represented in the state-space form as:

$$\begin{bmatrix}\dot{x}_1\\ \dot{x}_2\\ \dot{x}_3\\ \dot{x}_4\end{bmatrix} = \begin{bmatrix}0 & 0 & 1 & 0\\ 0 & 0 & 0 & 1\\ -\frac{\kappa}{m_1} & \frac{\kappa}{m_1} & 0 & 0\\ \frac{\kappa}{m_2} & -\frac{\kappa}{m_2} & 0 & 0\end{bmatrix}\begin{bmatrix}x_1\\ x_2\\ x_3\\ x_4\end{bmatrix} + \begin{bmatrix}0\\ 0\\ \frac{1}{m_1}\\ 0\end{bmatrix}u + \begin{bmatrix}0\\ 0\\ 0\\ \frac{1}{m_2}\end{bmatrix}w \tag{1.1}$$

$$y = x_2 + v \tag{1.2}$$

$$z = x_2 \tag{1.3}$$

where

x_1 = position of body 1
x_2 = position of body 2
x_3 = velocity of body 1
x_4 = velocity of body 2
u = control input
w = plant disturbance, a bounded function
y = sensor measurement
v = sensor noise
z = performance variable (output to be controlled).

It is assumed that $m_1 = m_2 = 1$ and $0.5 < \kappa < 2$. The nominal value of κ is 1. The problem is to design a constant gain linear feedback compensator of the form

$$\dot{x}_c = A_c x_c + B_c u \tag{1.4}$$

$$u = C_c x_c + D_c y \tag{1.5}$$

so that the closed-loop system achieves reasonable stability robustness. The stability robustness of the closed-loop system is the main objective of this paper, although we use a compensation different from (1.4), (1.5). Note that stability robustness is only one of the performance specifications proposed in Wie and Bernstein (1990).

2. System Description

We consider a class of systems modeled by:

$$\dot{x} = Ax + Bu + F(t,x,u,\delta,w) \tag{2.1}$$

$$y = Cx + v, \tag{2.2}$$

where

$x \in \mathbb{R}^n$ is the state vector

$u \in \mathbb{R}^m$ is the control input

$y \in \mathbb{R}^p$ is the sensor measurement

$\delta \in \mathbb{R}^s$ is the uncertain parameter

$w(t) \in \mathbb{R}^l$ is the plant disturbance

$v(t) \in \mathbb{R}^p$ is the sensor noise

F is a function that models the uncertainties and nonlinearities in the plant.

We will assume in this paper that:

A1. δ lies in a known compact subset of $\mathbb{R}^s$.
A2. v, w are bounded functions.
A3. The pair (A, B) is controllable.
A4. The pair (A, C) is observable.

An example of a dynamical system of this class is the plant described in the benchmark problem. Note that (1.1) can be written as

$$\dot{x} = \begin{bmatrix} 0 & 0 & 1 & 0 \\ 0 & 0 & 0 & 1 \\ -\kappa_o & \kappa_o & 0 & 0 \\ \kappa_o & -\kappa_o & 0 & 0 \end{bmatrix} x + \begin{bmatrix} 0 \\ 0 \\ 1 \\ 0 \end{bmatrix} u + \begin{bmatrix} 0 \\ 0 \\ (\kappa_o - \kappa)x_1 + (\kappa - \kappa_o)x_2 \\ (\kappa - \kappa_o)x_1 + (\kappa_o - \kappa)x_2 + w \end{bmatrix} \tag{2.3}$$

where κ_o is a nominal value and here we used the assumption that $m_1 = m_2 = 1$. Clearly (2.3) has the form (2.1) if $\delta = \kappa - \kappa_o$. It is easy to see that (2.3) satisfies assumption A1, A3, A4. We only need to investigate the boundedness of the uncertainty function in (2.3). This uncertainty function can be written as

$$\delta(x_1 - x_2) \begin{bmatrix} 0 \\ 0 \\ -1 \\ 1 \end{bmatrix} + \begin{bmatrix} 0 \\ 0 \\ 0 \\ 1 \end{bmatrix} w\,. \tag{2.4}$$

By assumption, w is a bounded function and $-0.5 \le \delta \le 1$. Thus condition A2 is satisfied. Furthermore, observe that $(x_1 - x_2)$ is the spring displacement and physical consideration dictates that $|x_1 - x_2|$ must be bounded. Therefore (2.4) is also bounded.

3. Basic Facts from the Theory of Variable Structure Systems

For the convenience of the reader and to fix our notation, we briefly review the basic facts that we need from the theory of variable structure systems. For a complete account of this theory, the reader can consult Utkin (1978). A state feedback Variable Structure Controller has a structure which changes in response to the changing state vector of the system. More specifically, the controller changes its structure according to the position of the state trajectory with respect to a chosen switching surface. The control is designed to force the state trajectory of the system onto the switching surface and to maintain it there. This is accomplished by a high speed switching law. The design of a variable structure controller (VSC) consists of two steps:

i) The design of the switching surface. The surface is chosen so that the system satisfies certain performance specifications, such as asymptotic stability, while on the surface.

ii) The design of the control strategy to steer the state trajectory to the switching surface.

Suppose $\{x \mid \sigma(x)=0\}$ is the chosen switching surface. In this paper, $\sigma(x)=Sx$ where $S \in \mathbb{R}^{mxn}$. We denote

$$\sigma(x) = \begin{bmatrix} \sigma_1(x) \\ \vdots \\ \sigma_m(x) \end{bmatrix} = \begin{bmatrix} s_1 x \\ \vdots \\ s_m x \end{bmatrix} \tag{3.1}$$

where $s_i \in \mathbb{R}^{1xn}$. We say that the system is in a sliding mode if $\sigma(x(t))=0$ for $t \geq t_o$, where $x(t)$ is the state trajectory and t_o is a specific time. It follows that in a sliding mode the velocity $\dot{x}$ is tangent to the switching surface. Equivalently,

$$(\nabla\sigma_i)^T \dot{x} = s_i\dot{x} = 0, \quad i = 1,...,m . \tag{3.2}$$

Hence

$$\frac{d}{dt}\sigma(x(t)) = S\dot{x} = 0 . \tag{3.3}$$

We can characterize the system in sliding mode by

$$\sigma(x(t)) = 0 \quad \text{and} \quad \dot{\sigma}(x(t)) = 0.$$

Consider the nominal plant

$$\dot{x} = Ax + Bu . \tag{3.4}$$

Combing (3.3), (3.4) we have

$$SAx + SBu = 0 . \tag{3.5}$$

If SB is nonsingular then (3.5) gives

$$u = -(SB)^{-1} SAx . \tag{3.6}$$

Substituting (3.6) into (3.4) yields

$$\dot{x} = [I_n - B(SB)^{-1}S] Ax . \tag{3.7}$$

The behavior of the system in sliding is therefore governed by

$$\begin{cases} \dot{x} = [I_n - B(SB)^{-1}S]Ax \\ Sx = 0 \end{cases} .$$

Note that while in sliding the (nominal) system is governed by a reduced set of differential equations.

4. Switching Surface Design

We will now describe a general procedure for designing a switching surface. The design is based on a procedure developed by El-Ghezawi et al (1983). See Utkin (1978) and DeCarlo et al (1988) for other methods.

Let $E = B(SB)^{-1}S$. Note that $E^2 = E$ and rank $E = m$ since SB is nonsingular. It follows that $(I_n - E)^2 = (I_n - E)$ and hence rank $(I_n - E) = n - m$. Therefore the matrix $[I_n - E]A$ has at most n−m nonzero eigenvalues. Since (A,B) is controllable, it follows that for any n complex numbers, symmetric with respect to the real axis, we can find a matrix K so that $A - BK$ has the given numbers as its eigenvalues. Our goal is to use this fact to find S so that

$$(I_n - E)A = A - B[(SB)^{-1}SA]$$

has n−m prescribed negative eigenvalues. We will then use the remaining degrees of freedom to partially shape the eigenstructure of $(I_n - E)A$ by prescribing n−m linearly independent eigenvectors $w_1, \ldots, w_{n-m}$. This is equivalent to

$$\{A - B[(SB)^{-1}SA]\}W = WJ \tag{4.1}$$

where

$$W = [w_1 \ldots w_{n-m}]$$

and

$$J = \text{diag}[\lambda_1, \ldots, \lambda_{n-m}] .$$

Observe that (4.1) implies the fact that the columns of $AW - WJ$ are in the range of B, that is,

$$\text{col}(AW - WJ) \subset \text{Range}(B) . \tag{4.2}$$

Note that we also require

$$SW = 0 . \tag{4.3}$$

Since SB is nonsingular, $SW = 0$ is equivalent to

$$\text{Range}(W) \cap \text{Range}(B) = \{0\} . \tag{4.4}$$

Hence $[W \vdots B]$ is invertible. The inverse $[W \vdots B]$ must have the form $\begin{bmatrix} W^g \\ B^g \end{bmatrix}$ where B^g, W^g denote left pseudoinverses of B, W. We have therefore

$$B^gB = I_m, \quad W^gB = 0, \quad B^gW = 0 .$$

If we let $S = B^g$, we have (4.3), $SB = I_m$, and (4.1).

We now use this procedure to design a switching surface for the nominal system of the benchmark problem. We have

$$A = \begin{bmatrix} 0 & 0 & 1 & 0 \\ 0 & 0 & 0 & 1 \\ -1 & 1 & 0 & 0 \\ 1 & -1 & 0 & 0 \end{bmatrix}, \quad B = \begin{bmatrix} 0 \\ 0 \\ 1 \\ 0 \end{bmatrix} . \tag{4.5}$$

To obtain an approximate settling time of 15 to 20 seconds we choose the prescribed eigenvalues to be $\lambda_1 = -0.2$, $\lambda_2 = -0.25$, $\lambda_3 = -0.3$. Hence $J = \text{diag}[-0.2, -0.25, -0.3]$. We now select the corresponding eigenvectors w_1, w_2, w_3 using (4.2). We have

$$AW - WJ = \begin{bmatrix} w_{31} + 0.2w_{11} & w_{32} + 0.25w_{12} & w_{33} + 0.3w_{13} \\ w_{41} + 0.2w_{21} & w_{42} + 0.25w_{22} & w_{43} + 0.3w_{23} \\ -w_{11} + w_{21} + 0.2w_{31} & -w_{12} + w_{22} + 0.25w_{32} & -w_{13} + w_{23} + 0.3w_{33} \\ w_{11} - w_{21} + 0.2w_{41} & w_{12} - w_{22} + 0.25w_{42} & w_{13} + w_{23} + 0.3w_{43} \end{bmatrix} .$$

Solving (4.2) we obtain the vectors

$$w_1 = a \begin{bmatrix} -5.2000 \\ -5.0000 \\ 1.0400 \\ 1.0000 \end{bmatrix}, \quad w_2 = b \begin{bmatrix} -4.2500 \\ -4.0000 \\ 1.0625 \\ 1.0000 \end{bmatrix}, \quad w_3 = c \begin{bmatrix} -3.6333 \\ -3.3333 \\ 1.0900 \\ 1.0000 \end{bmatrix}$$

where a,b,c are nonzero design parameters. We are now in a position to complete the design of the switching surface. We form the matrix $[W \vdots B]$ which can be written as

$$\begin{bmatrix} -5.2000 & -4.2500 & -3.6333 & 0 \\ -5.0000 & -4.0000 & -3.3333 & 0 \\ 1.0400 & 1.0625 & 1.0900 & 1 \\ 1.0000 & 1.0000 & 1.0000 & 0 \end{bmatrix} \begin{bmatrix} a & 0 & 0 & 0 \\ 0 & b & 0 & 0 \\ 0 & 0 & c & 0 \\ 0 & 0 & 0 & 1 \end{bmatrix} \tag{4.6}$$

Its inverse is

$$\begin{bmatrix} \frac{1}{a} & 0 & 0 & 0 \\ 0 & \frac{1}{b} & 0 & 0 \\ 0 & 0 & \frac{1}{c} & 0 \\ 0 & 0 & 0 & 1 \end{bmatrix} \begin{bmatrix} -40 & 37 & 0 & -22 \\ 100 & -94 & 0 & 50 \\ -60 & 57 & 0 & -27 \\ 0.75 & -0.735 & 1 & -0.815 \end{bmatrix}.$$

Thus $S = [0.75 \quad -0.735 \quad 1 \quad -0.815]$ and the switching surface is

$$0.75x_1 - 0.735x_2 + x_3 - 0.815x_4 = 0 .$$

5. State-Feedback Controller Design and Stability Analysis

Consider the uncertainty function $F(t,x,u,\delta,w)$ in (2.1). We assume that F can be decomposed as follows:

$$F(t,x,u,\delta,w) = f(t,x,\delta,w) + Bh(t,x,u,\delta,w) \tag{5.1}$$

with

$$f: \mathbb{R} \times \mathbb{R}^n \times \mathbb{R}^s \times \mathbb{R}^l \rightarrow (\text{Range } B)^\perp$$

and

$$h: \mathbb{R} \times \mathbb{R}^n \times \mathbb{R}^m \times \mathbb{R}^s \times \mathbb{R}^l \rightarrow \mathbb{R}^m$$

where $(\text{Range } B)^\perp$ denotes the orthogonal complement of Range B. Furthermore we assume that

$$\|f(t,x,\delta,w)\| \le \alpha_f \|x\| + \beta_f \tag{5.2}$$

and

$$\|h(t,x,u,\delta,w)\| \le \gamma_h \|u\| + \alpha_h \|x\| + \beta_h , \tag{5.3}$$

where α_f, β_f, γ_h, α_h, β_h are known nonnegative constants and $0 \le \gamma_h < 1/\sqrt{m}$. In this paper $\|\cdot\|$ denotes the Euclidean norm for vectors and the induced spectral norm for matrices. With these assumptions, we proceed with the design of a bounded controller. The controller will asymptotically stabilize or practically stabilize the plant (2.1) depending on the numerical values of the constants in (5.2), (5.3). Let $\sigma(x) = Sx = 0$ be the switching surface designed in Section 4 and let

$$\text{sgn}\, \sigma(x) = \begin{bmatrix} \text{sgn}\, \sigma_1(x) \\ \vdots \\ \text{sgn}\, \sigma_m(x) \end{bmatrix}.$$

We will now study the controller

$$u = -k \,\text{sgn}\, \sigma(x) , \tag{5.4}$$

where $k > 0$ is a design parameter. The closed-loop system (2.1), (5.4), (5.1) has the form

$$\dot{x} = Ax + B(-k \,\text{sgn}\, \sigma(x) + h) + f . \tag{5.5}$$

To facilitate the stability analysis of (5.5) we introduce the coordinate transformation, first used by Madani-Esfahani (see Madani-Esfahani et al (1988)). Let W, S be the matrices found in Section 4. Recall that S determines the switching surface and W the eigenstructure of $(I_n - E)A$. Let

$$M = \begin{bmatrix} W^g \\ S \end{bmatrix},$$

where W^g is a left pseudoinverse of W computed in Section 4. Note that $M^{-1} = [W \vdots B]$. Introduce the new coordinates

$$\hat{x} = Mx .$$

The closed-loop system (5.5) in the new coordinates has the form

$$\dot{\hat{x}} = MAM^{-1}\,\hat{x} + MB(-k \operatorname{sgn} \sigma + h) + Mf . \tag{5.6}$$

We write

$$MAM^{-1} = \begin{bmatrix} A_{11} & A_{12} \\ A_{21} & A_{22} \end{bmatrix},$$

where $A_{11} = W^gAW$, $A_{12} = W^gAB$, $A_{21} = SAW$, $A_{22} = SAB$. From Section 4, we have

$$MB = \begin{bmatrix} W^g \\ S \end{bmatrix} B = \begin{bmatrix} 0 \\ I_m \end{bmatrix} .$$

Let $z = W^gx$ and recall that $\sigma(x) = Sx$. Then (5.6) becomes

$$\dot{z} = A_{11}z + A_{12}\sigma + W^g\bar{f} \tag{5.7}$$

$$\dot{\sigma} = A_{21}z + A_{22}\sigma - k \operatorname{sgn} \sigma + \bar{h} + S\bar{f} \tag{5.8}$$

where

$$\bar{f}(t,z,\sigma,\delta,w) = f(t, Wz + B\sigma, \delta, w)$$

and

$$\bar{h}(t,z,\sigma,u,\delta,w) = h(t, Wz + B\sigma, u, \delta, w) .$$

From the assumptions (5.2) and (5.3) we have

$$\|\bar{f}\| \le \alpha_f\|W\|\,\|z\| + \alpha_f\|B\|\,\|\sigma\| + \beta_f \tag{5.9}$$

and

$$\|\bar{h}\| \le \gamma_h\|u\| + \alpha_h\|W\|\,\|z\| + \alpha_h\|B\|\,\|\sigma\| + \beta_h . \tag{5.10}$$

To analyze the stability of the closed-loop system (in the new coordinates) we need some preliminary computations. Suppose $\|\sigma\| \ne 0$. Then along the trajectory of the closed-loop system,

$$\frac{d\|\sigma\|}{dt} = \frac{\sigma^T\dot{\sigma}}{\|\sigma\|} = \frac{\sigma^T}{\|\sigma\|}\,(A_{21}z + A_{22}\sigma - k \operatorname{sgn} \sigma + \bar{h} + S\bar{f}) . \tag{5.11}$$

Let $a_{ij} = \|A\|$, $1 \le i, j \le 2$. Using (5.9) and (5.10) we have

$$\begin{aligned} \frac{d\|\sigma\|}{dt} &\le a_{21}\|z\| + a_{22}\|\sigma\| - k \sum_{i=1}^{m} \frac{|\sigma_i|}{\|\sigma\|} \\ &\quad + \gamma_h k\sqrt{m} + \alpha_h\|W\|\,\|z\| + \alpha_h\|B\|\,\|\sigma\| + \beta_h \\ &\quad + \alpha_f\,\|S\|\,\|W\|\,\|z\| + \alpha_f\,\|S\|\,\|B\|\,\|\sigma\| + \beta_f\|S\| \\ &\le \|z\|(a_{21} + \alpha_h\|W\| + \alpha_f\|S\|\,\|W\|) \\ &\quad + \|\sigma\|(a_{22} + \alpha_h\|B\| + \alpha_f\|S\|\,\|B\|) \\ &\quad + \beta_h + \beta_f\|S\| - k(1 - \gamma_h\sqrt{m}) . \end{aligned} \tag{5.12}$$

In the above manipulations we used the fact that $\sum_{i=1}^{m} |\sigma_i| \ge \|\sigma\|$ and $\|u\| = k\sqrt{m}$. With $\hat{a}_{21}$, $\hat{a}_{22}$, μ defined in the obvious way, (5.12) becomes

$$\frac{d\|\sigma\|}{dt} \leq \hat{a}_{21}\|z\| + \hat{a}_{22}\|\sigma\| - \mu \,. \tag{5.13}$$

Observe that from (4.1) and $W^g B = 0$ we have $A_{11} = W^g A\, W = J$.

Let $\quad \lambda = \min\{|\lambda_1|, \ldots, |\lambda_{n-m}|\}$.

We similarly evaluate for $\|z\| \neq 0$,

$$\frac{d\|z\|}{dt} = \frac{z^T \dot{z}}{\|z\|} = \frac{z^T}{\|z\|}(A_{11}z + A_{12}\sigma + W^g \bar{f})$$

$$\leq -\lambda\|z\| + a_{12}\|\sigma\| + \|W^g\|\,\|\bar{f}\| \,.$$

Using (5.9) we have for $\|z\| \neq 0$,

$$\frac{d\|z\|}{dt} \leq \|z\|(-\lambda + \alpha_f\|W\|\,\|W^g\|) + \|\sigma\|(a_{12} + \alpha_f\|B\|\,\|W^g\|) + \beta_f\,\|W^g\| \,.$$

$$= -\alpha'\,\|z\| + \hat{a}_{12}\|\sigma\| + \beta' \tag{5.14}$$

with α', $\hat{\alpha}_{12}$ and β' defined in the obvious way. In summary we have for $\|z\| \neq 0$, $\|\sigma\| \neq 0$,

$$\boxed{\begin{aligned} \frac{d\|\sigma\|}{dt} &\leq \hat{a}_{21}\,\|z\| + \hat{a}_{22}\|\sigma\| - \mu \\ \frac{d\|z\|}{dt} &\leq -\alpha'\|z\| + \hat{a}_{12}\|\sigma\| + \beta' \quad . \end{aligned}}$$

Recall that $\mu = k(1 - \gamma_h\sqrt{m}) - \beta_h - \beta_f\|S\|$ and $\alpha' = \lambda - \alpha_f\|W\|\,\|W^g\|$, where k, λ are design parameters. By adjusting k, λ we can make μ, α' as large as desired.

Theorem 1.

Let k, λ be chosen so that $\alpha' > 0$ and $\dfrac{\mu}{\hat{a}_{21}} > \dfrac{\beta'}{\alpha'}$. Let

$$\Sigma = \{(z,\sigma) \mid \|\sigma\| < \frac{\mu\alpha' - \hat{a}_{21}\beta'}{\hat{a}_{22}\alpha' + \hat{a}_{12}\hat{a}_{21}}, \quad \hat{a}_{21}\|z\| + \hat{a}_{22}\|\sigma\| < \mu - \epsilon_1\},$$

and let

$$\Gamma = \{(z,0) \mid \|z\| \leq \frac{\beta'}{\alpha'} + \epsilon_2\},$$

where ϵ_2 is any positive constant and $0 < \epsilon_1 < \mu$. Then a trajectory starting in Σ will hit Γ after a finite time and stay in Γ thereafter. That is, the closed-loop system is practically stable with respect to Γ. See Fig. 2 for an illustration of Theorem 1.

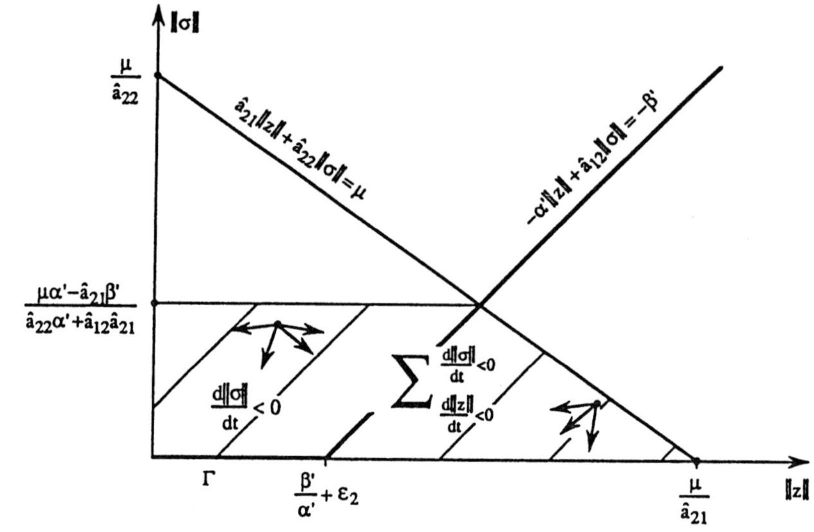

Fig. 2. Region of practical stability given in Theorem 1.

Proof.

It follows from the definition of Σ that $\frac{d||\sigma||}{dt} < -\epsilon_1$ for $(z,\sigma) \in \sum\backslash\{\sigma=0\}$ and $\frac{d||z||}{dt} < 0$ for $\{-\alpha'||z|| + \hat{a}_{12}||\sigma|| < -\beta'\}$. Therefore any trajectory starting in Σ will tend to Γ. See Fig. 2. We only need to verify that it will reach Γ in a finite time. Suppose the trajectory starts in $\Sigma\backslash\{\sigma=0\}$. Since $\frac{d||\sigma||}{dt} < -\epsilon_1$, the trajectory will hit $\{\sigma=0\}$ in no more than $\frac{||\sigma(t_0)||}{\epsilon_1}$ time units. If the trajectory starts in $\Sigma \cap \{\sigma=0\}$ then it will stay there for all subsequent time since $\frac{d||\sigma||}{dt} < -\epsilon_1$ in $\Sigma\backslash\{\sigma=0\}$ and $\frac{d||z||}{dt} < -\epsilon_2$ for $(z,\sigma) \in \{(z,0) \mid ||z|| > \frac{\beta'}{\alpha'} + \epsilon_2\} \cap \sum$. This implies that a trajectory starting in $\Sigma \cap \{\sigma=0\}$ will reach Γ in a finite time and that it stay in Γ thereafter. Combining these two facts completes the proof.

□

Remark 1.

Under the same assumptions on α_f, β_f, γ_h, σ_h, β_h, if an unbounded controller is allowed, then we can use

$$u = -(c_1||x|| + c_2)\text{sgn}\,\sigma(x)\,.$$

With this u we can find c_1, c_2 depending on α_f, β_f, γ_h, α_h, β_h so that the closed-loop systems is globally practically stable. This follows since $\frac{d||\sigma||}{dt}$ can be made negative everywhere along the trajectory of the system except $\{\sigma=0\}$ and the fact that $\frac{d||z||}{dt} < 0$ for $(z,\sigma) \in \{(z,0) \mid -\alpha'||z|| + \hat{a}_{12}||\sigma|| < -\beta'\}$.

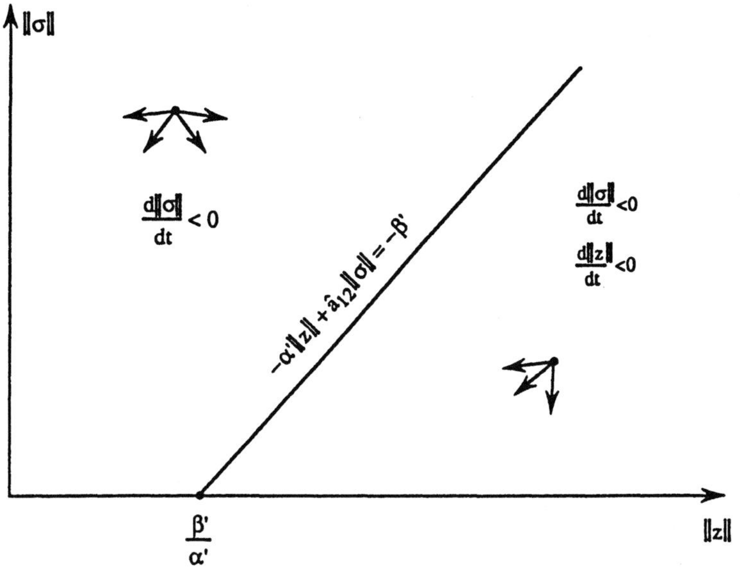

Fig. 3. Illustration of region of global practical stability for unbounded controller.

Remark 2.

In a plant where high speed switching, caused by the discontinuity of the controller (5.4), is undesirable, we can use the following continuous version. Let

$$u_i(x) = \begin{cases} -k \operatorname{sgn} \sigma_i(x) & \text{for } |\sigma_i(x)| \geq \delta_i \\ -k \dfrac{\sigma_i(x)}{\delta_i} & \text{for } |\sigma_i(x)| < \delta_i , \end{cases}$$

and let

$$u(x) = \begin{bmatrix} u_1(x) \\ \vdots \\ u_m(x) \end{bmatrix} .$$

The performance of this controller can be made arbitrarily close to the controller (5.4) we analyzed and the analysis is the same.

We now apply the analysis to construct a state feedback controller for the benchmark problem and estimate the region of practical stability for the closed-loop system driven by this controller. Recall that the uncertainty function of the benchmark problem is given by (2.4) and it can be written in the form required in (5.1) by letting

$$f = \delta(x_1 - x_2) \begin{bmatrix} 0 \\ 0 \\ 0 \\ 1 \end{bmatrix} + \begin{bmatrix} 0 \\ 0 \\ 0 \\ 1 \end{bmatrix} w$$

and

$$h = -\delta(x_1 - x_2) .$$

We need to estimate $||f||$ and $||h||$. Following the discussion at the end of Section 2 we assume that

$|x_1 - x_2|$ is bounded and hence $\alpha_f = \gamma_h = \alpha_h = 0$. Therefore $||f|| \le \beta_f$ and $||h|| \le \beta_h$, where β_f, β_h depend on the bounds of δ, w, $|x_1 - x_2|$. From Section 4, we choose

$$W^g = \begin{bmatrix} -1.3343 & 1.2342 & 0.0000 & -0.7339 \\ 3.3357 & -3.1356 & 0.0000 & 1.6679 \\ -2.0014 & 1.9013 & 0.0000 & -0.9006 \end{bmatrix} .$$

We obtained this W^g by setting $a = b = c = \dfrac{179.8722}{6}$. Note that 179.8722 is the spectral norm of W^g appearing in (4.6). We have

$$M = \begin{bmatrix} -1.3343 & 1.2342 & 0.0000 & -0.7339 \\ 3.3357 & -3.1356 & 0.0000 & 1.6679 \\ -2.0014 & 1.9013 & 0.0000 & -0.9006 \\ 0.7500 & -0.7350 & 1.0000 & -0.8150 \end{bmatrix}$$

and

$$MAM^{-1} = \begin{bmatrix} -0.2 & 0 & 0 & -1.3343 \\ 0 & -0.25 & 0 & 3.3357 \\ 0 & 0 & -0.3 & -2.0014 \\ 12.231 & 15.458 & 18.797 & 0.75 \end{bmatrix}, \quad MB = \begin{bmatrix} 0 \\ 0 \\ 0 \\ 1 \end{bmatrix} .$$

Simple computation shows that

$$||M|| = 6.0449 , \quad a_{12} = 4.1125 , \quad a_{21} = 27.2374 ,$$

$$||S|| = 1.6634 .$$

Thus $\hat{a}_{21} = a_{21} = 27.2374$,

$$\hat{a}_{22} = a_{22} = 0.75 , \; \alpha' = 0.2 ,$$

$$\hat{a}_{12} = a_{12} = 4.1125, \; \beta' = 6\beta_f ,$$

$$\mu = k - \beta_h - 1.6634\beta_f .$$

It follows that the regions Σ and Γ in Theorem 1 are

$$\Sigma = \{(z,\sigma) \mid ||\sigma|| < \frac{0.2\mu - 163.42\beta_f}{112.16} , \quad 27.24||z|| + 0.75||\sigma|| < \mu - \epsilon_1\}$$

and

$$\Gamma = \{(z,0) \mid ||z|| \le 30\beta_f + \epsilon_2\} .$$

6. Design of State Estimator

In Section 5 we synthesized a state-feedback controller. However in many practical situations the state vector of the plant is unavailable. This necessitates the construction of a state estimator, whose output is then used in place of the true state. In this section we assume

A5. F in (2.1) is a bounded function. That is, there is $\overline{\eta} > 0$ so that $||F|| \le \overline{\eta}$. We use the Luenberger (1971) state estimator of the form

$$\dot{\overline{x}} = (A - GC)\overline{x} + Bu + Gy . \qquad (6.1)$$

By the observability of (A,C) we can choose G so that A-GC is asymptotically stable with prescribed eigenvalues. Hence given Q, real symmetric positive-definite, there exists P, real symmetric positive-definite, so that

$$(A - GC)^T P + P(A - GC) = -Q \,. \tag{6.2}$$

The presence of uncertainties in (2.1) has a deteriorating effect on the quality of the state estimation. Let $e = \overline{x} - x$. Then

$$\dot{e} = (A - GC)e - F + Gv \,. \tag{6.3}$$

Let $V(e) = e^T P e$ be a Lyapunov function candidate for (6.3). Compute

$$\begin{aligned}\dot{V} &= 2\, e^T P \dot{e} \\ &= e^T [(A - GC)^T P + P(A - GC)]e - 2e^T PF + 2e^T PGv \\ &\leq -e^T Qe + 2||e||\ ||P||\ (||F|| + ||Gv||) \\ &\leq ||e||(-\lambda_{min}(Q)||e|| + 2\eta\lambda_{max}(P)) \,,\end{aligned}$$

where $\quad (||F|| + ||Gv||) \leq \eta \,.$

Note that $\dot{V}(e) < 0$ outside of the ball $\{e \mid ||e|| \leq \dfrac{2\eta\lambda_{max}(P)}{\lambda_{min}(Q)}\}$. Hence (6.3) is practically stable with respect to any ball with radius greater than

$$2\eta \frac{\lambda_{max}(P)}{\lambda_{min}(Q)} \sqrt{\frac{\lambda_{max}(P)}{\lambda_{min}(P)}} \,. \tag{6.4}$$

For the benchmark problem we choose the poles for the Luenberger estimator at $-0.5 \pm j$, $-1 \pm 2j$. The resulting gain matrix G is $[4.5,\ 3,\ 0,\ 6.25]^T$. Hence

$$A - GC = \begin{bmatrix} 0.00 & -4.50 & 1.00 & 0.00 \\ 0.00 & -3.00 & 0.00 & 1.00 \\ -1.00 & 1.00 & 0.00 & 0.00 \\ 1.00 & -7.25 & 0.00 & 0.00 \end{bmatrix} .$$

Solving the Lyapunov matrix equation (6.2) for $Q = -I_4$ yields

$$P = \begin{bmatrix} 2.0641 & -1.3462 & -0.5000 & -1.0000 \\ -1.3462 & 3.7276 & 1.0000 & -0.5000 \\ -0.5000 & 1.0000 & 2.0590 & -0.0051 \\ 1.0000 & -0.5000 & -0.0051 & 1.3410 \end{bmatrix}$$

We chose $Q = -I_4$ since Patel and Toda (1980) have shown that this choice of Q minimizes $\lambda_{max}(P)/\lambda_{min}(Q)$. The largest and smallest eigenvalues of P are $\lambda_{max}(P) = 4.9156$ and $\lambda_{min}(P) = 0.2096$. The ultimate bound (6.4) on the error of the estimator is given by 47.6102η.

7. Combined Estimator Controller Synthesis

In Section 5 we designed a controller assuming all the states are available for feedback and in Section 6 we designed an estimator for the states. In this section we implement the controller using the estimated states. The equations of the closed-loop system have the form: (see Fig. 4)

$$\dot{x} = Ax + Bu(\overline{x}) + F \tag{7.1}$$

$$y = Cx + v \tag{7.2}$$

$$\dot{\overline{x}} = (A - GC)\overline{x} + Bu + Gy \,. \tag{7.3}$$

We need to analyze the effect of using $\overline{x}$ in place of x on the stability of the closed-loop system.

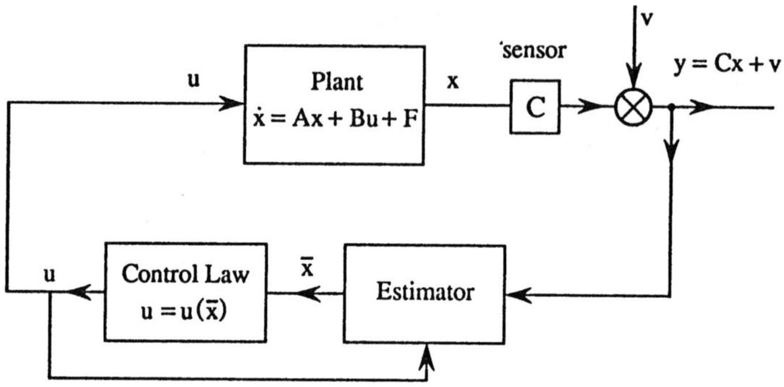

Fig. 4. Combined controller-estimator configuration.

We perform the analysis in the coordinate system introduced in Section 5. Using (5.8), and the same notation and method as in Section 5, we compute, for $||\sigma|| \neq 0$,

$$\frac{d||\sigma||}{dt} = \frac{\sigma^T \dot{\sigma}}{||\sigma||} = \frac{\sigma^T}{||\sigma||}\left(A_{21} z + A_{22}\sigma - k \operatorname{sgn} \sigma(x+e) + \bar{h} + S\bar{f}\right)$$

$$\leq \hat{a}_{21}||z|| + \hat{a}_{22}||\sigma|| + \beta_h + \beta_f||S||$$

$$+ k\,\gamma_h \sqrt{m} - k\,\frac{\sigma^T(x)}{||\sigma(x)||} \operatorname{sgn} \sigma(x+e)\,. \tag{7.4}$$

Observe that for $|\sigma_i| > |\hat{e}_{i+n-m}|$, $i = 1,\dots,m$, we have $\operatorname{sgn}\sigma(x+e) = \operatorname{sgn}\sigma(x)$ and hence (7.4) is the same as (5.12). Therefore we conclude that $\frac{d||\sigma||}{dt} < 0$ for $|\sigma_i| > |\hat{e}_{i+n-m}|$, $i = 1,\dots,m$. This leads to the following analogue of Theorem 1 whose proof is almost identical to that of Theorem 1.

Theorem 2.

Theorem 1 holds for the closed-loop system with the state estimator with

$$\tilde{\Gamma} = \left\{ (z,\sigma) \mid ||z|| \leq \frac{\beta'}{\alpha'} + \epsilon_2\,, \right.$$

$$\left. |\sigma_i| \leq 2\,\eta\,||s_i||\,\frac{\lambda_{max}(P)}{\lambda_{min}(Q)}\sqrt{\frac{\lambda_{max}(P)}{\lambda_{min}(P)}} + \epsilon_2,\ i = 1,\dots,m \right\},$$

and μ chosen so that $\tilde{\Gamma} \subset \Sigma$. That is, the closed-loop system with the estimator is practically stable with respect to $\tilde{\Gamma}$.

Note that the set $\tilde{\Gamma}$ is Γ with a "band" around it. See Fig. 5. So the effect of the estimator is the "fattening" of Γ in the σ-direction.

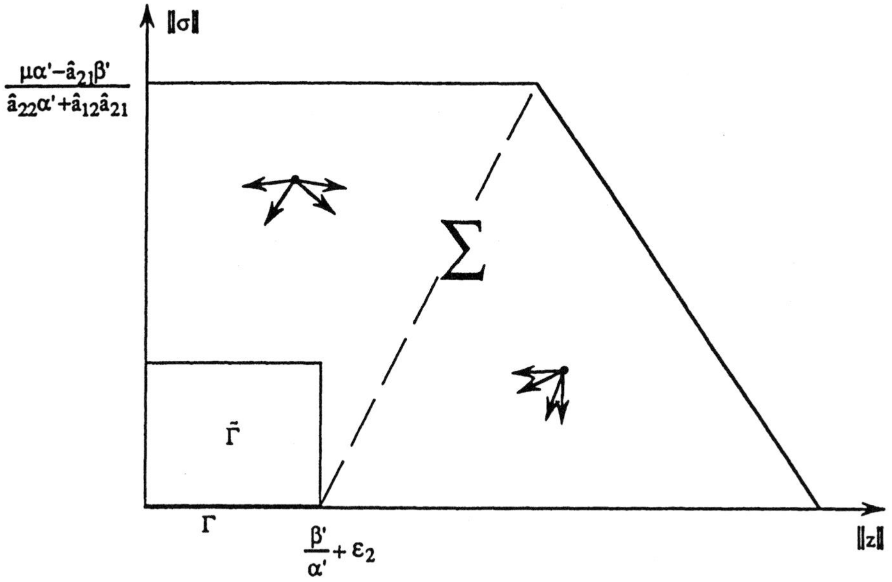

Fig. 5. Region of practical stability of the closed-loop system with the estimator.

For the benchmark problem the region of practical stability of the closed-loop system with the combined estimator-controller is given by

$$\tilde{\Gamma} = \{(z,\sigma) \mid \|z\| \leq 30\beta_f + \epsilon_2 , \quad |\sigma_i| \leq 79.1948\eta + \epsilon_2\} ,$$

where $\epsilon_2 > 0$ is arbitrary. As in Theorem 2, we need to choose μ so that $\tilde{\Gamma} \subset \Sigma$.

8. Conclusions

In this paper we proposed an output feedback control law to stabilize a class of nonlinear/uncertain dynamical systems. The proposed controller is simple to implement and the stability analysis of the closed-loop system can be performed by using the second method of Lyapunov. This approach provides explicit descriptions of the practical stability regions. We illustrated the results on a benchmark problem of Wie and Bernstein (1990).

References

Bondarev, A. G., S. A. Bondarev, N. E. Kostyleva, and V. I. Utkin, (1985) "Sliding modes in systems with asymptotic state observers," Automation and Remote Control, Vol. 46, No. 6, Pt. 1, pp. 679-684.

Corless, M. J., and G. Leitmann, (1981) "Continuous state feedback guaranteeing uniform ultimate boundedness for uncertain dynamical systems," IEEE Trans. Automat. Contr., Vol. AC-26, No. 5, pp. 1139-1144.

DeCarlo, R. A., S. H. Żak, and G. P. Matthews, (1988) "Variable structure control of non-linear multivariable systems: A tutorial," Proceedings of the IEEE, Vol. 76, No. 3, pp. 212-232.

El-Ghezawi, O. M. E., A. S. I. Zinober, and S. A. Billings, (1983) "Analysis and design of variable structure systems using a geometric approach," Int. J. Control, Vol. 38, No. 3, pp. 657-671.

Hui, S., and S. H. Żak, (1990) "Control and observation of uncertain systems: A variable structure systems approach," in Control and Dynamic Systems, Vol. 34, Part 1, Edited by C. T. Leondes, Academic Press, San Diego, pp. 175-204.

Lindorff, D. P., (1967), "Control of nonlinear multivariable systems," IEEE Trans. Automat. Contr., Vol. AC-12, No. 5, pp. 506-515.

Luenberger, D. G., (1971) "An introduction to observers," IEEE Trans. Automat. Contr., Vol. AC-16, No. 6, pp. 596-602.

Madani-Esfahani, S. M., S. Hui, and S. H. Zak, (1988), "On the estimation of sliding domains and stability regions of variable structure control systems," Proc. 26th Annual Allerton Conf. on Communication, Control, and Computing, Monticello, Illinois, pp. 518-527.

Patel, R. V., and M. Toda, (1980) "Quantitative measures of robustness for multivariable systems," Proc. Joint Automatic Control Conf. (JACC), San Francisco, CA, pp. TP8-A.

Ryan, E. P., and M. Corless, (1984) "Ultimate boundedness and asymptotic stability of a class of uncertain dynamical systems via continuous and discontinuous feedback control," IMA Journal of Mathematical Control & Information, Vol. 1, pp. 223-242.

Wie, B., and D. S. Bernstein, (1990) "A benchmark problem for robust control design," Proc. 1990 American Control Conf. (ACC), San Diego, CA, pp. 961-962.

Utkin, V. I., (1977) "Variable structure systems with sliding modes," IEEE Trans. Automat. Contr., Vol. AC-22, No. 2, pp. 212-222.

Utkin, V. I., (1978) "Sliding Modes and Their Application in Variable Structure Systems," Moscow, Mir Publishers.

OBSERVERS FOR CONTROLLERS WITH MATCHED UNCERTAINTY
IN A, B AND C MATRICES

Faryar Jabbari

Department of Mechanical Engineering
University of California
Irvine, CA 92717

ABSTRACT : In this paper, the effects of using observers on robust linear state feedback controllers are studied. The uncertainty, which can enter A, B or C matrices, is assumed to satisfy certain matching conditions. Lyapunov techniques are used to establish sufficient condition for stability for a given uncertainty bound. In particular, sufficient conditions are obtained that, if met, guarantee stabilization regardless of the size of the uncertainty.

1 Introduction

In this note, we study observer based stabilizing controllers for linear uncertain systems. In particular, we are interested in the problem of complete stabilizability; i.e., stabilizing the system for arbitrarily large amount of uncertainty. With full state feedback, it is well known that complete stabilizability is achievable only if the uncertainty satisfies the so called matching conditions. Consequently, we consider systems which satisfy matching conditions and investigate those properties of the nominal system that guarantee the stabilizability of the closed loop system via observer based controllers, for uncertainty arbitrary large but consistent with appropriate matching conditions.

Early work on robust stability problems with observers was presented in [1,2,3,4,5,6]. In [1] solvability of a related constrained Lyapunov equation studied, while in [2] certain inequalities based on the controller and observer gains are checked to determine stability. The structure of the uncertainty matrices ΔA and ΔB are used in [3] to obtain a similarity transformation. Observability of certain subsystems, in the transformed coordinates, implies the possibility of designing observers insensitive to uncertainty. The methods in [5,6] rely on two non-definite Riccati equations and certain relationships between their solutions. The approach used here is along the lines of [7] and

[8], where the uncertainty entered the A matrix only. We focus on developing conditions that concern the nominal system and are relatively easy to check (such as minimum-phase property or left invertibility). The overall approach here is based on the Lyapunov techniques so that the time domain framework is preserved, along with the ability to readily incorporate time varying uncertainty. As such, this method is a natural extension of standard results established via the Lyapunov approach. Lastly, we provide results that can also apply to systems that do not satisfy any additional conditions other than those used in the full state feedback problem.

While uncertainty is allowed to enter A, B and C matrices, the strongest results are obtained when either ΔB or ΔC is zero. In Section 2, it is shown that when ΔC is zero, high gain observers can be designed so that the closed system is stabilized. The matching conditions used are the well known conditions (e.g., $\Delta A = BD$). Also, the nominal system is required to be minimum phase and left invertible. In Section 3, we show that when ΔB is zero, high gain controllers can be designed so that the closed system is stabilized. The matching conditions used in this case are based on the C matrix (e.g., $\Delta A = FC$). These results are reminiscent of the results of [9], for the uncertainty entered the A matrix only, where input and output similarity are discussed, along with the connections to traditional results in the loop transfer recovery.

2 Preliminaries

We consider systems described by

$$\dot{x}(t) = [A + \Delta A]x(t) + [B + \Delta B]u(t) \quad , \quad y(t) = [C + \Delta C]x(t) \tag{1}$$

where $x(t) \in R^n$ is the state, $y(t) \in R^p$ is the measured output and $u(t) \in R^m$ is the control. The matrices A, B and C are the nominal system matrices and it is assumed that the nominal system is both observable and controllable. Throughout this section, we allow matrices ΔA, ΔB and ΔC be time varying; e.g., the system uncertainty matrix be of the form $\Delta A(r(t))$, where $r(t)$ is a vector in a compact set $\mathcal{R} \subset R^q$.

Here, we investigate the issue of observer design for the case when the entire state is not available for feedback . It is well known that when uncertainty satisfies the matching conditions, full state controllers exist that stabilize the system regardless of the size of ΔA, [7,8]. Our focus, therefore, is the possibility of stabilizing the system regardless of the size of ΔA when observers are used. When ΔB and ΔC are both zero, this problem becomes a special case of the results in [8], where the matching condition was not imposed on the uncertainty.

Consider the following structure for observer based controllers. The control law is of the form

$$u(t) = -\gamma_c B^T P_c z(t) \tag{2}$$

where $z(t)$ is from the (full order) observer equation

$$\dot{z}(t) = Az + Bu + \gamma_o P_o^{-1} C^T (y - Cz). \tag{3}$$

We now focus on choosing the positive definite matrices P_o and P_c as well as positive constants γ_o and γ_c such that the closed loop system is stabilized via (2) and (3). Defining the observer error

$$e(t) \triangleq x(t) - z(t) \tag{4}$$

and combining (1) and (3), we obtain

$$\dot{e}(t) = [A - \gamma_o P_o^{-1} C^T C + \gamma_c \Delta B B^T P_c] e(t) + [\Delta A - \gamma_o P_o^{-1} C^T \Delta C - \gamma_c \Delta B B^T P_c] x(t). \tag{5}$$

Using the control law of (2), equations (1) and (5) can be combined into

$$\begin{bmatrix} \dot{x} \\ \dot{e} \end{bmatrix} = \begin{bmatrix} A+\Delta A-\gamma_c BB^T P_c-\gamma_c \Delta BB^T P_c & \gamma_c BB^T P_c+\gamma_c \Delta BB^T P_c \\ \Delta A-\gamma_o P_o^{-1} C^T \Delta C-\gamma_c \Delta BB^T P_c & A-\gamma_o P_o^{-1} C^T C+\gamma_c \Delta BB^T P_c \end{bmatrix} \begin{bmatrix} x \\ e \end{bmatrix}. \tag{6}$$

For simplicity, we rewrite (6) as

$$\dot{\tilde{x}} = \tilde{A}\tilde{x} \tag{7}$$

with obvious definitions for $\tilde{A}$ and $\tilde{x}$. Introducing the Lyapunov function

$$V(x,e) = [x^T \;\; e^T] \begin{bmatrix} P_c & 0 \\ 0 & P_o \end{bmatrix} \begin{bmatrix} x \\ e \end{bmatrix} = \tilde{x}^T \tilde{P} \tilde{x} \tag{8}$$

for the system (6), we obtain, after standard manipulations, the following derivative of the Lyapunov function

$$\begin{aligned} \dot{V} = \tilde{x}^T & \begin{bmatrix} P_c A + A^T P_c - 2\gamma_c P_c BB^T P_c & \gamma_c P_c BB^T P_c \\ \gamma_c P_c BB^T P_c & P_o A + A^T P_o - 2\gamma_o C^T C \end{bmatrix} \tilde{x} \\ +\tilde{x}^T & \begin{bmatrix} P_c \Delta A + \Delta A^T P_c & \Delta A^T P_o - \gamma_o \Delta C^T C \\ P_o \Delta A - \gamma_o C^T \Delta C & 0 \end{bmatrix} \tilde{x} \\ +\tilde{x}^T & \begin{bmatrix} -\gamma_c P_c B \Delta B^T P_c - \gamma_c P_c \Delta BB^T P_c & -\gamma_c P_c B \Delta B^T P_o + \gamma_c P_c \Delta BB^T P_c \\ -\gamma_c P_o \Delta BB^T P_c + \gamma_c P_c B \Delta B^T P_c & \gamma_c P_c B \Delta B^T P_o + \gamma_c P_o \Delta BB^T P_c \end{bmatrix} \tilde{x}. \end{aligned} \tag{9}$$

Using standard techniques, it is straight forward to show that the closed loop system, (6), is asymptotically stable if appropriate positive definite matrices P_c , P_o and positive scalars γ_c, γ_o can be found that guarantee the Lyapunov derivative in (9) is negative. Obtaining these matrices and scalars is the focus of the following two sections. We will rely on the following matrix identities. For any two matrices of appropriate dimension X and Y

$$XY^T + YX^T \leq \frac{1}{a} XGX^T + aYG^{-1}Y^T \tag{10}$$

where a is any positive scalar and G is any positive definite matrix of appropriate dimension. Similarly

$$\begin{bmatrix} 0 & XY^T \\ YX^T & 0 \end{bmatrix} \leq \begin{bmatrix} \frac{1}{a} XGX^T & 0 \\ 0 & aYG^{-1}Y^T \end{bmatrix} \tag{11}$$

Finally, throughout this paper, by norm we mean the induced 2-norm or the maximum singular value;i.e., $\|E\| = \sigma_{max}(E) = \sqrt{\lambda_{max}(EE^T)}$.

3 No uncertainty in the C matrix; $\Delta C = 0$

In this section, we focus on the case where $\Delta C = 0$. For matching conditions, we use the well known form of

$$\Delta A(r) = B\,D(r(t)), \qquad \Delta B(s) = B\,E(s(t)), \tag{12}$$

where $r(t) \in \mathcal{R} \subset R^q$ and $s(t) \in \mathcal{S} \subset R^l$, with $\mathcal{R}$ and $\mathcal{S}$ compact sets. As usual, we require

$$\|E\| < 1, \qquad \forall s \in \mathcal{S}. \tag{13}$$

Naturally, there exists a positive scalar σ so that

$$EE^T < I, \quad E^TE < I\,, \quad 2I + E + E^T > \sigma I, \quad \sigma > 0 \tag{14}$$

Next, we use (10) and (11) to simplify the expression for the Lyapunov derivative in (9).

$$P_c\Delta A + \Delta A^T P_c \le \beta P_c BB^T P_c + \frac{1}{\beta} D^T D \tag{15}$$

$$\begin{bmatrix} 0 & \Delta A^T P_o \\ P_o\Delta A & 0 \end{bmatrix} = \begin{bmatrix} 0 & D^T B^T P_o \\ P_o BD & 0 \end{bmatrix} \le \begin{bmatrix} \frac{1}{\beta} D^T D & 0 \\ 0 & \beta P_o BB^T P_o \end{bmatrix} \tag{16}$$

$$\gamma_c \begin{bmatrix} 0 & P_c BB^T P_c \\ P_c BB^T P_c & 0 \end{bmatrix} \le \gamma_c \begin{bmatrix} \frac{1}{\beta} P_c BB^T P_c & 0 \\ 0 & \beta P_c BB^T P_c \end{bmatrix} \tag{17}$$

$$\gamma_c \begin{bmatrix} 0 & P_c B\Delta B^T P_o \\ P_o \Delta BB^T P_c & 0 \end{bmatrix} \le \gamma_c \begin{bmatrix} \frac{1}{\beta} P_c BB^T P_c & 0 \\ 0 & \beta P_o BB^T P_o \end{bmatrix} \tag{18}$$

$$\gamma_c \begin{bmatrix} 0 & P_c \Delta BB^T P_c \\ P_c B\Delta B^T P_c & 0 \end{bmatrix} \le \gamma_c \begin{bmatrix} \frac{1}{\beta} P_c BB^T P_c & 0 \\ 0 & \beta P_c BB^T P_c \end{bmatrix} \tag{19}$$

$$\gamma_c(P_cB\Delta B^TP_o+P_o\Delta BB^TP_c)\leq\gamma_c(P_oBB^TP_o+P_cBB^TP_c) \quad (20)$$

for any positive constant β. While different constants can be used in each inequality, using the same constant simplifies the development to follow. Using the above inequalities in (9) results in

$$\dot V\leq x^TM_1x+e^TM_2e \quad (21)$$

where

$$M_1=P_cA+A^TP_c+\frac{2}{\beta}D^TD-\gamma_cP_cB(2I+E+E^T-\frac{\beta}{\gamma_c}I-\frac{3}{\beta}I)B^TP_c \quad (22)$$

$$M_2=P_oA+A^TP_o-2\gamma_oC^TC+(\beta+\beta\gamma_c+\gamma_c)P_oBB^TP_o+(1+2\beta)\gamma_cP_cBB^TP_c. \quad (23)$$

In light of (21)-(23), we have the following preliminary lemma

Lemma 1 *If there exist suitable scalars γ_c, γ_o and β and positive definite matrices P_c and P_o such that M_1 and M_2 are negative definite, then* (2) *and* (3) *form a stabilizing control law for* (1), *with* $\Delta C=0$.

We now turn our attention to the main problem of interest, namely, choosing appropriate scalars and matrices that result in closed loop stability. Let σ be the positive scalar defined in (14). For this constant, consider the following two algebraic Riccati equations

$$PA+A^TP-\frac{9}{\sigma}PBB^TP+\frac{2\sigma}{9}\hat D+Q_1=0\quad,\quad Q_1>0 \quad (24)$$

$$S(A+\alpha I)^T+(A+\alpha I)S-\gamma_oSC^TCS+\frac{9(\sigma^2+3\sigma+27)}{\sigma^3}BB^T=0 \quad (25)$$

where γ_o, α and Q_1 are unspecified positive (definite) scalars and matrices. Also, along the lines of [10], $\hat D$ is chosen so that

$$\hat D\geq D^TD,\quad\forall\, r\in\mathcal{R}. \quad (26)$$

Note that controllability and observability of the nominal plant are sufficient conditions for both equations to possess positive definite solutions. (Indeed these condition can be weakened to appropriate stabilizability and detectability conditions, see [11]). We can state the following theorem

Theorem 1 *Assume the nominal system is both controllable and observable. If there exist positive scalars α and γ_o and positive matrix Q_1 such that the solutions of* (24) *and* (25) *satisfy*

$$R = -2\alpha S^{-1} - \gamma_o C^T C + \frac{27(\sigma + 18)}{\sigma^3} PBB^T P < 0 \tag{27}$$

then the control law of (2) *and* (3) *is stabilizing for P_c as the solution of* (24) *and P_o as the inverse of the solution of* (25).

Proof: For (22) and (23), choose

$$\beta = \frac{9}{\sigma} \longrightarrow \frac{3}{\beta} = \frac{\sigma}{3}, \; and \; \gamma_c = \frac{27}{\sigma^2} \longrightarrow \frac{\beta}{\gamma_c} = \frac{\sigma}{3}. \tag{28}$$

Letting $P_c = P$, and $P_o = S^{-1}$, where P and S are the positive definite solutions of (24) and (25), respectively, and using (28), it is easy to show that $M_1 < -Q_1$ and M_2 is the same as R of (27), which by the statement of the theorem implies M_2 is negative definite. The theorem follows by invoking Lemma 1. □

It is easy to show that for small enough ΔA and ΔB, i.e., $\hat{D}$ and $2 - \sigma$ small enough, the condition of the theorem is satisfied. Therefore, the theorem will produce a stabilizing control law for some nonzero amount of uncertainty. Also, through iterations on the design variables, estimates can be for amounts of uncertainty tolerated by the observer based controller. In general, σ and Q_1 are chosen as 'small' as possible. Iteration on the observer equation, by increasing the value of γ_o would produce nonincreasing solutions S, which in turn might allow larger amounts of uncertainty. In general, P_o cannot be made arbitrarily large. Consequently, this method may not result in a stabilizing controller for arbitrary magnitude of the uncertainty.

To strengthen this result, we need to determine conditions under which we can *guarantee* the existence of a stabilizing control law for *any* level of uncertainty satisfying (12)-(14). For this, we introduce additional assumptions on left invertibility and zero locations of the nominal transfer function. By zeros we mean the zeros of the numerator polynomials of the McMillan form of the transfer function, [12]. By left invertible we mean the transfer function $C(sI - A)^{-1}B$ has full column rank for almost all s and the rank of C is equal or greater than rank of B, [12,13].

Theorem 2 *Assume the nominal system is controllable and observable. If the transfer function $C(sI-A)^{-1}B$ has no zeros on the closed right half plane and is left invertible, then there exist an α and a γ_o^* such that for any $\gamma_o \geq \gamma_o^*$, the corresponding S (from* (25)*), results in a negative definite M_2 in* (23)*, and the control law of* (2)-(3) *is stabilizing for uncertainty bounds satisfying* (12)-(14).

Proof : The theorem follows from Theorem 1 and the results of [12] concerning the asymptotic behavior of (25). Let β, γ_c, P_c and P_o be as in Theorem 1. Since the transfer function $C(sI-A)^{-1}B$ has no zeros on the closed right half plane, there exists a $\alpha > 0$ such that $C(sI-A-\alpha I)^{-1}\hat{B}$ has no zeros on the open right half plane. Therefore, from the corollary to Theorem 2 of [12], the solution of (25) goes to zero as γ_o goes to infinity, $\lim_{\gamma_o \to \infty} S = 0$. As a result, by increasing γ_o, S becomes as close to zero as desired, and thus R (in (27)) becomes negative definite since σ and P_c are fixed and do not change with γ_o. The theorem follows from Lemma 2 and Theorem 1. □

In general, the closer σ gets closer to zero, and the larger the size of the ΔA, the larger will be the required γ_o.

Remark 1. Theorem 1 does not introduce any new assumption regarding the nominal system or the structure of the uncertainty, beyond those used in the full state problem. Theorem 2 introduces the conditions on zeros which are needed to ensure P_o can be made as large as required. If this assumption is not met, the following modification can be used: In equation (25) replace the term BB^T by $BB^T + NN^T$, where the matrix N is chosen such that $(A, [B\, N])$ is controllable and $C(sI-A)^{-1}[B\, N]$ is left invertible with zeros on the open left half plane. Since adding a column to B neither destroys controllability nor increases the number of zeros, Theorems 1 or 2 can now be invoked after replacing (27) with $-2\sigma S^{-1} - \gamma_c C^T C - \frac{9(\sigma^2+3\sigma+27)}{\sigma^3} S^{-1} N N^T S^{-1} + \frac{27(\sigma+18)}{\sigma^3} PBB^T P$. Note that this technique can be used in Theorem 2 only if $rank(C) > rank(B)$.

3.1 Example:

Consider the system

$$\dot{x}(t) = \begin{bmatrix} 0 & 1 \\ -4 + r(t) & 0 \end{bmatrix} x(t) + \begin{bmatrix} 0 \\ 1 + s(t) \end{bmatrix} u(t)$$

where $|r(t)| < 1$ and $|s(t)| < 0.1$. The measurement $y(t)$ is given by $y(t) = \begin{bmatrix} 1 & 0 \end{bmatrix} x(t)$. Since there are no zeros for this system, the results from Theorem 2 apply. The uncertainty in A is of the form

$$\Delta A = \begin{bmatrix} 0 & 0 \\ r(t) & 0 \end{bmatrix} = B \begin{bmatrix} r(t) & 0 \end{bmatrix} = BD$$

and the uncertainty in B is of the form

$$\Delta B = \begin{bmatrix} 0 \\ s(t) \end{bmatrix} = Bs(t) = BE.$$

To solve for P in equation (24) we used $\sigma = 1.8, \hat{D} = I$, and $Q_1 = 10^{-5} I$ resulting in the controller gains $K = -\gamma_c B^T P_c = \begin{bmatrix} -0.40441 & -2.6275 \end{bmatrix}$. To solve for S in equation (25), we tried $\gamma_o = 1$, but the resulting R in (27) is not negative definite. Increasing γ_o to 30,000, however, results in R being negative definite with the resulting observer gains $L = \gamma_o P_0^{-1} C^T = \begin{bmatrix} 72,626 & 1911 \end{bmatrix}^T$. These controller and observer gains result in a stable closed loop system for all allowable variations in $r(t)$ and $s(t)$.

4 No uncertainty in the B matrix; $\Delta B = 0$

In this section, we focus on the case where $\Delta B = 0$. For matching conditions, we use the following structure (recall the output similarity structure discussed in [9])

$$\Delta A(r) = G(r(t))\,C, \qquad \Delta C(s) = F(s(t))\,C, \tag{29}$$

where $r(t) \in \mathcal{R} \subset R^q$ and $s(t) \in \mathcal{S} \subset R^l$, with $\mathcal{R} and \mathcal{S}$ compact sets. Parallel to the previous section, we require

$$||F|| < 1, \qquad \forall s \in \mathcal{S}. \tag{30}$$

Equation (30) implies that there exists a positive scalar δ so that

$$FF^T < (1-\delta)I, \quad F^TF < (1-\delta)I \;, \; \delta > 0 \tag{31}$$

As before, (10) and (11) are used to simplify the expression for the Lyapunov derivative.

$$P_c\Delta A + \Delta A^T P_c = P_cGC + G^TC^TP_c \leq \beta_1 P_cGG^TP_c + \frac{1}{\beta_1}C^TC \tag{32}$$

$$\begin{bmatrix} 0 & \Delta A^T P_o \\ P_o\Delta A & 0 \end{bmatrix} = \begin{bmatrix} 0 & C^TG^TP_o \\ P_oGC & 0 \end{bmatrix} \leq \begin{bmatrix} \frac{1}{\beta_2}C^TC & 0 \\ 0 & \beta_2 P_oGG^TP_o \end{bmatrix} \tag{33}$$

$$\gamma_o \begin{bmatrix} 0 & \Delta C^T C \\ C^T\Delta C & 0 \end{bmatrix} = \gamma_o \begin{bmatrix} 0 & C^TFC \\ C^TFC & 0 \end{bmatrix} \leq \gamma_o \begin{bmatrix} \frac{1}{\beta_3}C^TF^TFC & 0 \\ 0 & \beta_3 C^TC \end{bmatrix} \tag{34}$$

Using the above inequalities in (9) results in

$$\dot{V} \leq x^TN_1x + e^TN_2e \tag{35}$$

where

$$N_1 = P_cA + A^TP_c - \gamma_c P_cBB^TP_C + (\frac{1}{\beta_1} + \frac{1}{\beta_2} + \frac{\gamma_o(1-\delta)}{\beta_3})C^TC + \beta_1 P_cGG^TP_c \quad (36)$$

$$N_2 = P_oA + A^TP_o - 2\gamma_o C^TC + \beta_3\gamma_o C^TC + \beta_2 P_oGG^TP_o + \gamma_c P_cBB^TP_c. \quad (37)$$

In light of (35)-(37), we have the following preliminary lemma

Lemma 2 *If there exist suitable scalars γ_c, γ_o and β_1-β_3 and positive definite matrices P_c and P_o such that N_1 and N_2 are negative definite, then* (2) *and* (3) *form a stabilizing control law for* (1), *with $\Delta B = 0$.*

Let δ be the positive scalar defined in (31). For this constant, consider the following two algebraic Riccati equations

$$P(A+\alpha I) + (A+\alpha I)^TP - \gamma_c PBB^TP + \gamma_o(1-\frac{\delta}{2})C^TC = 0 \quad (38)$$

$$SA^T + AS - \frac{\delta\gamma_o}{2}SC^TCS + \frac{4}{\delta\gamma_o}\hat{G} + Q_2 = 0, \quad Q_2 > 0. \quad (39)$$

where γ_c, α and Q_2 are unspecified positive (definite) scalars and matrices. Again, $\hat{G}$ is chosen so that

$$\hat{G} \geq G^TG, \ \forall\, r \in \mathcal{R}. \quad (40)$$

Theorem 3 *Assume the nominal system is both controllable and observable. If there exist positive scalars α and γ_c and positive matrix Q_2 such that the solutions of* (38) *and* (39) *satisfy*

$$R_1 = -2\alpha P_c + \frac{4}{\delta\gamma_o}P_c\hat{G}P_c = -P_c[2\alpha I - \frac{4}{\delta\gamma_o}\hat{G}P_c] < 0 \quad (41)$$

and

$$R_2 = -(1-\frac{\delta}{2})\gamma_o C^TC + \gamma_c P_cBB^TP_c - P_oQ_2P_o < 0 \quad (42)$$

then the control law of (2) *and* (3) *is stabilizing for P_c as the solution of* (38) *and P_o as the inverse of the solution of* (39).

Proof: For (36) and (37), choose

$$\beta_1 = \beta_2 = \frac{4}{\delta\gamma_o}, \quad \beta_3 = 1 \tag{43}$$

Letting $P_c = P$, and $P_o = S^{-1}$, where P and S are the positive definite solutions of (38) and (39), respectively, and using (43), it is easy to show that $N_1 \leq R_1$ and $N_2 \leq R_2$, which by the statement of the theorem are negative definite. The theorem follows by invoking Lemma 2. □

As before, it is easy to show that for small enough ΔA and ΔC, the condition of the theorem is satisfied, and the theorem can be used to estimate the allowable uncertainty level. This method, however, may not result in a stabilizing controller for arbitrary magnitude of the uncertainty. To strengthen this result, we introduce additional assumptions.

Theorem 4 *Assume the nominal system is controllable and observable. If the transfer function* $C(sI - A)^{-1}B$ *has no zeros on the closed right half plane and is right invertible, then there exist an* α *and a* γ_c^* *such that for any* $\gamma_c \geq \gamma_c^*$*, the corresponding* P *(from* (38)*), results in a negative definite* N_1 *and* N_2*, and the control law of* (2)-(3) *is stabilizing for uncertainty bounds satisfying* (29)-(31).

Proof: The theorem follows from Theorem 3 and the asymptotic behavior of (38). Let β_1-β_3 be as in (43), and P_o be the solution of (39) for some γ_o. Since the transfer function $C(sI - A)^{-1}B$ has no zeros on the closed right half plane, there exists a $\alpha > 0$ such that $C(sI - A - \alpha I)^{-1}B$ has no zeros on the open right half plane. Therefore, the solution of (38) goes to zero as γ_c goes to infinity, $\lim_{\gamma_c \to \infty} P = 0$, which implies the term in the bracket in (41) will be positive definite for large enough γ_c. Also, $\lim_{\gamma_c \to \infty} PBB^TP = (1 - \frac{\delta}{2})C^TC$, which implies that term in brackets in (42) can be made arbitrary small (recall that γ_o and P_o are fixed). The theorem follows from Lemma 2 and Theorem 3. □.

Remark 2. Note that for Theorem 4, the number of actuators is required to be equal or larger than the number of sensors. A square, invertible and minimum phase nominal plant, which is both controllable and observable, meets the conditions of both Theorem 2 and Theorem 4.

4.1 Example:

Consider a two-mass spring system of the form

$$\dot{x}(t) = \begin{bmatrix} 0 & 1 & 0 & 0 \\ -1-r(t) & 0 & 1+r(t) & 0 \\ 0 & 0 & 0 & 1 \\ 1+r(t) & 0 & -1-r(t) & 0 \end{bmatrix} x(t) + \begin{bmatrix} 0 & 0 \\ 1 & 0 \\ 0 & 0 \\ 0 & 1 \end{bmatrix} u(t).$$

The outputs are given by

$$y(t) = \begin{bmatrix} 1+s_1(t) & 0 & 0 & 0 \\ 0 & 0 & 1+s_2(t) & 0 \end{bmatrix} x(t)$$

where $|r(t)| < 0.5, |s_1(t)| < 0.2$, and $|s_2(t)| < 0.2$. For this system there are no transmission zeros and, hence, the results from Theorem 4 apply. The uncertainty in A is of the form

$$\Delta A = \begin{bmatrix} 0 & 0 & 0 & 0 \\ -r(t) & 0 & r(t) & 0 \\ 0 & 0 & 0 & 0 \\ r(t) & 0 & -r(t) & 0 \end{bmatrix} = \begin{bmatrix} 0 & 0 \\ -r(t) & r(t) \\ 0 & 0 \\ r(t) & -r(t) \end{bmatrix} \begin{bmatrix} 1 & 0 & 0 & 0 \\ 0 & 0 & 1 & 0 \end{bmatrix} = GC$$

and the uncertainty in C is of the form

$$\Delta C = \begin{bmatrix} s_1(t) & 0 & 0 & 0 \\ 0 & 0 & s_2(t) & 0 \end{bmatrix} = \begin{bmatrix} s_1(t) & 0 \\ 0 & s_2(t) \end{bmatrix} C = FC.$$

The uncertainties in A and C lead to the use of $\hat{G} = I$ in (40) and $\delta = 0.96$ in (31). With $\gamma_c = 10^5, \gamma_o = 100, \alpha = 1$, and $Q_2 = I$, we obtain P and S in (38) and (39) which result negative definite R_1 and R_2. The resulting controller and observer gains are

$$K = \begin{bmatrix} -2347.9 & -69.533 & -1.0139 & -0.014795 \\ -1.0139 & -0.014795 & -2347.9 & -69.583 \end{bmatrix}, \; L = \begin{bmatrix} 16.458 & 0.22714 \\ 12.937 & 1.7944 \\ 0.22714 & 16.458 \\ 1.7944 & 12.937 \end{bmatrix}.$$

With these gains, the closed loop system is stable for all allowable parameter variations.

References

[1] A.R Galimidi and B.R. Barmish, "The constrained Lyapunov problem and its application to robust output stabilization," *IEEE Transactions on Automatic Control*, vol. 31, pp. 410–419, 1986.

[2] B.R. Barmish and A.R. Galimidi, "Robustness of Luenberger observers: linear systems stabilized via nonlinear control," *Automatica*, vol. 22, pp. 413–423, 1986.

[3] V.W. Breinl and P.C. Muller, "Ein parameterunempfindlicher zustandsbobacher und seine anwendung bei einem tragregelsystem eines magnetschwebefahrzeugs," *Regelungstechnik*, vol. 30, pp. 403–412, 1982.

[4] V.W. Breinl and G. Leitmann, "State feedback for uncertain dynamical systems," *Applied Mathematics and Computation*, vol. 22, pp. 65–87, 1987.

[5] I.R. Petersen, "A Riccati equation approach to the design of stabilizing controllers and observers for a class of uncertain linear systems," *IEEE Transactions on Automatic Control*, vol. 30, pp. 904-907, September 1985.

[6] I.R. Petersen and C.V. Hollot, "High gain observers applied to problems in stabilization of uncertain linear systems, disturbance attenuation and H_∞ optimization," *International of Adaptive Control and Signal Processing*, vol. 2, pp. 347–369, 1988.

[7] F. Jabbari and W.E. Schmitendorf, "Robust linear controllers using observers," 1990. To appear in *IEEE Trans. on Automatic Control*.

[8] F. Jabbari and W.E. Schmitendorf, "Effects of observers on stabilization of linear uncertain system," . Proceedings of 1991 *ACC*.

[9] M. Tahk and J.L. Speyer, "Modeling of parameter variations and asymptotic LQG synthesis," *IEEE Transactions on Automatic Control*, vol. 32, pp. 793–801, September 1987.

[10] F. Jabbari and W.E. Schmitendorf, "A non-iterative method for design of linear robust controllers," *IEEE Transactions on Automatic Control*, vol. 35, pp. 2954–957, August 1990.

[11] H. Kano, "Existence condition of positive definite solutions for algebraic matrix Riccati equations," *Automatica*, vol. 23, pp. 393–397, June 1987.

[12] B.A. Francis, "The optimal linear quadratic time invariant regulator with cheap control," *IEEE Transactions on Automatic Control*, vol. 24, pp. 616–621, August 1979.

[13] T. Kailath, *Linear Systems*. New Jersey: Prentice-Hall, 1980.

[14] R.J. Veillette, J.V. Medanic, and W.R. Perkins, "Robust stabilization and disturbance rejection for systems with structured uncertainty," in 28th *IEEE Conference on Decision and Control*, (Tampa, Florida), pp. 936–941, December 1989.

PARAMETRIC ROBUST CONTROL
BY QUANTITATIVE FEEDBACK THEORY

Osita Nwokah
School of Mechanical Engineering
Purdue University
West Lafayette, Indiana 47907

Suhada Jayasuriya
Department of Mechanical Engineering
Texas A&M University
College Station, Texas 77843-3123

Yossi Chait
Department of Mechanical Engineering
University of Massachusetts
Amherst, MA 01003

Keywords: Parametric uncertainty, QFT, robust control.

ABSTRACT

The problem of performance robustness, especially in the face of significant parametric uncertainty, has been increasingly recognized as a predominant issue of engineering significance in many design applications. Quantitative feedback theory (QFT) is very effective for dealing with this class of problems even when there exist hard constraints on closed loop response. In this paper, SISO-QFT is viewed formally as a sensitivity constrained multi objective optimization problem which can be used to set up a constrained H^{∞} minimization problem whose solution provides an initial guess at the QFT solution. In contrast to the more recent robust control methods where phase uncertainty information is often neglected, the direct use of parametric uncertainty and phase information in QFT results in a significant reduction in the cost of feedback. An example involving a standard problem is included for completeness.

List of Symbols Used.

L_∞: Banach space of essentially bounded Baire functions

H^∞: Banach space of bounded analytic functions.

RH^∞: Banach space of bounded analytic functions with elements from the ring of stable, proper real rational functions.

Unit of RH^∞: An element of RH^∞ whose inverse $\in RH^\infty$.

e,r: relative degree of transfer function

SISO: single input, single output

MIMO: multi-input, multi-output

ω, λ: radian frequency

Ω: compact parameter space with elements α.

QFT: Quantitative Feedback Theory

1. Introduction

The last decade has witnessed a steady and growing research effort in robust control; see for example the recent book [9]. The majority of this effort has been devoted to systems that are assumed to have unstructured uncertainty. This allows such problems to be transformed into a form where the small gain theorem [10] and powerful recent mathematical techniques from functional analysis and operator theory [11,12] can be successfully employed for system analysis and synthesis. However many problems of practical interest appear as models with both large parametric uncertainty and high frequency non-parametric uncertainty. Typical examples include flight control and tubomachinery control over a flight envelope parametrized by power level, height and mach number, as well as general automotive engine control problems. All these problems yield a collection of linear models obtained by a linearization of a parametrically dependent nonlinear differential equation set about a finite number of different operating points. This problem class is often endowed with hard stability and performance constraints such as on rise time and overshoot. This problem class also requires plant uncertainty to be expressed as variations in both gain and phase. Traditional control of this problem has relied on gain-scheduling or the on-line switching of controllers designed for models obtained at the different operating points; as and when due. The design of the switching logic and some resulting stability problems are nontrivial. When feasible; parametric robust control makes gain scheduling redundant. A non-parametric description of uncertainty is of course possible but will almost always result in loss of phase information. This often leads to higher bandwidth controllers. The quantitative feedback theory (QFT) robust control methodology introduced by Horowitz [13] is perhaps the only known technique that considers both large parametric uncertainty and phase information simultaneously. The major pay-off is the ability to satisfy both robust stability and multi-objective hard performance constraints, with the minimum possible cost of feedback. The downside is that the method though systematic and powerful in the hands of an experienced control engineer has not until recently lent itself easily to formal mathematization as in the more recent paradigms such as H^∞ control and μ-synthesis. The present effort is an attempt to bridge the gap. The QFT problem can be posed as a formal sensitivity constrained optimization problem [14] which reduces to the problem statements in H^∞-control when the hard performance constraints and parametric uncertainty descriptions are relaxed. Consequently the newer methods may be viewed as restricted quantitative feedback design methods [18]. Therefore when uncertainty is independent of phase, QFT and H^∞ should give identical results for equivalent specifications.

This paper is divided into six sections of which this is the first one. In section two, we develop a formal statement of the QFT problem and convert it to a sensitivity constrained optimization problem. In section three we use the Krylov-Karhunen-Wilcox (KKW) [3] representation to convert the QFT problem into an H^∞ problem, whose solution provides an initial solution to the QFT problem. In section four we translate these requirements into the classical QFT format and draw some comparisons with H^∞ control. In section five we give a nontrivial example with corresponding H^∞ solution. This example clearly demonstrates the higher controller bandwidth which is called for whenever there is a loss of phase information. Finally in section six we give some concluding remarks.

2. *Quantitative Feedback Theory (The SISO Case)*

The QFT problem can be stated as follows: There is given an uncertain family of finite dimensional linear time invariant plants:

$$P = P_L(\alpha, s) + P_H(s) . \tag{1}$$

Here $P_L(\alpha, s)$ represents the low to mid frequency plant-set with parametric uncertainity $\alpha \in \Omega$, where Ω is a compact parameter space and $\alpha \in R^p$ is a p-vector of uncertain parameters. $P_H(s)$ represents the high frequency uncertain plant-set model to account for such unknowns as unmodelled dynamics, measurement errors and unknown parameter variations. If the relative degree of $P_L(\alpha, s)$ is r $\forall\ \alpha \in \Omega$, then $P_H(s)$ is defined by:

$$| P_H(i\omega)| \leq \frac{k_L(\alpha)}{\omega^r}, \quad \forall\ \omega \geq \omega_h , \tag{2}$$

where:

$$\text{(i)}\ k_L(\alpha) = \lim_{|s| \to \infty} s^r \cdot P_L(\alpha, s) , \tag{3}$$

and $k_L(\alpha)$ is called the plant high frequency gain set.

(ii) ω_h is an effective high frequency from which P is structurally unconstrained and effectively allowed to roll off at an arbitrary rate. This is called the Horowitz frequency. It is only from this frequency that P is assumed to have arbitrary phase uncertainity.

There is also given an ideal target closed loop transfer function $T_o(s)$ and an ideal disturbance response transfer function $T_D^o(s)$. The high frequency plant $P_H(s)$ puts a stability margin constraint on the controlled system. Write

$$T(\alpha, s) = H(\alpha, s) \cdot F(s) \tag{4}$$

$$H(\alpha, s) = \frac{L(\alpha, s)}{1 + L(\alpha, s)} . \tag{5}$$

$$L(\alpha, s) = P_L(\alpha, s)\, G(s) . \tag{6}$$

The QFT problem is to find an admissible pair of strictly proper rational and preferably stable transfer functions (G(s), F(s)) in the 2 degree of freedom arrangement shown in Fig. 1 such that the following conditions are satisfied with some measure of optimality.

$$\text{a) } T(\alpha,s) \text{ is stable}, \ \forall\, \alpha \in \Omega .$$

$$\text{b) } \max_{\alpha\in\Omega} | T(\alpha,s) - T_o(s)| \le \delta_T(s), \ \forall\, s \tag{7}$$

$$\text{c) } \max_{\alpha\in\Omega} | T_D(\alpha,s)| \le | T_D^o(s)| = \delta_D(s), \ \forall\, s ; \tag{8}$$

where $\delta_T(s)$ and $\delta_D(s)$ are prespecified. For physical systems for which $r \ge 2$, we assume that for $s = i\omega$, $\lim_{\omega\to\infty} \delta_T(\omega) = \infty$ and $\lim_{\omega\to\infty} |T_D(\alpha,i\omega)| = \lim_{\omega\to\infty} |G_w(i\omega)|$. This reflects the fact that feedback is only effective over a finite frequency band in either reducing sensitivity to plant uncertainty or in disturbance attenuation. Note that $\delta_T(\omega) \ge 0$, and $\delta_D(\omega) \ge 0 \ \forall\, \omega \in [0 \ \infty)$, because of the parametric uncertainty $\alpha \in \Omega$. It is further assumed that *P* satisfies the following conditions:

(i) *P* is path connected.

That *P* is connected implies that $\dfrac{k_L(\alpha)}{k_L(\alpha_o)} > 0 \ \forall\, \alpha_o, \alpha \in \Omega$, [21].

(ii) Every $P \in$ *P* has the same number of non-minimum phase poles and zeros.

(iii) Internal stability dictates that all unstable zeros of P_L^o also appear in $T_o(s)$.

Under the above assumptions the QFT problem can be converted to a sensitivity constrained optimization problem as follows:

From Bodes' sensitivity relation:

$$\frac{T(\alpha,s) - T_o(s)}{T_o(s)} = S(\alpha,s) \cdot \frac{P_L(\alpha,s) - P_L^o(s)}{P_L^o(s)} \tag{9}$$

where

$$S(\alpha,s) = \frac{1}{1 + L(\alpha,s)} . \tag{10}$$

Therefore:

$$\max_{\alpha\in\Omega}|T(\alpha,s)-T_o(s)| = \max_{\alpha\in\Omega}|T_o(s)S(\alpha,s)\frac{P_L-P_L^o}{P_L^o}|\ ,\forall\ s \tag{11}$$

Let

$$\max_{\alpha\in\Omega}|\frac{P_L-P_L^o}{P_L^o}| \le \delta_G(s)\ ,\ \forall\ s\ , \tag{12}$$

where $\delta_G(s) \ge 0$. Now put $s = i\omega$,

Then the tracking constraint reduces to:

$$\max_{\alpha\in\Omega}|S(\alpha,i\omega)| \le \frac{\delta_T(\omega)}{\delta_G(\omega)|T_o(i\omega)|}\ ,\ \forall\ \omega \ge 0\ . \tag{13}$$

Recall that:

$$T_D(\alpha,s) = \frac{G_w(s)}{1+L(\alpha,s)} = S(\alpha,s)\cdot G_w(s) \tag{14}$$

where $G_w(s)$ is the disturbance transfer function. Consequently the constraint $\max_{\alpha\in\Omega}|T_D(\alpha,s)| \le |T_D^o(s)|$ is satisfied if:

$$\max_{\alpha\in\Omega}|S(\alpha,i\omega)| \le |\frac{T_D^o(i\omega)}{G_w(i\omega)}|\ ,\ \forall\ \omega \ge 0\ . \tag{15}$$

Finally, it can be shown [1] that the stability margin constraint can always be transformed to a sensitivity constraint of the form:

$$\max_{\alpha\in\Omega}|S(\alpha,i\omega)| \le \gamma(\omega)\ ,\ \forall\ \omega \ge 0\ . \tag{16}$$

Usually $\gamma(\omega) \equiv 1 + M_p$, where M_p is the maximum closed loop amplification. We observe further that for physical systems having loop transmission functions $L(\alpha,s)$ with relative degree $r \ge 2$, $\lim_{\omega\to\infty}|S(\alpha,i\omega)| = 1$, $\forall\ \alpha \in \Omega$.

Hence all the QFT constraints are simultaneously satisfied if:

$$\text{(i)} \quad \max_{\alpha\in\Omega}|S(\alpha,i\omega)| \le \min\left\{\frac{\delta_T}{\delta_G|T_o|}, \frac{|T_D^o|}{|G_w|}, \gamma\right\}, \ \forall\, \omega \ge 0 .$$
$$\text{(ii)} \quad \lim_{\omega\to\infty}|S(\alpha,i\omega)| = 1 \ \forall\, \alpha \in \Omega . \qquad (17)$$

Since (17) is a multi-objective measure, any finite number of performance or stability specifications which can be suitably expressed as sensitivity inequalities can be accommodated. It is desired to satisfy the QFT constraints with the minimum cost of feedback. Let G represent the set of all stabilizing controllers for P. Bode's law of conservation of attenuation [8] states that for any sensitivity function whose plant satisfies certain relative degree conditions:

$$\int_0^\infty \log|S(i\omega)|\, d\omega = \int_0^{\omega_c} \log|S(i\omega)|\, d\omega + \int_{\omega_c}^\infty \log|S(i\omega)|\, d\omega = C \ge 0$$

where C is a real non-negative constant that depends entirely on the unstable poles in the plant, and ω_c is the minimum finite frequency at which $|S(i\omega)| = 1$. The first integral on the RHS of the above equation is the area of sensitivity decrease (i.e. the benefit of feedback), while the second integral represents the area of sensitivity increase (or the cost of feedback). Similar expressions can be developed for $T_o(s)$ with the corresponding non-minimum phase zeros in $L_o(s)$. It is because of these integral constraints that we strive to avoid the use of non-minimum phase and/or unstable controllers in $L_o(s)$ if at all possible. Since the cost of feedback is given by the area of sensitivity increase, we may pose the QFT problem as follows [5]:

$$\min_{G\in G}\left[\max_{\alpha\in\Omega}\int_{\omega c}^{\infty} \log|S(\alpha,i\omega)|\,d\omega\right] \qquad (18)$$

subject to the sensitivity constraint:

$$\begin{aligned}&\text{(i)} \quad \max_{\alpha\in\Omega}|\,S(\alpha,i\omega)|\le \min\left\{\frac{\delta_T}{\delta_G|T_o|},\ \frac{|T_D^o|}{|G_w|},\ \gamma\right\},\ \forall\ \omega\ge 0,\\&\text{(ii)} \quad \lim_{\omega\to\infty}|S(\alpha,i\omega)| = 1\ \forall\ \alpha\in\Omega\ ,\end{aligned} \tag{19}$$

where ω_c is the sensitivity cut-off frequency of the sensitivity function. If S_{opt} is the optimal sensitivity function given by

$$S_{opt} = \frac{1}{1+P_oG_{opt}}\ , \tag{20}$$

then the solution to the above optimization problem must satisfy:

$$|\,S_{opt}(i\omega)| = \begin{cases}\min\left\{\dfrac{\delta_T}{\delta_G|T_o|},\ \dfrac{|T_D^o|}{|G_w|},\ \gamma\right\} & \forall\ \omega\le\omega_h\ ,\\ 1, \quad \forall\ \omega>\omega h\end{cases} \tag{21}$$

provided G is internally stabilizing. In other words, if an optimal sensitivity function exists, it lies on the sensitivity boundary:

$$M(\omega):=\begin{cases}\min\left\{\dfrac{\delta_T}{\delta_G|T_o|},\ \dfrac{|T_D^o|}{|G_w|},\ \gamma\right\} & \forall\ \omega\le\omega_h\ ,\\ 1, \quad \forall\ \omega>\omega h\end{cases} \tag{22}$$

where ω_h is the frequency at which the performance boundary crosses the stability boundary as shown in Fig. 2. For an alternative discussion of this conclusion (see Horowitz and Gera 1980) [2]. Note that in (22) above: $\delta_T\ge 0, |T_D^o|\ge 0, \gamma>0$, from the nature of the specifications (7), (8), and (16). Consequently for physical systems, $M(\omega)\ge 0$. Note that (19) - (21) imply that $|M^{-1}(\omega)S(\alpha,i\omega)|\le 1\ \forall\ \alpha\in\Omega$. One of the most difficult problems in QFT loop shaping is design initialization. This difficulty is overcome by first regarding the QFT problem as an H^∞ problem. Consider the plant set P(s).

Theorem 3.1 [19]

Assume that all plants in the family:

$$\mathcal{P}_L = \left\{ P_L(\alpha, s): \max_{\alpha \in \Omega} \left| \frac{P_L(\alpha, s) - P_L^o(s)}{P_L^o(s)} \right| \leq \delta_G(s) \right\}, \ \forall s, \tag{23}$$

have the same number of RHP poles. Let $W(s) \in RH^\infty$ satisfy: $M^{-1}(\omega) \leq |W|, \forall s$. Then the closed loop system will meet the performance specification:

$$\max_{\alpha \in \Omega} \| W(s) S(\alpha, s) \|_\infty < 1 \tag{24}$$

if and only if:

$$\text{(i)} \qquad H_o = \frac{P_o G_H}{1 + P_o G_H} \text{ is stable}, \tag{25}$$

and

$$\text{(ii)} \qquad \left\{ \delta_G(s) |H_o| + |M^{-1}(\omega) S_H(s)| \right\} < 1, \ \forall s, \tag{26}$$

where

$$S_H = \frac{1}{1 + P_o G_H}, \tag{27}$$

and G_H is the H^∞ controller. Constructive necessary and sufficient conditions for the existence of G_H satisfying the above theorem conditions are not known. However by generating suitable RH^∞ weighting functions W(s) and V(s), the following H^∞ minimization problem can be set up:

$$\min_{G_H \in \mathcal{G}_{\Pi}} \left\{ \sup_{\omega} [\|WS\|_\infty + \|VH\|_\infty] \right\} \tag{28}$$

In standard H^∞ control, the choice of W and V is qualitative. We can make it quantitative by considering the bounds given in the Theorem, as follows: Select $V(s) \in RH^\infty$ and satisfying

$$\delta_G(\omega) \leq |V(i\omega)| \quad \forall \omega \geq 0 \tag{29}$$

Assume that $\delta_G(\omega) \in L_\infty$ and satisfies

$$\int_{-\infty}^{\infty} \log \delta_G(\omega)d\omega > -\infty . \tag{30}$$

Then it is known that the function

$$\tilde{V}(s) := \exp\left[\frac{1}{\pi}\int_{-\infty}^{\infty} \frac{1-i\lambda s}{s-i\lambda}\,\frac{\log \delta_G(\lambda)}{1+\lambda^2}\,d\lambda\right] \tag{31}$$

as well as $\tilde{V}^{-1}(s)$ are both in H^∞ [3,6]; such that

$$|\tilde{V}(i\omega)| = \delta_G(\omega)$$

almost everywhere. Next approximate $\tilde{V}(s)$ uniformly by an RH^∞ function $V(s)$ such that:

$$|\tilde{V}(i\omega)| \le |V(i\omega)| \quad \forall\, \omega \ge 0 \tag{32}$$

Now the sensitivity specifications (22) and (24) => $\|M^{-1}(\omega)S(\alpha, i\omega)\|_\infty \le 1\ \forall\, \alpha \in \Omega$. Define $M^{-1}(\omega) := \hat{M}(\omega)$. Similarly, assume that both $\hat{M}(\omega) \in L_\infty$ and $\dfrac{\delta_G|T_o|}{\delta_T} \in L_\infty$ and satisfy:

$$\int_{-\infty}^{\infty} \log \hat{M}(\omega)d\omega > -\infty \tag{33}$$

and

$$\int_{-\infty}^{\infty} \log \frac{\delta_G|T_o|}{\delta_T}\,d\omega > -\infty . \tag{34}$$

so that

$$\tilde{W}(s) := \exp\left[\frac{1}{\pi}\int_{-\infty}^{\infty} \frac{1-i\lambda s}{s-i\lambda}\,\frac{\log \hat{M}(\lambda)}{1+\lambda^2}\,d\lambda\right] \tag{35}$$

and $\tilde{W}^{-1}(s) \in H^\infty$. Proceed to carry out a uniform approximation to $\tilde{W}(s)$ by $W(s)$. Now insert $W(s)$ and $V(s)$ at the appropriate place in the H^∞ minimization problem (28), and use either the polynomial method of Kwakernaak [22] or the broadband matching method of Verma and Jonchere [23] to obtain an appropriate controller G_H. Alternatively solve the simpler H^∞ problem [12]:

$$\min_{G \in \mathcal{G}} \left\| \frac{WS}{VT} \right\|_{\infty} = \mu . \qquad (36)$$

Indeed $\mu < 1/2$ is a sufficient condition for the original H^{∞} problem to be solved [11]. Once G_H is determined, we draw the graph of $L_o(i\omega) = P_o G_H(i\omega)$ on the Nichols chart. This forms the initializing loop transmission function for the QFT optimization algorithm. On the same Nichols chart is superimposed the standard QFT performance and stability boundaries obtained from the templates [14]. The over design in the H^{∞} solution will usually be graphically apparent. The QFT optimization routine strives to reduce the gain-bandwidth area of L_{opt} by moving $L_o(i\omega)$ towards the boundaries at every frequency.

3. The QFT Optimization Algorithm

As is clear from the foregoing developments a complete analytical solution of the problem appears hopelessly complicated. The graphical solution of the problem (if and when one exists) is a major contribution of QFT. In classical QFT the above design problem is solved almost entirely graphically on the Nichols chart through the following procedure:

1. For $P_L(\alpha, i\omega)$ at some frequency $\omega \le \omega_h$, the mapping $P_L(\alpha, i\omega) : \Omega \rightarrow R(g, \phi, \omega)$ generates a compact set $R(g, \phi, \omega)$ in the gain-phase plane called the plant **template** at that frequency.

2. By use of $R(g, \phi, \omega)$, the performance specifications $\left\{ \delta_T(\omega) , \delta_D(\omega) \right\}$ are mapped into $B_p(\omega)$ which is a constraint on the loop transmission function $L(\alpha, i\omega)$ at that frequency, according to:

$$R(g, \phi, \omega) : \min \left\{ \delta_T(\omega) , \delta_D(\omega) \right\} \rightarrow B_p(\omega) .$$

3. The high frequency plant $P_H(i\omega)$ maps the high frequency stability constraint $\gamma(\omega)$ into a stability boundary according to:

$$P_H(i\omega) : \gamma(\omega) \rightarrow B_s(\omega)\ , \text{ for } \omega \geq \omega_h\ .$$

When $\omega_h \rightarrow \infty$, $B_s(\omega) \rightarrow B_s(\infty)$ which is called the universal high frequency boundary.

4. The stability specification is given as constraints $B(g_m, \phi_m)$ on the gain margin g_m and phase margin ϕ_m around the critical point on the Nichols chart.

5. The union of $B(g_m, \phi_m)$ and $B_s(\infty)$ given by:

$$B_\infty(g_m, \phi_m) = B_s(\infty) \cup B(g_m, \phi_m)$$

generates a contour around the critical point (0 dB, -180°) called the universal high frequency contour [U - contour].

6. Finally the design constraints on $L(\alpha, i\omega)$ are given by $B_\infty(g_m, \phi_m) \cup \left\{B_p(\omega)\right\}$.

Let $L_o(i\omega)$ be the nominal loop transmission function determined in section 3, where $L_o(i\omega) = P_L^o(i\omega)\, G_H(i\omega)$. Any $L_o(i\omega)$ which is QFT admissible must satisfy:

$$|\, L_o(i\omega)| \geq B_p(\omega)\ ,\ |\, L_o(i\omega)| \geq \min_{\alpha \in \Omega} |\, L(\alpha, i\omega)|\ ,\ \ \forall\ \omega \geq 0\ , \omega \in [0\ \omega_h]\ ;$$

and:

$$|\, L_o(i\omega)| \text{ nonmember } B_\infty(g_m, \phi_m)\ \forall\ \omega > \omega_h\ .$$

Subject to these constraints, $L_o(i\omega)$ can be rolled off as rapidly as desired by the addition of extra poles after $\omega = \omega_h$. The object of the optimization is to reduce over-design, in particular to reduce controller bandwidth.

An optimal loop transmission function L_{opt} must satisfy the constraints with the minimum possible gain at every frequency [2]. Consequently $|\, L_{opt}| = \phi(\omega) \in B_p(\omega)$ for all $\omega \leq \omega_h$. This optimum (if it exists) can be shown to be unique but may not be real rational. The QFT loop shaping methodology reduces to several CAD techniques for obtaining finite order real rational sub-optimal loop transmission functions which approach L_{opt} as closely as desired. A constrained nonlinear programming optimization design software has recently been developed to solve this problem [4].

Note that, $L_o(i\omega)$ cannot penetrate the U-contour at any frequency and at a given frequency ω, must have a gain $|L_o(i\omega|$ which lies on or above the corresponding frequency boundary $B_p(\omega)$. The basic idea is to modify the loop shape of $L_o(i\omega)$ to L_{opt}, such that at each frequency, $|L_{opt}|$ lies on the corresponding $B_p(\omega)$. Whether one uses the sensitivity formulation or the classical formulation, one ends up with a unique optimal H^∞ sensitivity function S_{opt}. In general, $S_{opt} \notin RH^\infty$. Note that S_{opt} is given by: $S_{opt} = (1 + L_{opt})^{-1}$. In either case one must then find suitable real rational approximation and interpolation algorithms which yield an $S_u \in RH^\infty$ satisfying:

$$\| S_{opt} - S_u \|_\infty < \varepsilon \text{ for any given } \varepsilon > 0 \text{, provided } S_{opt} \text{ is continuous on the boundary.}$$

In this respect QFT is no different from the more recent robust control methodologies such as H^∞ - optimization and μ-synthesis, but predates these by over two decades. The major difference lies in the way classical QFT arrives at S_{opt} by making graphical use of the phase uncertainty information contained in both $R(g,\phi,\omega)$ and $S(\alpha,i\omega)$. For example, in the more recent methodologies, the above rational approximation provides an adequate solution to the H^∞ optimization problem. In QFT however, one must check that $|S_u(i\omega)|$ is not only a good approximation to $M(\omega)$ at every frequency but also that the phase of $S_u(i\omega)$ is a good approximation to that of $A(i\omega)\,M(i\omega)$ at every frequency. Satisfaction of the above H^∞ norm conditions in general does not guarantee the latter. The performance and stability boundaries

$B_p(\omega)$ and $B_s(\infty)$ in QFT provide a graphical methodology for extracting from the set of all S_u which satisfy the given norm condition, that which also satisfies the additional phase requirement.

4. Design Example

Here we present an example which details how uncertainty representation and the difference in the subsequent design philosophies can produce vastly different costs of feedback. The example is a standard problem for control system design [15]. Both the H^∞ design techniques and the QFT method were applied to this example. Both methods were able to meet the design specifications, with QFT doing slightly better in meeting the performance specifications. However the difference in the cost of feedback between the QFT and the H^∞ solutions is quite large with QFT producing a significantly smaller cost of feedback than the corresponding H^∞ solution. This is shown dramatically by comparing the respective controller gains and bandwidths for the H^∞ solution and QFT solution as done in Fig. 4. The H^∞ problem was solved first and then used as the initial design for the QFT optimization. After 3 iterations, the controller shown in Fig. 5 was obtained.

Example 1: [15]

Consider a plant transfer function:

$$P(s, k, a) = \frac{ka}{s(s+a)}, \quad k \in [1,10] \; a \in [1,10]. \tag{37}$$

Here, $\alpha = [k, a]^T \in \Omega \subset IR^2$. It is desired to find a suitable controller G(s) such that:

$$\max_{\alpha \in \Omega} | S(i\omega, \alpha)| \leq 0.089 \, |i\omega|^2 \quad \forall \, \omega \geq 0 \tag{38}$$

and

$$\max_{\alpha\in\Omega} |\, H(i\omega, \alpha)| \le \frac{2.5}{|\, i\omega|^2}\ , \quad \forall\, \omega \ge 0 \tag{39}$$

By choosing the nominal model as

$$P_o(s) = \frac{10}{s(0.1s+1)} \tag{40}$$

i.e. with k and a at their maximum values, both QFT and H_∞ were able to solve the problem. For this model, the unstructured H^∞ perturbation is

$$\Delta_m(s) = \frac{0.9\left[\dfrac{i\omega}{.91} + 1\right]}{\left[\dfrac{i\omega}{1.0} + 1\right]}\ . \tag{41}$$

Whereas for QFT, any member of P(s, k, a) is an admissible nominal model, for H^∞, one must use the "maximum" plant as given by $P_o(s)$, or a specific model for which the H^∞ weighting functions yield an admissible solution for otherwise $|\, \Delta_m(0)| > 1$. But for robust stability: $|\, H(i\omega)| < \dfrac{1}{|\, \Delta_m(i\omega)|}$; which would then imply that $|\, H(0)| < 1$, thus contradicting the performance requirement following from $S(i\omega) + H(i\omega) \equiv 1$, that $H(0) = 1$. For details, see [15].

Using the same nominal models, the H^∞ solution with the weights: $\left|\begin{matrix} W_1(\omega)\ S(i\omega) \\ W_3(\omega)\ H(i\omega) \end{matrix}\right|$, where $W_1(\omega) = \dfrac{1}{.089|\, (i\omega)|^2}$, and $W_3(\omega) = \dfrac{1}{\Delta_m(i\omega)}$, is given by:

$$G_H = 22123.81\left[\frac{(s+3E-7)(s+1)(s+1.54)(s+10)}{(s+8E-10)(s+1E-7)(s+2)(s+4498)}\right]. \tag{42}$$

Using this as the initial controller, produced the following QFT design after a three iterate optimization:

$$G_Q = 3.837\left[\frac{(s+1)(\frac{s}{1.54}+2)(\frac{s}{10}+1)}{s(\frac{s}{100}+1)(\frac{s}{750}+1)}\right] \tag{43}$$

Examination of figures 3 and 4 clearly show that

$$\int_{o}^{\omega c} \log | S_H^{-1} | \, d\omega \geq \int_{o}^{\omega c} \log | S_Q^{-1} | \, d\omega . \tag{44}$$

This implies that the QFT cost of feedback is much less than the corresponding H^∞ cost of feedback when both address the same nominal models and the same performance specifications as expected, where S_H and S_Q are respectively the H^∞ and QFT sensitivity functions.

5. Conclusions

Quantitative feedback theory is a very useful robust design methodology whenever large parametric uncertainty and hard constraints on closed loop response are indicated. Until recently however, the technique relied almost entirely on semi-analytical and graphical methods, making comparison with H^∞ control at best indirect. By using H^∞ control as an initial trial design however, the QFT methodology can be systematized. The formalization of the QFT process such as is presented here makes comparison with H^∞ and μ–synthesis straightforward. The result of the non inclusion of phase information in these methods is the inevitability of a higher cost of feedback, as shown by the examples. The extension of these ideas to multivariable systems is not difficult [18]. Investigations into the adaptation of this approach to nonlinear control design along the lines originally suggested by Horowitz is in progress.

6. References

1. D'Azzo, J. J., Houpis, C. H., Linear Control System Analysis and Design, 5th Edition, McGraw Hill, New York, 1988.
2. Gera, A., Horowitz, I. M., Optimization of the Loop Transfer Function, International J. Control 31, 389-398, 1980.
3. Robinson, E. A., Random Wavelets and Cybernetic Systems, Hafner Publishers, New York 1962.

4. Thompson, D. F., Optimal and Sub-optimal Loop Shaping in Quantitative Feedback Theory, Ph. D. Thesis, Purdue University, August 1990.

5. Nwokah, O.D.I., Thompson, D. F., Perez, R. A., On the Existence of QFT Controllers, ASME-WAM, Dallas, TX, 1990.

6. Hoffman, K., Banach Spaces of Analytic Functions, Prentice-Hall, New Jersey, 1962.

7. Lurie, B.J., Feedback Maximization, Artech House Inc., Dedham, Massachusetts, 1986.

8. Freudenberg, J.S., Looze, D.P., Frequency Domain Properties of Scalar and Multivariable Feedback Systems, Springer-Verlag, New York, 1988.

9. Dorato, P., Yedavali, R. (Eds.), Recent Advances in Robust Control, IEEE Press, New York, 1990.

10. Zames, G., Feedback and Optimal Sensitivity: Model Reference Transformation, Multiplicative Semi Norms, and Approximate inverses, IEEE Trans. Autom Control, AC-26, April 1981.

11. Francis, B.A., A Course in H^∞ Control Theory, Lecture Notes in Control and Information Sciences, No. , Springer-Verlag, New York, 1987.

12. Maciejowski, J., Multivariable Feedback Design, Addison-Wesley, Reading, MA, 1989.

13. Horowitz, I.M., Synthesis of Feedback Systems, Academic Press, Orlando, FL, 1963.

14. Jayasuriya, S., Nwokah, O.D.I., Yaniv, O., The Benchmark Problem Solution by Quantitative Feedback Theory, Proc. ACC, Boston, MA, 1991. Also submitted to AIAA Journal of Guidance and Control.

15. Chait, Y., Hollot, C.V., A Comparison Between H^∞ Methods and QFT for a SISO Plant with Both Plant Uncertainty and Performance Specifications, ASME Winter Annual Meeting, Dallas, TX, November 1990.

16. Betzold, R.W., MIMO Flight Control Design with Highly Uncertain Parameters: Application to the C-135 Aircraft, M.S. Thesis, Dept. of Electrical Engineering, Air Force Institute of Technology, AFIT/GE/EE/83D-11, WPAFB, OH 1983.

17. Thompson, D.F., Analytical Loop Shaping Methods in Quantitative Feedback Theory, ASME-Winter Annual Meeting, Dallas, Texas, November 1990.

18. Perez, R.A.,Nwokah, O.D.I., Thompson, D.F., Almost Decoupling by Quantitative Feedback Theory, submitted to the American Control Conference, Boston, MA, 1991. Also submitted to the ASME Journal of Dynamic Systems and Control.

19. Morari, M., Zafiriou, E., Robust Process Control, Prentice-Hall, NJ, 1989.

20. Thompson, D.F., Nwokah, O.D.I., Frequency Response Specifications and Sensitivity Functions in Quantitative Feedback Theory, ACC, Boston, 1991, Also submitted to ASME J. Dynamic Systems Measurement and Control.

21. Nwokah, O.D.I., Strong Robustness In Uncertain Multivariable Systems, IEEE CDC, Austin, TX, December 1988.

22. Kwakernaak, H. Minimax Frequency Domain Performance and Robustness Optimization of Linear Feedback Systems IEEE Trans. Autom. Control AC-30, 994-1004, 1985.

23. Verma, M. Jonckheere, E. L^∞ Compensation with Mixed Sensitivity as a Broadband Matching Problem, Systems and Control Letters, 4, 125-129, 1984.

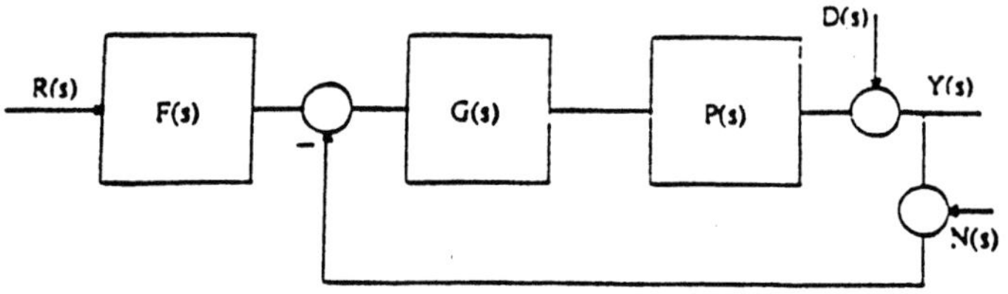

Fig. 1: The Standard two-degree-of-freedom feedback structure.

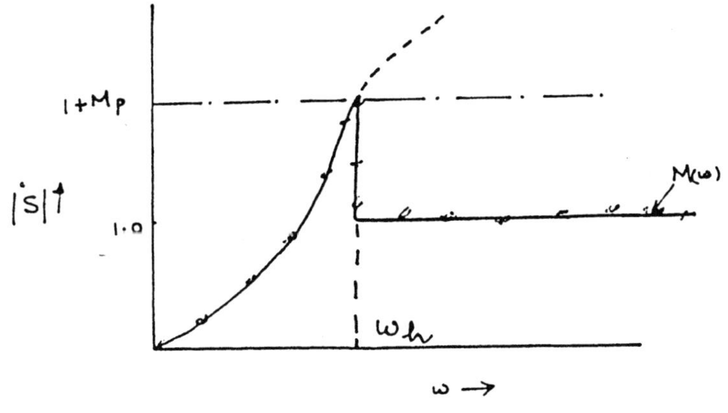

Fig. 2: The QFT performance and stability boundaries.

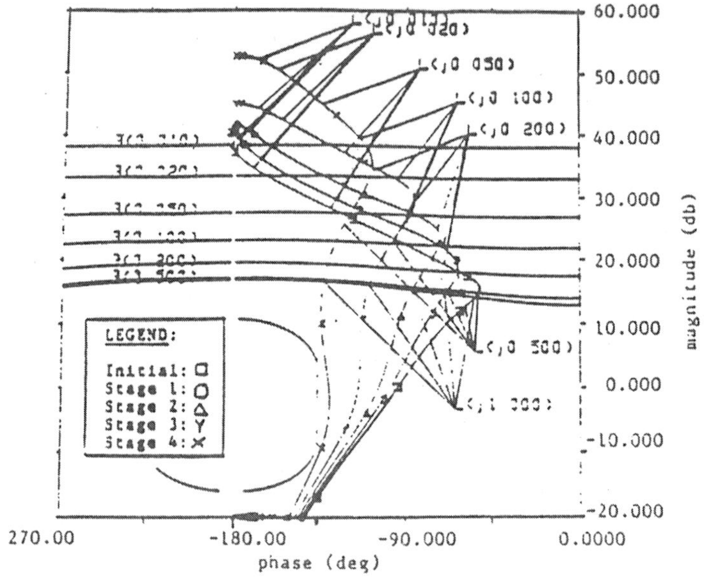

Fig. 3: Typical controller optimization from initial H$^\infty$ controller.

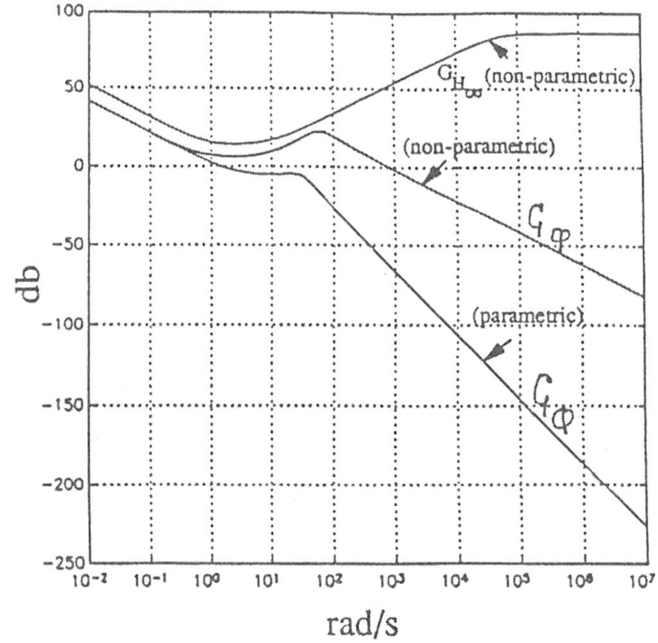

Fig. 4: Frequency response of H$^\infty$ and QFT controllers, showing bandwidth differences.

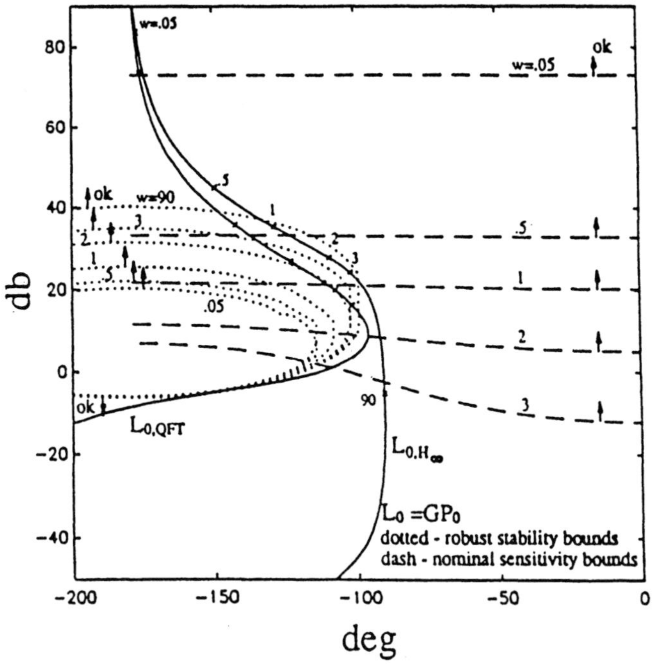

Fig. 5: Optimization of initial H^∞ controller for bandwidth and gain reduction.

ON ROBUSTNESS OF SYSTEMS WITH STRUCTURED UNCERTAINTIES

A. OLAS*

Abstract - We consider the robust stability problem for nominally linear system with nonlinear, structured perturbations. The system is of the form

$$\dot{x} = A_N x + \sum_{j=1}^{q} p_j A_j x$$

The Lyapunov direct method has been often utilized to determine the bounds for nonlinear, time-dependent functions p_j which can be tolerated by a stable nominal system. In most cases quadratic forms are used either as components of vector Lyapunov function or as a function itself. The resulting estimates are usually conservative. As it is known, often the conservatism is due to the fact that a quadratic form is used as a Lyapunov function. To reduce the conservatism of the bounds we propose to use a piecewise quadratic Lyapunov function. An example demonstrates application of the proposed method.

I. INTRODUCTION

The design of robust controllers for multivariable linear systems remains one of the important issues of modern technology, see Davison in [1], Desoer et all. in [2], and the Siljak review in [3].

Recently a significant research has taken place in applications of Lyapunov's direct method in robust design; it is prompted by the fact that this approach can easily accommodate nonlinear and time-varying perturbations. Using Lyapunov's direct method Siljak in [4] and Patel et al. in [5] established the procedures to calculate quantitative measures of robustness, i.e., the bounds on perturbations such that for the perturbations so bounded the system remains stable. Yedavalli and Liang in [6] using transformations obtained improvements in bounds estimation. Results of [5] and [6] are based on the selection of some quadratic form as the Lyapunov function. As it is known this approach has its limitations; Becker and Grimm in [7] proved that applying the small gain theorem provides the stability bound which cannot be improved by state transformations. Radziszewski in [8] examining two-dimensional structurally-perturbed system discussed quadratic forms as the class of Lyapunov function candidates and determined the best Lyapunov function

*A. Olas is with the Department of Mechanical Engineering, Oregon State University, Corvallis, OR 97331, USA.

in this class. He found that robustness bounds obtained using this function were worse than those obtained by other methods, applicable for two-dimensional system. Siljak in [3] shows that the bound estimation strongly depends on selection of the system state space (250% improvement of the estimation). Further, he suggests use of vector Lyapunov functions to reduce conservatism of robustness estimation and provides an example of application. In this paper piecewise quadratic Lyapunov functions are proposed to improve the bound estimation. The procedure proposed may be treated as a natural extension of existing procedure.

II. PROBLEM STATEMENT AND DEFINITIONS

Consider a nominally linear system with structured perturbation

$$\dot{x} = A_N x + \sum_{j=1}^{q} p_j A_j x \tag{1}$$

with $p_j = p_j(x,t)$ and A_N having negative real parts of its eigenvalues. Let $\mathbf{p} \triangleq [p_1,\ldots,p_q]^T$. We define the general robust stability problem for (1) as the assignment to determine the set R belonging to the parameter space R^q such that iff $\mathbf{p}(x,t) \in R$ for all x,t then the trivial solution $x = 0$ of the system (1) is stable in the sense of Lyapunov. Most often a reduced problem is discussed when instead of the set R a parallelopiped or a ball embedded in R is to be determined, see Siljak's review in [3] for details. We limit our interest to Lyapunov direct method approach. We denote

$$\sum_{j=1}^{q} p_j A_j = A_U(p)$$

Let Π denote a parallelopiped in R^q

$$\Pi = \{p \in R^q : p^- \le p \le p^+\}$$

with $p^1, p^2, ..$ being the 2^q vertices of Π. Usually some quadratic form

$$V(x) = x^T S x \tag{2}$$

where S is a symmetric, positive-definite matrix, is selected as a Lyapunov function candidate.

The derivative $dV(x)/dt$ of this function along the solutions of (1) is a quadratic form of x and linearly depends on parameters $p_i(t,x)$. Thus when searching for the solution of robust stability problem of (1) in a form of a parallelopiped $\Pi \in R^q$ the 2^q quadratic forms

$$QF_j \triangleq x^T\left(A_N^S + SA_N + A_U^T(p^j)S + SA_U(p^j)\right)x\,, \qquad j = 1,\dots,2^q \tag{3}$$

generated by 2^q vertices $p^1,p^2,\dots$ of the parallelopiped Π are considered (see for instance the paper [9] by Herisberger and Belanger). If the forms QF_j are non-positive for all x then Π is a solution to robust stability problem. In such a case we say that V(x) guarantees the solution Π (to robust stability problem).

We shall start our consideration with the assumption that the quadratic Lyapunov function (2) and resulting from its application parallelopiped Π are known.

Of our interest is the case when the parallelopiped Π is maximal (with respect to applied Lyapunov function) i.e., if enlarging it by decreasing a single bound p_j^- or increasing a single bound p_j^+ causes at least one of the forms QF_i to attain a positive value for some x. Such a definition implies that if the parallelopiped is maximal then for each p_j^-, p_j^+ there is at least one vertex p^i such that the corresponding form QF_i attains the value zero at some point $x \neq 0$.

In the paper we shall use the piecewise quadratic forms as Lyapunov functions to solve the robust stability problem for the system (1). The proposed procedure is a natural extension of the procedure described above. We shall show that under some assumptions less conservative bounds (i.e. larger parallelopiped) are obtained when using piecewise quadratic functions.

III. PIECEWISE QUADRATIC FORMS

Let $F_1,\dots,F_k$ be nonempty, open, and disjoint sets of R^n. Denote by F_i^* the closure of the set F_i and assume that the union of the closures F_i^* fills entire R^n. Let $V_1,\dots,V_k$ be quadratic forms

$$V_i(x) = x^T S_i x$$

with symmetric matrices S_i. We call the function $V^{(p)}: R^n \to R^n$, such that

$$V^{(p)}(x) = V_i(x) \text{ for } x \in F_i$$

a piecewise quadratic form (PQF) if it is continuous. Further we call the set F_i^* the i-th set of definition of $V^{(p)}(x)$. If PQF is formed by k quadratic forms then we call it k-piece PQF. The sum of two PQF's is a PQF, and a quadratic form is a single-piece PQF. It is easy to observe that the function $V^{(p)}(x)$ is locally lipschitzian in x, but it is not differentiable. The case of such a Lyapunov function was considered by Rouche et al. in [10]. They considered the system

$$\dot{x} = f(t,x) \tag{4}$$

with assumptions on f which are fulfilled by the system (1). They utilized an upper right Dini derivative of the continuous function V(t,x) along the solutions of the system (4)

$$D^+V(t,x) = \limsup_{h \to 0+} \frac{1}{h} [V(t+h,x+hf(t,x))-V(t,x)] \tag{5}$$

The function a: $R^+ \to R^+$ belongs to the class $\mathfrak{K}$ in the sense of Hahn [11], if it is continuous, strictly increasing with a(0) = 0. Let Ω be a domain in R^n, containing 0. We formulate following

Conclusion 1. By virtue of the Theorem 6.2 in [10] the zero solution of the system (1) is stable in the sense of Lyapunov if for some function $a \in \mathfrak{K}$ and every $(t,x) \in I \times \Omega$ the following conditions are fulfilled

(i) $V^{(p)}(x) \geq a(\|x\|)$

(ii) $D^+ V^{(p)}(x) \leq 0$

■

Using definition (5) it is easy to relate the Dini derivative $D^+V^{(p)}(x)$ to the derivatives $dV_i(x)/dt$ of the quadratic form V_i along the solution of a differential equation (1). If $x \in F_i$ then

$$D^+V^{(p)}(x) = dV_i(x)/dt$$

If x is the point belonging to the product of l closures $F^*_{i1},\dots,F^*_{il}$, $l \leq k$ then

$$D^+V^{(p)}(x) \leq \max_{ij} [dV_{i1}(x)/dt,\dots,dV_{il}(x)/dt] \tag{6}$$

IV. ROBUST STABILITY PROBLEM USING PQF

A convenient way of analyzing sign properties of the derivatives $dV_i(x)/dt$ for $x \neq 0$ is to consider the derivatives on the unit sphere Φ_1 in R^n. The relation

$$dV_i(x)/dt \leq 0 \text{ for } x \in F_i^* \cap \Phi_1$$

implies that

$$dV_i(x)/dt \leq 0 \text{ for } x \in F_i^*$$

Denote

$$\rho = \max_{l \in \{1,\dots,k\}} \max_{x \in F_l^* \cap \Phi_1} [dV_l(x)/dt] \tag{7}$$

Then if $x \in F_l^*$, $i = 1,\dots,k$ then

$$dV_l(x)/dt \le \rho \; \|x\| \tag{8}$$

Inserting the relation (8) in the expression (6) yields

$$D^+V^{(p)}(x) < \rho \; \|x\| \text{ for } x \in R^n$$

Relating the discussed properties of PQF to the conditions (i) - (ii) of the Conclusion 1 we formulate

Conclusion 2. Let $V^{(p)}(x)$ be a PQF. The zero solution of the system (1) is stable in the sense of Lyapunov if the following conditions are fulfilled:

(i) $\max_{x \in \Phi_1} V^{(p)}(x) > 0$

(ii) $\max_{l \in \{1,\dots,k\}} \max_{x \in F_l^* \cap \Phi_1} [dV_l(x)/dt] \le 0$

■

As was the case for quadratic Lyapunov function, the derivatives $dV^{(p)}(x)/dt$ depend linearly on parameters $p_i(t,x)$. As previously we search for a solution to a robust stability problem in a form of a parallelopiped $\Pi \in R^q$, with vertices p^j, $j = 1,\dots,2^q$. We introduce $k \bullet (2^q)$ quadratic forms

$$QF_j^{(l)} \triangleq x^T(A_N^T S_l + S_l A_N + A_U^T(p^j) S_l + S_l A_U(p^j))\, x$$

where $i = 1,\dots,k$; $j = 1,\dots,2^q$. In the conclusion of the above consideration we formulate Theorem 1. Let $V^{(p)}(x)$ be a k-piece PQF and $\Pi \in R^q$ a parallelopiped with vertices p^j, $j = 1,\dots,2^q$. If

(i) $\max_{x \in \Phi_1} V^{(p)}(x) > 0$

(ii) $QF_j^{(l)} \le 0$ for $x \in F_l^* \cap \Phi_l$; $\quad i = 1,\dots k$; $j = 1,\dots,2^q$

then a parallelopiped Π is a solution to robust stability problem of (1). ■

Some Definitions. We redefine the concept of maximal parallelopiped Π (with respect to applied Lyapunov function) saying that it is maximal if enlarging it by decreasing a single bound p_s^- or increasing a single bound p_s^+ causes at least one of the forms $QF_j^{(i)}$, $i = 1,\dots,k$; $j = 1,\dots 2^q$, to attain a positive value for some $x \in F_i^* \cap \Phi_1$. Such a definition implies that if the parallelopiped is maximal then for each p_s^-, p_s^+ there is at least one vertex p^j and the form $QF_j^{(i)}$ such that it attains the value zero at some point $x \in F_i^* \cap \Phi_1$.

The bound p_s^- (p_s^+) is called active on the vertex p^j if decreasing the single bound p_s (increasing the single bound p_s^+) makes at least one of the forms $QF_j^{(i)}$, $i = 1,\ldots,k$, attain a positive value at some point $x \in F_i^* \cap \Phi_1$.

If all the bounds are inactive on the vertex p^j then we call the vertex p^j inactive.

We call a point $\varsigma \in F_i^* \cap \Phi_1$ a root of the form $QF_j^{(i)}$ if $QF_j^{(i)}(\varsigma) = 0$. We denote the set of all roots of all the forms $QF_j^{(i)}$ by Ψ.

Let the parallelopiped Π be given and let $V_1^{(p)}(x)$, $V_2^{(p)}(x)$ be two piecewise quadratic Lyapunov functions. Further let Ψ_1, Ψ_2 be the sets of the roots of the forms $QF_j^{(i)}$ of correspondingly $V_1^{(p)}(x)$, $V_2^{(p)}(x)$. We call the function $V_1^{(p)}(x)$ better than the function $V_2^{(p)}(x)$ (with respect to the parallelopiped Π) if either

(i) $V_1^{(p)}(x)$ guarantees the solution Π, while $V_2^{(p)}(x)$ does not,

or

(ii) both functions guarantee the solution Π, and the set Ψ_1 is smaller than Ψ_2, i.e.

$$\psi_1 \subset \psi_2 \quad \text{and} \quad \psi_1 \neq \psi_2$$

V. CONSTRUCTION OF PIECEWISE QUADRATIC FORM

Of our interest will be a simple method of generating piecewise quadratic forms based on two planes belonging to R^n. Let two non-zero vectors γ_1, γ_2 be given. The vectors define two planes $(\gamma_1)^T x = 0$, $(\gamma_2)^T x = 0$ in R^n, such that they contain the point 0 of R^n. Let $V(x)$ be an arbitrary quadratic form on R^n. Define by $F_1 \in R^n$ the set such that if $x \in F_1$ then

$$(\gamma_1)^T x \geq 0, \quad (\gamma_2)^T x \leq 0$$

Introduced notation is sufficient to formulate the Lemma 1.

Lemma 1. The function

$$V^{(p)}(x) = \begin{cases} V(x) + \lambda \left[x^T \gamma_1 (\gamma_2)^T x\right] & \text{for } x \in F_1 \\ V(x) & \text{for } x \in R^n \backslash F_1 \end{cases}$$

is a PQF for any values of a constant λ.

To prove the Lemma it is enough to observe that the boundary of F_1 consists of the two previously introduced planes. ■

Denote

$$V_1(x) = \lambda\, x^T \gamma_1 (\gamma_2)^T x$$

Let η be an arbitrary constant, non-zero n-dimensional vector. Select an arbitrary $x = x^* \in R^n$. We have the following expression for $\mathrm{grad}V_1(x)$.

$$\mathrm{grad}V_1(x) = 2\lambda\, \gamma_1 (\gamma_2)^T x$$

Lemma 2. If the product

$$(\eta)^T \mathrm{grad}V_1(x^*) = 0$$

then there are arbitrarily small changes of the vectors γ_1, γ_2 such that

$$(\eta)^T \mathrm{grad}V_1(x^*) \neq 0$$

For proof of the Lemma see Appendix.

Select a nonempty subset F of Φ_1. Define two following quadratic forms on F:

(i) a form $x^T S_1 x$ such that zero is its largest and single eigenvalue, and zero is attained at some point $\eta \in F$, thus being the maximum of the form,

(ii) a form $x^T S_2 x$ such that

$$\eta^T S_2 \eta < 0 \tag{9}$$

Lemma 3. Under assumptions made, there is a sufficiently small $\lambda > 0$ such that

$$x^T S_1 x + \lambda\, x^T S_2 x < 0 \text{ for } x \in F$$

For proof of the Lemma 3 see Appendix.

Let $\zeta_1,\ldots\zeta_m$ be a finite set of points on Φ_1, such that $\zeta_1 \neq \zeta_m$.

Lemma 4. There exist a non-zero PQF $V_1^{(p)}(x)$ and a closed set $F_1 \in R^n$ such that

(i) $\zeta_1 \in F_1$

(ii) $\zeta_i \notin F_1$, $i = 2,\dots,m$

(iii) $V_1^{(p)}(x) = 0$ for $x \in R^n \setminus F_1$.

Sketching the proof observe first that it is always possible to draw such a plane through points ζ_1 and 0 that it does not contain any other points ζ_i. The points ζ_i are then not closer to the plane than some minimal distance. Thus during a small positive or negative rotation of the plane about some in-plane line perpendicular to ζ_1 no point ζ_i, $i = 1,\dots,m$ is incorporated into the plane. Writing the equations of rotated planes by

$$g^{+T} x = 0 , \quad g^{-T} x = 0$$

where g^+ and g^- are such that $g^{+T} \zeta_1 > 0$ and $g^{-T} \zeta_1 < 0$ and defining the set F_1 as the set of x such that

$$g^{+T} x \geq 0 , \quad g^{-T} x \leq 0$$

we conclude that $\zeta_1 \in F_1$ and $\zeta_i \notin F_1$. The function

$$V_1^{(p)} = \begin{cases} x^T g^+ g^{-T} x & \text{for } x \in F_1 \\ 0 & \text{for } x \in R^n \backslash F_1 \end{cases}$$

fulfills the assumption (iii), which ends the proof. ■

VI. MAIN RESULT

We return to the problem statement of the Sec. II. We assume that for a given system (1) and quadratic Lyapunov function (2) the maximal parallelopiped Π has been determined thus by a definition the 2^q quadratic forms QF_j are nonpositive. Considering the set Ψ of the roots of the forms QF_j we observe that generally the structure of the set may be quite complex. However the most common situation when investigating the robust stability problem is when matrices of some of the forms QF_j have single zero eigenvalues. Then the set Ψ contains the eigenvectors of these matrices. To apply piecewise quadratic function to robust stability problem we consider the case when there is a finite number of the roots of the forms QF_j and they are isolated points of Φ_1, which seems to be the most often case in applications. However weaker assumptions may be used. Further we assume that the roots are not equilibrium points of the system (1); thus the right hand side of the equation (1) at the roots of QF_j is not zero.

Theorem 2. If

(i) $V(x)$ guarantees the solution Π to robust stability problem,

(ii) there is a finite number of roots of the forms QF_j, $j = 1,...,2^q$, and the roots are isolated points of Φ_1,

(iii) there is a root ζ_r, $1 \leq r \leq m$ such that it is not an equilibrium point of the system (1) and at this root only one of the forms QF_j is equal zero,

then there exists a two-piece piecewise Lyapunov function $V^{(p)}(x)$ better than $V(x)$.

For proof of the Th. 2 see Appendix.

Remark 1. If the vertex p^h is the only vertex on which some bound p_j^- (p_j^+) is active then application of piecewise quadratic Lyapunov function makes possible the determination of the new parallelopiped Π_1, containing but larger than the previous one.

Remark 2. The function $V(x)$ of the Th. 2 has been defined as a quadratic Lyapunov function. With similar assumptions a more general theorem utilizing a piecewise quadratic function may be proved.

Remark 3. Simple algorithm to design a piecewise quadratic Lyapunov function will be the subject of the next paper.

VII. EXAMPLE

Consider the robust stability problem for the two-dimensional system

$$\dot{x} = A_N x + p_1(x,t)A_1 x , \qquad x \in R^2 \tag{10}$$

where $p_1(x,t)$ is a scalar function, and

$$A_N = \begin{bmatrix} 0 & 1 \\ -1 & -1 \end{bmatrix} , \qquad A_1 = \begin{bmatrix} 0 & 0 \\ -1 & 0 \end{bmatrix}$$

In Radziszewski's paper [8] the quadratic Lyapunov function

$$V = x_1^2 + x_1 x_2 + x_2^2$$

was used to analyze robust stability.

The resulting vertices of the one-dimensional maximal parallelopiped Π were $p^1 = -\sqrt{3}/2$, $p^2 = \sqrt{3}/2$, thus stability of the trivial solution of (1) was ensured if

$$p_1^- = -\sqrt{3}/2 \le p_1(x,t) \le \sqrt{3}/2 = p_1^+ \quad \text{for all } x,t$$

Simultaneously, Radziszewski proved that the no quadratic Lyapunov function exists such that its application to the robust stability problem of (13) yields the larger parallelopiped Π_1 containing Π.

To enlarge the parallelopiped Π we shall apply the piecewise quadratic Lyapunov function. Testing the assumptions of the Th. 2 we find that the forms QF_j, $j = 1,2$ associated with the vertices p^1, p^2 have two roots $\zeta_1 = [0.5907, -0.8069]^T$, $\zeta_2 = [0.939, 0.344]^T$. All the assumptions of the Th. 2 are fulfilled. Since the algorithm to build a piecewise quadratic Lyapunov function is not ready we arbitrarily test the function

$$V_1^{(p)}(x) = -\varepsilon x_1 x_2$$

on the set $x_1x_2 < 0$, finding that its derivative along solutions of (13) is negative at the point ζ_1. Also the function

$$V_2^{(p)}(x) = \varepsilon\, x_1 x_2$$

treated on the set $x_1x_2 \ge 0$ has negative derivative at the point ζ_2. Combining the function V(x) with functions $V_1^{(p)}(x)$ and $V_2^{(p)}(x)$ and testing for the value of ε such that assumptions of the Th. 2 are fulfilled leads to the piecewise quadratic Lyapunov function

$$V = \begin{cases} x_1^2 + 1.5\, x x_2 + x_2^2 & \text{for } x_1\, x_2 \le 0 \\ x_1^2 + 0.5\, x_1\, x_2 + x_2^2 & \text{for } x_1\, x_2 > 0 \end{cases}$$

The vertices of the new enlarged parallelopiped Π_1 are $p^1 = -.94$, $p^2 = .94$
and the system (13) is stable if

$$p_1^- = -.94 \le p_1(x,t) \le .94 = p_1^+$$

which represents 8.5% improvement. The bound p_1^- is only 6% less than the value $p_1^- = -1$ which makes the system unstable.

VIII. CONCLUSION

The paper shows applicability of piecewise quadratic Lyapunov functions to robust stability problem of nominally linear systems with structured perturbations. The approach proposed is a natural extension of the traditional procedure utilizing quadratic Lyapunov functions. It seems that a simple numerical algorithm may be built to systematize the proposed approach.

APPENDIX

Proof of the Lemma 2.

Lemma 2. If the product

$$(\eta)^T \, \mathrm{grad}V_1 \, (x^*) = 0$$

then there are arbitrarily small changes of the vectors γ_1, γ_2 such that

$$(\eta)^T \, \mathrm{grad}V_1 \, (x^*) \neq 0$$

Proof. Let η_i and x_j^* be non-zero components of correspondingly η and x^*. Change the i-th component of γ_1 by $\Delta\gamma_{1i}$. The resulting change of the product

$$(\eta)^T \, \mathrm{grad}V_1 \, (x^*) \tag{11}$$

is equal to

$$\lambda \, \Delta\gamma_i \, \eta_i \left(\gamma_{21} \, x_1^* + \dots \gamma_{2n} \, x_n^*\right) \tag{12}$$

If (12) is equal zero then change γ_2 by $\Delta\gamma_{2j}$. The corresponding change of the product (11) is of the form

$$2\lambda \left(\eta_1\gamma_1 + \dots + \eta_n \, \gamma_{1n}\right) \Delta\gamma_{2j} \, x_j^*$$

If this expression is also equal to zero then introduce the changes $\Delta\gamma_{1i}$, $\Delta\gamma_{2j}$ simultaneously. The corresponding change of the product (11) is equal to

$$2\lambda \, \eta_i \, \Delta\gamma_{2j} \, x_j^*$$

and is not equal zero, which ends the proof. ■

Proof of the Lemma 3.

Lemma 3. Under assumptions made there is a sufficiently small $\lambda > 0$ such that

$$x^T S_1 x + \lambda x^T S_2 x < \text{ for } x \in F$$

Proof. By virtue of (12) there is a neighborhood $N \in F$ such that

$$x^T S_2 x < 0 \text{ for } x \in N$$

Therefore for any $x \in N$ and any $\lambda > 0$ we have

$$x^T S_1 x + \lambda x^T S_2 x < 0$$

Denote by $-M_1 < 0$ the upper bound of $x^T S_1 x$ on $F \backslash N$, and by $M_2 > 0$ the upper bound of $x^T S_2 x$ on $F \backslash N$ (for $M_2 \leq 0$ the problem is trivial). Select $\lambda = 0.5(M_1/M_2)$ and estimate the upper bound on $F \backslash N$

$$x^T S_1 x + \lambda x^T S_2 x < -M_1 + 0.5(M_1/M_2) M_2 < 0$$

which ends the proof. ■

Proof of the Th. 2.

Theorem 2. If

(i) $V(x)$ guarantees the solution II to robust stability problem,

(ii) there is a finite number of roots of the forms QF_j, $j = 1,\dots,2^q$, and the roots are isolated points of Φ_1,

(iii) there is a root ζ_r, $1 \leq r \leq m$ such that it is not an equilibrium point of the system (1) and at this root only one of the forms QF_j is equal zero,

then there exists a two-piece piecewise Lyapunov function $V^{(p)}(x)$ better than $V(x)$.

Proof. By virtue of the Lemma 4 there exist the non-zero $V_1^{(p)}(x)$ and a set F_1 such that the point $\zeta_r \in F_1$ and the remaining points of the set Ψ do not belong to F_1.

By virtue of the assumption the vector of the right hand-side of (1) at the point ζ_r is not zero. Therefore, either the product of this vector and the vector of the gradient of the function $V_1^{(p)}$ at the point ζ_r is non-zero or by virtue of the Lemma 2 the product may be made non-zero by means of the arbitrarily small changes of the function $V_1^{(p)}(x)$. But this means that the derivative of the function along the solution of (1) at the point ζ_r is not equal zero after the changes. Further, the point ς_r still remains in the set F_1 and the other points of Ψ outside of F_1 when executing such small changes. Assume that such changes were made if necessary and let $V_1^{(p)}(x)$ represent now the adjusted function.

Since the sign of the value of the derivative of $V_1^{(p)}(x)$ at the point ς_r may be changed by manipulating the sign of the function we assume that the sign of the derivative of the adjusted function is negative at ς_r.

Consider the function

$$V^{(p)}(x) = \begin{cases} V(x) + \lambda\, V_1^{(p)}(x) & \text{for } x \in F_1 \\ V(x) & \text{for } x \in R^n \backslash F_1 \blacktriangle F_2^* \end{cases}$$

It is easy to see that selecting sufficiently small λ the condition (i) of the Th. 1 is fulfilled. Since on the set F_2 the function $V^{(p)}(x)$ is identical with $V(x)$ the corresponding quadratic forms $QF_j^{(2)}$, $j = 1,2,\dots,2^q$ fulfill on this set condition (ii) of the Th. 1. Moreover, the forms attain the zero values at no other points than the points at which the forms QF_j attain this value, namely $\Psi\backslash\varsigma_r$.

On the set F_1 by virtue of the definition of the function $V^{(p)}(x)$ we have

$$QF_j^{(1)} = QF_j + QF_j^{*(1)}$$

where $QF_j^{*(1)}$ is a quadratic form resulting from entering the vertex p^j into the expression for the derivative of the function $V_1^{(p)}(x)$ along solutions of (1). According to the assumptions all but one (which we denote QF_h) forms QF_j, $j = 1,2,\dots,2^q$, are negative on the set $F_1 \cap \Phi_1$. Therefore, selecting sufficiently small λ it is possible to have all the forms $QF_j^{(1)}$, $j \neq h$, negative on the set $F_1 \cap \Phi_1$. The form $QF_h^{*(1)}$ is negative at the point ς_r and the form $QF_h^{(1)}$ is nonpositive on the set $F_1 \cap \Phi_1$, and attains the value zero only at the point ς_r. Thus, the assumptions of the Lemma 3 are fulfilled. Considering the set $F_1 \cap \Phi_1$ we find that for sufficiently small λ all the forms $QF_j^{(1)}$ are negative on this set. We conclude that the function $V^{(p)}(x)$ fully satisfies both assumptions of the Th. 1 and therefore $V^{(p)}(x)$ guarantees the solution II to robust stability problem. Also we have

$$\psi_1 = \psi\backslash\zeta_r$$

which proves the Theorem. ■

REFERENCES

1. E.J. Davison, "The Robust Control of a Servo-Mechanism Problem for Linear Time-Invariant Multivariable Systems," *IEEE Trans. Automat. Contr.*, Vol. AC-21, pp. 25-34, February 1976.
2. C.A. Desoer, F.M. Callier, and W.S. Chan, "Robustness of Stability Conditions for Linear Time-Invariant Feedback Systems," *IEEE Trans. Automat. Contr.*, Vol. AC-22, pp. 586-590, August 1977.
3. D. Siljak, "Parameter Space Methods for Robust Control Design: A Guided Tour," *IEEE Trans. Automat. Contr.*, Vol. AC-34, pp. 674-688, July 1989.
4. D. Siljak, *Large Scale Dynamic Systems: Stability and Structure*, North-Holland, Amsterdam, The Netherlands, 1978.
5. R.V. Patel and M. Toda, "Quantitative Measures of Robustness of Multivariable Systems," *Proc. JACC*, San Francisco, TP8-A, 1980.
6. R.K. Yedavalli, Z. Liang, "Reduced Conservatism in Stability Robustness Bounds by State Transformation," *IEEE Trans. Automat. Contr.*, Vol. AC-31, pp. 863-866, 1986.
7. N. Becker, W.M. Grimm, "Comments on Reduced Conservatism in Stability Robustness Bounds by State Transformations," *IEEE Trans. Automat. Contr.*, Vol. AC-33, pp. 223-224, 1988.
8. B. Radziszewski, "O. Najlepszej Funkcji Lapunowa," IFTR Reports, No. 26, 1977.
9. H.P. Harisberger and P.R. Belanger, "Regulators for Linear, Time-Invariant Plants with Uncertain Parameters," *IEEE Trans. Automat. Contr.*, Vol. AC-21, pp. 705-708, 1976.
10. N. Rouche, P. Habets, and M. Laloy, *Stability Theory by Liapunov's Direct Method*, Springer-Verlag, New York, 1977.
11. W. Hahn, *Stability of Motion*, Springer-Verlag, New York, 1967.

DISCRETE ADAPTIVE ALGORITHMS AND NEURAL NETWORK CONTROLLERS FOR NONLINEAR DYNAMICS

M.B. PSZCZEL AND T. PAYNE

Weapons System Research Laboratory,
Surveillance Research Laboratory,
Defence Science and Technology Organisation,
P.O. Box 1700, South Australia 5108

ABSTRACT. Authors consider nonlinear models applicable to analysis of motion of robotic manipulator and/or flexible space structures. The aim of the paper is to derive an adaptive class of stabilizing controllers via Liapunov direct method and compare them with the control obtained through the design of a suitable neural network. Similarities and differences and advantages/defficiencies of both approaches are shown, but at this stage analysis is mainly theoretical.

1. Modelling of flexible structures.

Modelling and control of flexible structures such as a robotic manipulator arm or large space structures (viz. space station) should take into account problems of quickly varying parameters (uncertainty of entries of inertia matrix, Coriolis forces, and spring characteristics), necessity of hybrid modelling (rigid and elastic dynamics), and retention in the model of nonlinear functions determining the dynamics of structures.

The approach presented here for modelling of a manipulator with flexible links will closely follow that described by A. Truckenbrodt [18]. For simplicity only the flexibility w.r.t. one chosen plane will be later assumed in the model.

With ν_i, κ_i denoting planar bendings of link (i), and using the Ritz-Kantorovitch series expansion we have :

$$\bar{\nu}_i(x_i,t) = \sum_{j=1}^{n} \nu_i^j(x_i)\nu_i^j(t) \tag{1.1}$$

The exact solution is expected as $n \to \infty$. Practically n is taken large enough to physically justify the linearization of the process. The lenght of step subdividing the link may change in order to fulfill the assumed tolerance condition following linearization.

Denoting :

$$\eta_i = [\nu_i(x_i,t), \kappa_i(x_i,t), \zeta_i(x_i,t)]^T$$

where $\zeta_i(x_i,t)$ denotes torsion of the i-th link, the following hybrid model may be obtained for flexible structures (cf. [14] for derivation) :

$$\begin{bmatrix} A & A_c \\ A_c^T & A_\eta \end{bmatrix} \begin{bmatrix} \ddot{q} \\ \ddot{\eta} \end{bmatrix} + \begin{bmatrix} C(q,\dot{q},w) \\ C(\eta,\dot{\eta},w) \end{bmatrix} + \begin{bmatrix} 0 & C_c \\ 0 & C_\eta \end{bmatrix} \begin{bmatrix} \dot{q} \\ \dot{\eta} \end{bmatrix} + \begin{bmatrix} G(q,w) \\ G_\eta(\eta,w) \end{bmatrix} + \tag{1.2}$$

$$+ \begin{bmatrix} 0 & P_c \\ 0 & P_{q,\eta} \end{bmatrix} \begin{bmatrix} q \\ \eta \end{bmatrix} = \begin{bmatrix} Q^F \\ 0 \end{bmatrix} + \begin{bmatrix} Q^D(q,\dot{q},w) \\ Q_\eta(\eta,\dot{\eta},w) \end{bmatrix} + \begin{bmatrix} 0 & D_c \\ 0 & D_\eta \end{bmatrix} \begin{bmatrix} \dot{q} \\ \dot{\eta} \end{bmatrix}$$

where $A_\eta(\eta,w)$, $C(\eta,\dot{\eta},w)$, $G_\eta(\eta,w)$, $Q_\eta(\eta,\dot{\eta},w)$ are elastic correspondents of $A(q,w)$ $C(q,\dot{q},w)$ (matrix of Coriolis and gyroscopic forces of rigid model), $G(q,w)$ (matrix of all potential forces of rigid structure) and $Q^D(q,\dot{q},w)$ (matrix of damping forces), while $A_c(q,\eta,w)$, $C_c(q,\dot{q},\eta,\dot{\eta},w)$,

$C_\eta(q,\dot q,\eta,\dot\eta,\ w)$ $P_c(q,\eta,w)$, $P_{q,\eta}(q,\eta,w)$, $D_c(q,\dot q,\eta,\dot\eta,w)$, $D_\eta(q,\dot q,\eta,\dot\eta,w)$ are matrices coupling the elastic and joint coordinates.
The inertially decoupled hybrid model may be written as

$$\begin{bmatrix}\ddot q\\ \ddot\eta\end{bmatrix}+\begin{bmatrix}\Gamma^1\\ \Gamma^2\end{bmatrix}+\begin{bmatrix}\Pi^1\\ \Pi^2\end{bmatrix}=\begin{bmatrix}u\\ u_c\end{bmatrix}+\begin{bmatrix}\Phi^1\\ \Phi^2\end{bmatrix} \tag{1.3}$$

The procedure for the inertial decoupling may be described as follows. We let :

$$\hat A=\begin{bmatrix}A & A_c\\ A_c^T & A_\eta\end{bmatrix}$$

Now matrix $\hat A^{-1}$ is divided into four blocks dimensionally corresponding to A, A_c, and A_η. They will be denoted simply as $\hat A_i^{-1}$ $i=1,\dots,4$ where

$$\begin{aligned}
\hat A_1^{-1}&=I(A-A_cA_\eta^{-1}A_c)^{-1}\\
\hat A_2^{-1}&=-I(A-A_cA_\eta^{-1}A_c)^{-1}A_cA_\eta^{-1}\\
\hat A_3^{-1}&=I(A_cA^{-1}A_c-A_\eta)^{-1}A_cA_\eta^{-1}\triangleq[\hat A_2^{-1}]^T\\
\hat A_4^{-1}&=-I(A_cA^{-1}A_c-A_\eta)^{-1}
\end{aligned}$$

The new numerical characteristics for all other forces are determined in the following ways :

$$\begin{bmatrix}\Gamma^1\\ \Gamma^2\end{bmatrix}=\begin{bmatrix}\hat A_1^{-1}[C_{q.\dot q,w}+C_c\dot\eta]+\hat A_2^{-1}[C_{\eta,\dot\eta,w}+C_\eta\dot\eta]\\ \hat A_3^{-1}[C_{q,\dot q,w}+C_c\dot\eta]+\hat A_4^{-1}[C_{\eta,\dot\eta,w}+C_\eta\dot\eta]\end{bmatrix}$$

defines the centrifugal and Coriolis characteristics of the decoupled, hybrid model,

$$\begin{bmatrix}\Pi^1\\ \Pi^2\end{bmatrix}=\begin{bmatrix}\hat A_1^{-1}[G(q,w)+P_c\eta]+\hat A_2^{-1}[G_\eta(\eta,w)+P_{q,\eta}\eta]\\ \hat A_3^{-1}[G(q,w)+P_c\eta]+\hat A_4^{-1}[G_\eta(\eta,w)+P_{q,\eta}\eta]\end{bmatrix}$$

determines the new spring/gravity characteristics,

$$\begin{bmatrix}\Phi^1\\ \Phi^2\end{bmatrix}=\begin{bmatrix}\hat A_1^{-1}[Q^D(q,\dot q,w)+D_c\dot\eta]+\hat A_2^{-1}[Q_\eta(\eta,\dot\eta,w)+D_\eta\dot\eta]\\ \hat A_3^{-1}[Q^D(q,\dot q,w)+D_c\dot\eta]+\hat A_4^{-1}[Q_\eta(\eta,\dot\eta,w)+D_\eta\dot\eta]\end{bmatrix}$$

defines the dissipative forces and

$$\begin{bmatrix}\hat A_1^{-1}Q^F\\ \hat A_3^{-1}Q^F\end{bmatrix}=\begin{bmatrix}u\\ u_c\end{bmatrix}$$

determines the new control vector.
Introducing the vector of state-space variables x represented by $(x_1,x_2,\dots,x_n)$, where $n=2N$ rigid model and $n=4N$ for hybrid (elastic) structure the inertially decoupled, state-space version of the (1.1.8)-(1.1.11) may now be presented. For the convenience of presentation the vector of state-space variables will be subdivided in the following way :

$$[x_1,\dots,x_N,x_{N+1},\dots,x_n]^T=[x_{r1},x_{r2}]^T$$

in case of rigid structure

$$[x_1,\dots,x_N,x_{N+1},\dots,x_{2N},x_{2N+1},\dots,x_{3N},x_{3N+1},\dots,x_n]^T=$$
$$[x_{r1},x_{r2},x_{e1},x_{e2}]^T$$

in case of hybrid model.

The newly introduced variables are chosen in such a way that x_{r1}, x_{e1} correspond to generalized displacements q, η while x_{r2}, x_{e2} correspond to generalized velocities $\dot{q}$, $\dot{\eta}$. Then the state-space representation may be written as :

$$\begin{bmatrix} \dot{x}_{r1} \\ \dot{x}_{r2} \end{bmatrix} = \begin{bmatrix} x_{r2} \\ \mathrm{u} - (\Gamma - \Phi + \Xi + \Psi) \end{bmatrix} \tag{1.4}$$

for the model of the rigid body and

$$\begin{bmatrix} \dot{x}_{r1} \\ \dot{x}_{r2} \\ \dot{x}_{e1} \\ \dot{x}_{e2} \end{bmatrix} = \begin{bmatrix} x_{r2} \\ u - (\Gamma^1 + \Pi^1 - \Phi^1) \\ x_{e2} \\ u_c - (\Gamma^2 + \Pi^2 - \Phi^2) \end{bmatrix} \tag{1.5}$$

for the hybrid model (1.3).

Without two-fold definition of x, either of these two state-space forms are equivalent to a system

$$\dot{x} = \tilde{f}(x, u, w) \tag{1.6}$$

For a general case, with the forces characteristics on the right-hand-sides of (1.4), (1.5) explicitly depending upon t we have

$$\dot{x} = f(x, u, w, t) \tag{1.7}$$

In the above $x \in R^n$, $u \in R^m$, $w \in R^s$, $t \in R^+$, $\tilde{f} \,:\, R^n \times R^m \times R^s \to R^n$ and $f : R^n \times R^m \times R^s \times R_0^+ \to R^n$ is such as to generate unique solutions in Δ and $\Delta \times R$ respectively. Whenever the control vector and/or the vector of disturbances are not specified (1.6), (1.7) change to contingent forms :

$$\dot{x} \in \{\tilde{f}(x, u, w) \;|\; u \in U,\; w \in W\}, \tag{1.8}$$

$$\dot{x} \in \{f(x, u, w, t) \;|\; u \in U, w \in W\} \tag{1.9}$$

respectively. The equations (1.6), (1.7) are called *selector equations* for (1.6), (1.7), with $\tilde{f}$, f being the selector of the vector fields (orientors) which they produce in (1.8), (1.9). Under suitable conditions, see Fillipov [4], (1.9) has absolutely continous solutions $k(x^0, t_0, u, \cdot) \,:\, R^+ \to \Delta$ and moreover there are functions $u(\cdot)$, $w(\cdot)$ such that these solutions satisfy (1.9). Given (x^0, t_0) the class of such solutions is denoted as $\mathcal{K}(x^0, t_0)$.
With the selectors to (1.8)-(1.9) being nonlinear and coupled we have now to solve the problem of nonlinear identification and design of robust controllers. The theory of adaptive identification and control, based on model reference approach, is discussed in the sections dealing with the stabilization, boundedness and reduced order stabilization for hybrid systems. We then reformulate the problem in terms of neural networks and present a possible solution.

2. Continuous and discrete systems.

2.1. Stabilization and boundedness. Continuous systems.

The properties of general nonlinear control systems discussed here are derived via the Liapunov direct method. The literature on the subject is now rather extensive (for review see Skowronski [15]). The results given rely on classical texts (Yoshizawa [18], Cesari [2], Hahn [6], Zubov [20],

Krassovski [9]) but have been extended (cf. Pszczel [14]) Theorems concerning partial stability refer to papers published in the seventies and early eighties by Oziraner and Rumiantsev [10]-[12]. Referring to (1.6), (1.7), we shall mean either the Lagrangian, or the Hamiltonian representation. The equilibria of (1.6) represent rest position of the manipulator which means $x(t) \equiv 0$ and extends cylindrically (for all $t_0 \in R$) into $x^e \times R \subset R^{2n+1}$ for the nonautonomous systems (1.7). In turn the equilibria of a system with noise will be these of (1.3) but subject to ranging in W. Thus we shall have a set of q^e, p^e specified by $p = 0$, $Q^P(q,w) = 0$, $w \in W$, or more precisely and generally defined as follows. Given specific $\tilde{w} \in W$, $\tilde{f}(x^e, 0, \tilde{w}) = 0$ has finite sequence of isolated roots x^{ei}, $i = 1, 2, \dots, k$, (see Section 1.4) each surrounded by the i-th equilibrium range

$$X^{ei} = \{x^{ei} \mid f(x^{ei}, 0, w) = 0,\ w \in W\}\ i = 1, 2, \dots, k$$

Then consider $X^{ei} \times R \subset R^{2N+1}$ and let $X^{ei}(t)$ be a t-section of it. Since the union of invariant sets is invariant, the set $X^{ei} \times R$ is invariant, and under the class $\mathcal{K}(X^{ei}, R)$ is minimal invariant. Let

$$\rho(x(t), X^{ei}(t)) \triangleq \inf \rho(x(t), x^{ei}(t)) \mid x^{ei}(t) \in X^{ei}(t)$$

be the instantenous distance of the solution from X^{ei}.

Definition 2.1.1. *The set $X^{ei} \times R$ is strongly stabilizable under motions of (1.9), iff there is a control program $\mathcal{U}(\cdot)$ such that for each $\epsilon > 0$, $u \in U(x)$, $x^0 \in \Delta_H^i$, $t_0 \in R$, $k(x^0, t_0, u, \cdot) \in \mathcal{K}(x^0, t_0)$ there is $\delta > 0$ such that $\rho(x^0, X^{ei}(t_0)) < \delta$ implies*

$$\rho(k(x^0, t_0, u, t), X^{ei}(t)) < \epsilon, \quad \forall t \geq t_0$$

When δ does not depend upon x^0, t_0, $k(\cdot)$ the stability is x^0-, t_0-, $k(\cdot)$- uniform respectively.

Definition 2.1.2. *Suppose $X^{ei} \times R$ is strongly stabilizable. Then iff in addition there is $\delta_0 > 0$ such that $\rho(x^0, X^{ei}(t_0)) < \delta_0$ $x^0 \in \Delta_{AS} \subset \Delta_H^i$, $t_0 \in R$ implies*

$$\rho(k(x^0, t_0, u, t), X^{ei}(t)) \to 0$$

as $t \to \infty$, then $X^e \times R$ is strongly asymptotically stabilizable and Δ_{AS} is the region of such stabilizability.

The region Δ_{AS} is open and contains at least some neighbourhood of X^{ei} where there may not be another X^{ej} $i \neq j$. Thus X^{ei} is isolated. In particular, if we specify $w = \tilde{w}$ in W, X^{ei} becomes a singleton $\{\tilde{x}^{ei}\}$, defined as one of the equilibria of (1.3), or of

$$f(x^e, 0, \tilde{w}) = 0. \tag{2.1.1}$$

For the case of autonomous (1.6) the Definitions 2.1.1, 2.1.2 simplify relating to $\Delta \subset R^{2N}$ enclosing X^{ei} and trajectories $k(x^0, u, \cdot) \in \mathcal{K}(x^0)$ instead of $\Delta \times R$ enclosing $X^{ei} \times R$ and motions $k(x^0, t_0, u, \cdot) \in \mathcal{K}(x^0, t_0)$. The distance $\rho(x^0, X^{ei}) < \delta \Rightarrow \rho(k(x^0, u, t), X^{ei}) < \epsilon$ and $\rho(x^0, X^{ei}) < \delta_0 \Rightarrow \rho(k(x^0, u, t), X^{ei}) \to 0$, $t \to \infty$, respectively.
Observe that the concept of X^{ei} (in terms defined here) is local. For stability, we want to keep the energo-motions below secure energy levels thus estimating the corresponding behaviour we will need to use the reference to an energy surface $\tilde{\mathcal{H}}$ which is defined by

$$\tilde{H}(x) \triangleq \inf_{w,\ t} H(x, w, t) \mid w \in W,\ t \in R^+$$

Let $w = w^*$ be the extremizing w. Then

$$\tilde{H}(x) \leq H(x, w^*, t), \ \forall t \in R^+ \tag{2.1.2}$$

as desired. Then also by (2.1.1), we reduce X^{ei} to the singleton x_*^{ei}, and may study the $\tilde{H}(x)$-energy cups Z_H^i about it.

Theorem 2.1.1. *The set $X^{ei} \times R$ is strongly stabilizable under (1.7) on some $\Delta^i_H \times R$ enclosing $X^{ei} \times R$, if the corresponding hamiltonian $H(x, w, t)$ of the system generates suitable cup of $\tilde{H}(x)$ about X^{ei} on Δ^i_H and there is a control program $U(x)$ defined on Δ^i_H such that for each $u \in U(x)$ on Δ^i_H,*

$$\tilde{L}(x,t) = \langle \nabla \tilde{H}(x), f(x,u,w,t) \rangle \leq 0, \ \forall w \in W \tag{2.1.3}$$

Proof follows from the fact that (2.1.3) renders to the energy flow nonaccumulative everywhere on Δ^i_H, thus implying that each energy level Z_c is crossed by the motion either staying on it or entering its interior, tending towards the equilibrium x^{ei}_*.

Theorem 2.1.2. *Under the condition of Theorem 2.1.1 if for each $u \in U(x)$ on Δ^i_H there is $c(r) > 0$ continously increasing and such that*

$$\tilde{L}(x,u,w,t) = \langle \nabla \tilde{H}(x), f(x,u,w,t) \rangle \leq -c(\|x\|), \ \forall w \in W, \tag{2.1.4}$$

then $X^{ei} \times R$ is strongly asymptotically stabilizable in Δ^i_H

Proof follows by the same argument, but with the adjustment, that (2.1.4) gives dissipative flow i.e. only strict entry by the motions concerned when crossing any level Z_c in Δ^i_H.
The control program in question might be then found from the following.

Corollary 2.1.2. *Given $(x,t) \in \Delta^i_H \times R$, if there is a pair u^*, w^* such that*

$$L(x,t,u^*,w^*) = \min_u \max_w \tilde{L}(x,t,u,w) < -c(\|x\|) \tag{2.1.5}$$

then condition (2.1.4) is met with $u^* \in U$.

The proof follows from the fact that

$$\min_u \max_w \tilde{L}(x,t,u,w) \geq \min_u \tilde{L}(x,t,u,w), \ \forall w \in W \tag{2.1.6}$$

Consider now the product system of two (1.7) :

$$\begin{cases} \dot{x} \in \{f(x,t,u,w) \mid u \in U(x), \ w \in W\} \\ \dot{x}_m \in \{f_m(x_m,t,u_m,) \mid u_m \in U(x) \ \} \end{cases} \tag{2.1.7}$$

with solutions $k(x^0,t_0,u,t), k_m(x^0_m,t_0,u_m,t)$, $t \geq t_0$ generating a vector $(x(t),x_m(t)) \in \Delta \times \Delta$ and with two Hamiltonians $H(x,w,t)$, $H_m(x_m,t)$ which form the product Hamiltonian

$$H(x,x_m,w,t) = H(x,w,t) + H(x_m,t)$$

We can apply now the same reasoning to $H(x,x_m,w,t)$ as we did for $H(x,w,t)$ and introduce a product energy frame of reference $\tilde{H}(x,x_m)$, then applying Theorems 2.1.1, 2.2.2 to the product system (2.1.7) in $\Delta^i_H \times \Delta^i_{H_m}$ about the equilibrium (x^{ei}, x^{ei}_m).

Theorem 2.1.3. *The set $X^{ei} \times X^{ei}_m \times R$ is strongly asymptotically stabilizable - under the product system (2.1.7) on some $\Delta^i_H \times \Delta^i_{H_m} \times R$ enclosing the corresponding equilibria, formed by the product $\tilde{H}(x,x_m)$ if there is a control program $U(x,x_m)$ defined on $\Delta^i_H \times \Delta^i_{H_m}$ and that for each $u \in U(x,x_m)$ there is a constant $c > 0$ such that*

$$\langle \nabla \tilde{H}(x,x_m), (f,f_m) \rangle \leq -c \tag{2.1.8}$$

for all $w \in W$.

Proof follows by the same argument as for Theorem 2.1.2. We may obviously have a similar Corollary, see (2.1.6), generating a desired controller.
Stabilization in real time requires ultimate boundedness rather than asymptotic stability. Since our discussion will refer to various regions of the energy surface, it is convenient to use general definitions and suficient conditions first, before applying them in particular problems.

Definition 2.1.3. *The motions of (1.9) are strongly Lagrange stable (SLS) in a given bounded set Δ_0 in Δ if there is a control program $U(x)$ such that $(x^0, t_0) \in \Delta_0 \times R$, $k(x^0, t_0, u, \cdot) \in \mathcal{K}(x^0, t_0)$ implies*

$$k(x^0, t_0, u, t) \in \Delta_0, \quad \forall t \geq t_0$$

Consider a neighbourhood $N(\partial\Delta_0)$ of the boundary $\partial\Delta_0$ and let $N_S = N(\partial\Delta_0) \cap \text{int}\Delta_0$. Then introduce a test function $V(\cdot) : N_S \to R$ with $v_0 \triangleq \inf V(x)$ for $x \in \partial\Delta_0$. The following has been proved in Skowronski [15].

Theorem 2.1.4. *Definition 2.1.3 holds if given $\Delta_0 \subset \Delta$, there is $U(x)$ defined on Δ_0 and a C^1-function $V(x)$ such that for all $x \in N_S$,*

(i) $V(x) \leq v_0$

(ii) for each $u \in U(x)$

$$\langle \nabla V(x), f(x, t, u, w) \rangle < 0 \tag{2.1.9}$$

Proof follows immediatly from contradiction between (i) and (ii) generated when any motion $k(\cdot) \in \mathcal{K}(x^0, t_0)$, $x^0 \in \Delta_0$, crosses $\partial\Delta_0$.

Definition 2.1.4. *The motions of (1.9) are strongly uniformly ultimately bounded (SUUB) iff there is a control program $U(x)$ and a constant $B > 0$ and for some $\Delta_0 \in \Delta$ there is $T_B(\Delta_0) < \infty$ such that $(x^0, t_0) \in \Delta_0 \times R$, $k(x^0, t_0, u, t) \in \Delta_B = \{x \in \Delta \mid \|x\| < B\}$, $\forall\, t \geq t_0 + T_B$. (Here $\|\cdot\|$ denotes any norm in R^n.)*

The following sufficient conditions are an extension of these proved by Yoshizawa [18].

Theorem 2.1.5. *Definition 2.1.4 holds if there are Δ_0, Δ_B, T_B, $U(x)$ of this Definition together with a C^1-function $V(\cdot) : \Delta_r \times R \to R$,*

$\Delta_r = \{x \in \Delta \mid \|x\| > r\}$, $r > 0$ may be large, such that

(i) $a(\|x\|) \leq V(x,t) \leq b(\|x\|)$ where $a(\rho)$, $b(\rho)$ continuously increasing and $a(\rho) \to \infty$ as $\rho \to \infty$;

(ii) for each $u \in U$,

$$\frac{\partial V}{\partial t} + \langle \nabla V(x,t), f(x,t,u,w) \rangle \leq -c(\|x\|), \ \forall w \in W \tag{2.1.10}$$

where $c(\rho) > 0$ continous.

The proof follows by the same argument as in [18], with the obvious adjustment for the class $\mathcal{K}(x^0, t_0)$ and with the suitable choice of $r > 0$ such that $r \leq B$ i.e. $\Delta_r \supset \partial\Delta_B$.
Both, Definition 2.1.4 and Theorem 2.1.5 open the option that the bound $B > 0$ (or the set Δ_B), the set of initial conditions Δ_0 and the real time interval $T_B < \infty$ may be stipulated. Skowronski [15], calls the case of stipulated Δ_0, Δ_B ***real time attraction*** provided $\Delta_B \subset \Delta_0$, but Δ_B may be an arbitrary open set. Then Δ_B is the ***real time attractor*** and the maximal Δ_0 (or the union of all Δ_0 possible) is called the region of such attraction. We shall adapt these names for SUUB. When $\Delta_0 = \Delta_B$ we have $T_B = 0$ and SUUB becomes SLS. In turn when $T_B < \infty$ is stipulated, the SUUB becomes SUUB ***during*** T_B. In case of real time attraction during T_B, Theorem 2.1.5 has to be adjusted (see Skowronski [16]). We have then the following theorem.

Theorem 2.1.6. *Given Δ_0, a set $\Delta_B \subset \Delta_0 \subset \Delta$ is a strong real time attractor for motions of (1.9) from Δ_0, if Theorem 2.1.4 holds and moreover there is a C^1-function $V(\cdot) : D \to R$, D (open) $\supset \overline{C\Delta_B}$, $D \cup \{0\} = \emptyset$, $C\Delta_B = \Delta_0 - \Delta_B$, such that*

(i) $V_B(x) \leq v_B^+$, $\forall x \in \partial\Delta_B$

(ii) $0 \leq V_B(x) \leq v_B^+$,
where $v_B^+ = \sup V_B(x) \mid x \in \partial\Delta_B$, $v_B^+ > 0$

(iii) given $U(x)$ of Theorem 2.1.4, for each $u \in U(x)$ there is a constant $T_B < \infty$ such that

$$\langle \nabla V_B(x), f(x,t,u,w) \rangle \leq -\frac{v_B^+}{T_B}, \ \forall w \in W \tag{2.1.11}$$

Proof. By Theorem 2.1.4 we have SLS in Δ_0. Consider $k(x^0,t_0,u,\cdot) \in \mathcal{K}(x^0,t_0)$, $(x^0,t_0) \in C\Delta_B \times R$. Integrating (2.1.11) along such motion, we obtain the estimate

$$t \le t_0 + \frac{V(x^0) - V(x)}{v_B^+} T_B.$$

By (ii), $V_B(x) \ge 0$ and $V_B(x) - v_B^+ \le 0$, hence $V_B(x^0) - V_B(x) \le v_B^+$. Thus for $t \ge t_0 + T_B$ the motion must be out of $C\Delta_B$. Since Δ_0 is SLS, it must enter Δ_B. By (ii) and (iii) it may not return, since upon re-entry to $C\Delta_B$, these two conditions contradict. □

Corollary 2.1.6. *Given $x \in D$, if there is a pair u^*, w^* such that*

$$L(x,t,u^*,w^*) = \min_u \max_w L(x,t,u,w) \le -\frac{v_0 - v_B}{T_B} \tag{2.1.12}$$

then condition (iii) of Theorem 2.1.6 is met with $u \in U$.

Proof follows by the same argument as for Corrolary 2.1.2.
Now, when we would want to stabilize, or bound motions below some energy levels we may let $V(\cdot)$ of Theorems 2.1.4 - 2.1.6, be defined by $\tilde{H}(x)$ of (1.9) and we may have the conditions of these theorems satisfied, depending on where we are on the corresponding energy surface $\tilde{\mathcal{H}}$, and whether the control program can secure a dissipative field on the corresponding region of $\tilde{\mathcal{H}}$. On the other hand if we wanted to bound the motions above some energy levels, say in a neighbourhood of an energy threshold, we must let $V(\cdot)$ of these theorems be defined by $-\tilde{H}(x)$, with the same comments. Obviously then the contradicting inequalities expressed in terms of such $\tilde{H}(x)$ will be inverted.

2.2 Stabilization of hybrid systems.

The discussion about stabilization in the previous section referred to the system (1.9), which covers also the case of rigid-flexible motion, described by the selector equation (1.7). For a large class of flexible manipulators the number of model DOF's i.e. the dimension of the vector $\eta(t)$ is at best uncertain and so is the dynamics in the second part of (1.5). It is advisable to secure the stability of the system ignoring $\eta(t)$, as long as we know that it is bounded.
In general, the partial stability means stability with respect to some components of the state vector only, while the other are assumed bounded. Given below the definitions and suficient conditions extend the results discussed by Rumiantsev and Oziraner [12].
Let us represent the state vector $x(t)$ of (1.7) as

$$x(t) \triangleq [y(t), z(t)] \text{ with } y(t) = [y_1(t), \dots, y_k(t)]^T \triangleq [x_1(t), \dots, x_k(t)]^T$$

and

$$z(t) = [z_1(t), \dots, z_l(t)] \triangleq [x_{k+1}, \dots, x_{2N}(t)]^T, \; l = 2N - k$$

Assuming that the motions of (1.7) are SLS in some Δ_0 i.e. there are suitable constants $M, K < \infty$ such that

$$\|y(t)\| < M \;, \quad \|z(t)\| < K, \; \forall t \ge t_0 \tag{2.2.1}$$

Definition 2.2.1. *The set $X^{ei} \times R$ is strongly stabilizable under motions of (1.7) relative to $y(t)$, briefly strongly y-stabilizable iff there is $U(x)$ such that for each $\epsilon > 0$, $u \in U(x)$, $x^0 \in \Delta_H^i$, $t_0 \in R$, $k(x^0,t_0,u,\cdot) \in \mathcal{K}(x^0,t_0)$ there is $\delta > 0$ such that $\rho(x^0, X^{ei}) < \delta$ implies $\rho\left([y(t),z(t)], X^{ei}\right) < \epsilon$, $\forall z(t)$, $t_0 \ge t_0$*

This stability is automatically z-uniform. When δ does not depend upon x^0, t_0, $k(\cdot)$ it is also x^0-, t_0- , $k(\cdot)$-uniform.

Definition 2.2.2. *The y-stability of Definition 2.1.5 becomes asymptotic, iff in addition there is $\delta_0 > 0$ such that $\rho(x^0, X^{ei}) < \delta_0$, $x^0 \in \Delta_{AS} \subset \Delta^i_H$ implies $\rho\left([y(t), z(t)], X^{ei}\right) \to 0$, as $t \to \infty$, $\forall z(t), t \geq t_0$.*

The notions of uniformity are the same, Δ_{AS} is the region of strong asymptotic y-stabilizability. The following theorem is an immediate extension of a theorem proved by Rumiantsev-Oziraner quoted above.

Theorem 2.2.1. *Definition 2.1.6 holds if there is $U(x)$ defined on Δ^i_H and a C^1-function $V(\cdot)$: $\Delta^i_H \times R \to R$ such that*
(i) $a(\|y\|) \leq V(x,t) \leq b(\rho)$
(ii) for all $u \in U(x)$

$$\frac{\partial V}{\partial t} + \langle \nabla V(x,t), f(x,t,u,w) \rangle \leq -c(\rho) \qquad \forall w \in W \tag{2.2.2}$$

where $a(\cdot)$, $b(\cdot)$, $c(\cdot)$ as in Theorem 2.1.5, $a(0) = 0$, and $\rho = \left(\sum_{i=1}^{s} x_i^2\right)^{1/2}$, $k \leq s \leq n$.

Proof follows by the same argument as in Oziraner [12].
Turning now to (1.2.9), we can let $k = N$, thus specifying $y(t) \triangleq (q(t)\dot{q}(t))$, while $z(t) = (\eta(t), \dot{\eta}(t))$. Using Theorem 2.1.4, we let $V(x) = \tilde{H}(x)$ to have (i) implied by (1.4.7) and (ii) implied by the suitable energy characteristics (1.4.3), provided the system is dissipative. Hence we obtain SLS for the entire system (1.2.9) which generates (2.2.1). Then, we use Theorem 2.2.1 and again let $V(x,t) \equiv \tilde{H}(x)$. Since $\tilde{H}(y) \leq \tilde{H}(x)$ and $\tilde{H}(x) \leq \hat{H}(x) = \sup_{w,t} H(x,w,t)$, we have condition (i) satisfied while condition (ii) is again implied by the dissipativness of energy flow on Δ^i_H, if applicable. Hence we obtain

Theorem 2.2.2. *When (1.9) with the selector (1.7) is dissipative on some local Δ^i_H about X^{ei}, the latter set is strongly asymptotically stable irrespective of the dimension and behaviour of $(\eta, \dot{\eta})$, $t \geq t_0$.*

For real time attraction and tracking with respect to part of the state components, the following concept is useful.

Definition 2.2.3. *The motions of (1.7a) are strongly uniformly bounded with respect to part of the state variables (i.e. part of the state variables are SUUB) iff there is $U(x)$ and a constant $B > 0$, and for some $\Delta_0 \in \Delta$ there is $T_B(\Delta_0) < \infty$ such that $(x^0, t_0) \in \Delta_0 \times R$, $k(x^0, t_0, u, \cdot) \in \mathcal{K}(x^0, t_0)$ implies $k(y^0, z, t_0, u, t) \in \Delta_B = \{(y,z) \in \Delta \mid \|y\| < B\}$, $\forall t \geq t_0 + T_B$, $\forall z$.*

The sufficient conditions can be obtained from Theorem 2.1.5 with (i), (ii) replaced by

$$(i)' \quad a(\|y\|) \leq V(x,t) \leq b(\|y\|), \ \forall x \in \Delta$$
$$(ii)' \quad \frac{\partial V}{\partial t} + \langle \nabla V(y,z,t), f(y,z,t,u,w) \rangle \leq -c(\|y\|)$$

for all $w \in W$, $z \in \Delta$, which imply (i), (ii).

2.3 Relative dynamics.

The relative dynamic behaviour of the two systems forms the theory for our model reference adaptive control of manipulators, so we may introduce it now in the general format applicable to rigid and hybrid modelling. We want to estimate the relative behaviour of two control systems in terms of their motions (states) and parameters. Let us consider the product system as in (2.1.7) but make the first system dependant upon adjustable parameters $\lambda^1(t) = (\lambda^1_1(t), \ldots, \lambda^1_l(t)) \in \Lambda \subset R^l$, where Λ is a bounded set, while the second, call it target system will have the parameters fixed λ^2 =const$\in \Lambda \subset R^l$. Moreover one of the control vectors, say u^2, is given. The choice of which λ is fixed and which u is given may be reversed without consequence to our argument. Thus we have

$$\begin{cases} \dot{x}^1 \in \{f^1(x^1,t,u^1,\lambda^1,w^1) \mid u^1 \in U(x),\ w \in W\} \\ \dot{x}^2 \in \{f^2(x^2,t,u^2,\lambda^2,w^2) \mid w^2 \in W\} \end{cases} \tag{2.3.1}$$

with $x^1(t)$, $x^2(t) \in \Delta \subset R^n$, λ^1, $\lambda^2 \in \Lambda \subset R^l$. We want a convergence in real time of $x^1(t) \to x^2(t)$, $\lambda^1(t) \to \lambda^2$, the second state and parameters represent a target dynamical system. The convergence should occur on a specific subset $\Delta_0 \subset \Delta$ and with a stipulated accuracy, perhaps achieved within a stipulated time $T_c < \infty$.

The first way of doing this that comes to mind is the classical linear technique (see next Chapter) of substracting $x^1 - x^2 \triangleq e(t)$, $\dot{x}^1 - \dot{x}^2 = \dot{e}(t)$ and writing (2.3.1) in terms of an error equation

$$\dot{e} \in \{(f^1 - f^2)(e,u^1,u^2,\lambda^1,\lambda^2,t,x^1,x^2) \mid u^1 \in U(x),\ w \in W\} \tag{2.3.2}$$

This technique may work for linear systems, as the wide literature quoted on MRAC shows, it may work for a large class of nonlinear systems like Hamiltonian systems, see Flashner-Skowronski [5], but it does not work in general and the extensions provided so far do not work in case of x^1 and x^2 having different dimensions. The latter is essential when we want to reduce dynamics. We thus use the product state method introduced by Skowronski [16].

Let $X(t) = (x^1(t), x^2(t)) \in \Delta \times \Delta \triangleq \Delta^2 \subset R^{2n}$ and $F \triangleq (f^1, f^2) \in R^{2n}$ tangent to $X(t)$, while $\alpha(t) \triangleq \lambda^1(t) - \lambda^2$ implying $\dot{\alpha} = \dot{\lambda}^1$.

Then the system (2.3.1) becomes

$$(\dot{X}, \dot{\alpha}) \in \{F(X,\alpha,t,u^1,\lambda^1,\lambda^2,w) \mid u^1 \in U(x),\ w \in W\}$$

with motions $k(X^0,\alpha^0,t_0,u^1,\cdot) \in \mathcal{K}(X^0,\alpha_0) \in \Delta^2 \times \Lambda$ for almost all $t \geq t_0$. Our objective can be specified as follows. Let $\Delta_0^2 \triangleq \Delta_0 \times \Delta_0$ be a subset of Δ^2 and define

$$M_\epsilon \triangleq \{(X,\alpha) \in \Delta_0^2 \times \Lambda \mid \|x^1 - x^2\| < \epsilon,\ \|\alpha\| < \epsilon\} \tag{2.3.3}$$

Definition 2.3.1. *The system 1 of (2.3.1) tracks the target system 2 in real time and with precision $\epsilon > 0$ on a given subset Δ_0 of Δ, iff there is a control program $U(x,\alpha)$ and a time interval $T < \infty$, such that $(X^0,\alpha^0) \in \Delta_0^2$, $k(X^0,\alpha^0,t_0,\cdot) \in \mathcal{K}(X^0,\alpha^0)$ implies $k(X^0,\alpha^0,t_0,t) \in M_\epsilon$, $\forall t \geq t_0 + T$.*

To produce sufficient conditions for the objective we use Theorems 2.1.4, and 2.1.6 applied to the vector (X,α) instead of x. We re-define : $\partial(\Delta_0^2 \times \Lambda)$ and $N_S^2 \triangleq N^2[\partial(\Delta_0^2 \times \Lambda)] \cap \text{int}\ \Delta_0^2$ generating $V_S(\cdot)\ :\ N_S^2 \to R$ with $v_s \triangleq \inf V(X,\alpha) \mid (X,\alpha) \in \partial(\Delta_0^2 \times \Lambda)$. Moreover. $CM_\epsilon \triangleq (\Delta_0^2 \times \Lambda) - M_\epsilon$, open $D^2 \supset \overline{CM_\epsilon}$ and $V_\epsilon(\cdot)\ :\ D^2 \to R$ with $v_\epsilon = \inf V(X,\alpha) \mid (X,\alpha) \in \partial M_\epsilon \cap \overline{CM_\epsilon}$.

Theorem 2.3.1. *Definition 2.1.3 holds, if given Δ_0, ϵ, there is $U^1(X,\alpha)$ and a C^1-functions $V_S(\cdot)$, $V_\epsilon(\cdot)$ defined above such that*

(i) $V_S(X < \alpha) \leq v_s$, $(X,\alpha) \in N_S$

(ii) $\forall u \in U(X,\alpha)$ $\langle \nabla V_S(x,\alpha), F(X,\alpha,t,u^1,\lambda^1,\lambda^2,w)\rangle < 0$, $\quad \forall w \in W$

(iii) $0 \leq V_\epsilon(X,\alpha) \leq v_\epsilon^+$, $v_\epsilon^+ > 0$, $\forall (X,\alpha) \in CM_\epsilon$

(iv) $V_\epsilon(X,\alpha) \leq v_\epsilon^-$, $\forall (X,\alpha) \in CM_\epsilon$

(v) *for each $u \in U(X,\alpha)$, there is $T < \infty$ such that*

$$\langle \nabla V_\epsilon(X,\alpha), F(X,\alpha,t,u^1,\lambda^1,\lambda^2,w)\rangle \leq -\frac{v_\epsilon^+}{T} \tag{2.3.4}$$

for all $w \in W$. When T is stipulated, we obtain stipulated time tracking.

Proof follows immediately from Theorems 2.1.4 and 2.1.6.

The following test function may now be introduced :

$$V_S = H^2(x^1) + H^2(x^2) + a\alpha, \tag{2.3.5}$$

$$V_\epsilon = \begin{cases} |H^2(x^1) - H^2(x^2)| + a\alpha, \ (X,\alpha) \in CM_\epsilon, \\ a\alpha, \ (X,\alpha) \in M_\epsilon, \end{cases} \tag{2.3.6}$$

where $a = (\text{sign}\ \alpha_1, \ldots, \text{sign}\ \alpha_l)$ and $\alpha_i \neq 0$, as then the adjustment of parameters is redundant. Note that $H^2(x^1)$ is the energy function of the target system $H^2(\cdot)$ considered along motions of the tracking system. It is of interest to choose Δ_0 either within some local Δ_H^i or encompassing all Δ_H^i's i.e. enclosing Δ_L. In either case we have N_S filled with the energy levels with positive bound satisfying condition (i). The character of $H^2(\cdot)$ as a square form also secures (iii) and (iv). It remains to check (ii) and (v). Differentiating (2.3.5), (2.3.6),

$$\dot{V}_S = \dot{H}^2(x^1) + \dot{H}^2(x^2) + a\dot{\alpha} \tag{2.3.7}$$

$$\dot{V}_\epsilon = \begin{cases} \dot{H}^2(x^1) - \dot{H}^2(x^2) + a\dot{\alpha}, \ \delta H^2 \geq 0 \\ \dot{H}^2(x^2) - \dot{H}^2(x^1) + a\dot{\alpha}, \ \delta H^2 < 0 \\ a\dot{\alpha}, \ (X,\alpha) \in M_\epsilon \end{cases} \tag{2.3.8}$$

where $\delta H^2 = H^2(x^1) - H^2(x^2)$, and

$$\dot{H}^2(x^1) = \langle \nabla H^2(x^1), f^1(x^1, u^1, \lambda^1, w) \rangle, \tag{2.3.9}$$

$$\dot{H}^2(x^2) = \langle \nabla H^2(x^2), f^2(x^2, u^2, \lambda^2, w) \rangle. \tag{2.3.10}$$

The following two condition secure (ii), (v) in the obvious way :

Control Condition

$$\min_{u^1} \max_{w} \langle \nabla H^2(x^1), f^1(x^1, t, u^1, \lambda^1, w) \rangle = \dot{H}^2(x^2), \ \delta H^2 \geq 0 \tag{2.3.11a}$$

$$\max_{u^1} \min_{w} \langle \nabla H^2(x^1), f^1(x^1, t, u^1, \lambda^1, w) \rangle = \dot{H}^2(x^2), \ \delta H^2 < 0, \tag{2.3.11b}$$

Adaptation Condition

$$a\dot{\alpha} \leq -\left(\frac{v_\epsilon^+}{T} + 2|\dot{H}^2(x^2)| \right), \ \alpha \neq 0. \tag{2.3.12}$$

The latter is implied in turn by the adaptive laws :

$$\dot{\alpha}_i \text{sign}\ \alpha_i = -\left(\frac{v_\epsilon^+}{T} + 2\left|\frac{\partial H^2}{\partial x_i^2} f_i^2\right| \right), \ i = 1, \ldots, l = n \tag{2.3.13}$$

or adjusted by partial sums of components, when $l < n$. The case $l > n$ should be avoided when selecting adjustable parameters.
The control program $U(X, \alpha)$ is then selected from (2.3.11a)-(2.3.11b), but it may be done only after functions $f^1(\cdot)$, $f^2(\cdot)$ are specified, i.e. in case studies.

2.4 Discrete systems.

We consider now the discrete equivalent of the dynamical (1.7), namely

$$x(k+1) = f(x(k), t_0, t_k, u(k+1)) \tag{2.4.1}$$

where $f(\cdot)$ is the previously defined nonlinear function. We may treat (2.4.1) as a discrete mapping and analyze its global behaviour (regions of attraction) using theory of cell to cell mapping (Hsu [7]). Stability conditions similar to those formulated in 2.1-2.3 can also be presented for (2.4.7).

Obviously those conditions also apply to the augmented system in its discrete version. Instead of the continuous test function with $H(x)$ and the gradient $\nabla H(x)$ we will use a discrete Liapunov (test) function $H(x(k))$ and the increment defined as

$$\Delta H(x(k)) = H(x(k+1)) - H(x(k)) \tag{2.4.2}$$

Similarly as in Theorem 2.1.3, for the asymptotic stability of the discrete system it is required that

$$\Delta H(x(k)) \leq -c, \ \forall k \in N \tag{2.4.3}$$

With the above changes conditions (2.3.11)-(2.3.12) may be adjusted to :
Discrete Control Condition

$$\min_{u^1} \max_{w} \langle \Delta H^2(x^1(k)), f^1(x^1(k), t_k, u^1(k+1), \lambda^1(k), w)\rangle = \Delta H^2(x^2(k)), \ \delta H^2 \geq 0 \tag{2.4.4}$$

$$\max_{u^1} \min_{w} \langle \Delta H^2(x^1(k)), f^1(x^1(k), t_k, u^1(k+1), \lambda^1(k), w)\rangle = \Delta H^2(x^2(k)), \ \delta H^2 < 0, \tag{2.4.5}$$

Discrete Adaptation Condition

$$a\Delta\alpha(k) \leq -\left(\frac{v_\epsilon^+}{T} + 2|\Delta H^2(x^2(k))|\right), \ \alpha(k) \neq 0 \tag{2.4.6}$$

3. Neural networks framework for an adaptive control systems.

Instead of a single-processor type implementation of the adaptive control problem as formulated in 2.3, we now consider an implementation of a control system in the form of a neural network. The application of this techniques to nonlinear control has already been discussed in numerous papers (viz [8], [19] and others). Among particular methods suitable for tackling the problem are Hopfield model, Kohonen self-organizing feature maps and backpropagation. An example of network controller is presented in the following section.

4. Flexible Multiple Link. Manipulator Control Problem.

A possible neural solution is presented which should minimise the training required and ensure stability after operational installation to obtain the optimum response.
Work by several researchers, Kawato et al. [8], Chen and Pao [3] and others using an approach referred to by Psaltis et al. [13] as specialized learning have demonstrated the success of neural controllers on rigid systems. However it is pointed out in V. Zeman et al.[19] that this approach is probably unsuitable for flexible manipulators since these systems are only marginally stable, and during the early stages of learning, adaption of the neural net will only reinforce the oscillatory tendency of the plant. To avoid this problem V. Zeman et al.[20] adopt the method of generalized learning. In this paper a different approach is considered.
The problem is simplified to an N link manipulator which is assumed to operate in a plane, so that only bending in one dimension need be considered. The control problem is divided into two parts, with the Lagrange coordinate q representing the angle at each joint, and η representing the flex in the structure as discussed in the first section (see (1.2)). The rigid problem involves developing a controller to determine the control force u so that $|q^* - q| < \epsilon$ for $t > t_M$ where the desired path is represented by q^*. This design assumes that there is no flex in the structure. This is done using conventional control techniques, and can be derived from knowledge of the physical system. Due to the unconsidered flex in the system, the true position of the manipulator $\tilde{z}$ will be different to that required (z represents the position in global coordinates) and can be represented by

$$\tilde{z} = f(x^r, x^e) \tag{4.1}$$

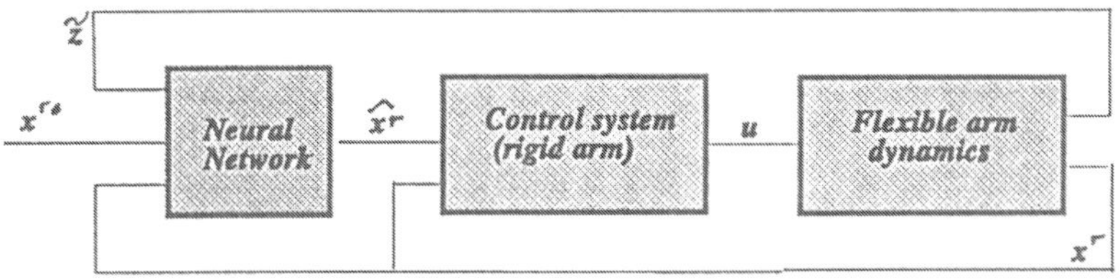

FIGURE 1. Control Structure and Feedback

where

$$x^r \triangleq [q, \dot{q}] \tag{4.2}$$

$$x^e \triangleq [\eta, \dot{\eta}] \tag{4.3}$$

The control problems can be restated in a first order difference representation,

$$x_{n+1}^r = F(x_n^r, u_{n+1}) \tag{4.4}$$

$$x_{n+1}^e = G(x_n^r, x_n^e, u_{n+1}) \tag{4.5}$$

and the rigid controller can be specified as

$$u_{n+1} = g(x_{n+1}^{r*}, x_n^r) \tag{4.7}$$

The rigid controller need not be discrete. The modification required to the control law, due to the existence of flexibility is provided by a feed forward neural controller which modifies x^{r*} to $\widehat{x^r}$ which is the reference seen by the rigid controller, so that the resultant positioning of the manipulator by the rigid controller is as close as possible to the reference demand. ie.

$$\widetilde{z_{n+1}} = f(F(x_n^r, g(\widehat{x_{n+1}^r}, x_n^r)), x_{n+1}^e) \tag{4.8}$$

This structure has the advantage that it can be added to existing systems which ignore the flexibility of their structure. Errors in the rigid controller can also be compensated for by the neural controller so it is not necessary for the rigid controller to be adaptive.

To implement the controller, feedback is required. The rigid controller has x^r available. These signals may be continuous or discrete depending on the type of controller used. The neural controller requires knowledge (particularly during training) of both x^r and x^e, however it is not practical to measure x^e so the neural controller must make use of the location (in global coordinates) of some significant link(s). For example, it may be only the end point of the gripper which is needed to be positioned accurately. For this application it will be assumed that only a subset of the angles between links and a global coordinate axis are known.

The reference path consists of a discrete sequence of x_n^{r*}s. Since the input is discrete in time and the values can be discretized the network depicted will act merely as a mapping from the reference vector to a vector which will produce a result close to the original demand. It is possible for a neural network to learn this mapping function, however this would require a large number of neurons, it is more appropriate to attempt to generate a network whose dynamic response approximates the required function. The input-output relationship should be the same, but the first technique relies on recognition while the second uses modeling.

The neural solution to this problem will require a substantial amount of training. If the controller was to be performed entirely with the neural network (no rigid controller), then for much of this time, the system performance would be largely useless, and there is no guarantee that a suitable stable solution would ever be found. However using the structure presented here, if the neural network is initially configured so that

$$\widehat{x^r} = x^{r*} + C \tag{4.9}$$

ie. the neural controller corrects for a shift in equilibria due to the gravity and unmodelled dynamics at that point. Then with the manipulator allowed to settle to a stable position this value of the neural network is correct. For points which are yet to be trained the responses of the neural network are adjusted to maintain piece-wise continuity. The training of the network will occur gradually by training regions in the phase space which are located at an increasing distance from this initial point. Physically, this could mean, that the system would be operational over a full range of motion, however only at a low speed. As the system was trained the speed of operation could slowly be increased, until full speed was obtained.
This technique assumes that the transfer function provided by the neural network is continuous. However this may not be adequate if the links are capable of oscillating in modes other than the first order. In this case there needs to be a discontinuity in the network's response as links change modes of oscillation. This is a much more difficult problem. It may be possible to provide the network with a signal which represents the oscillation mode of each link, in which case the system can be considered (and trained) as being made up of a number of networks which are selected for the appropriate mode structure of the flex. If this signal is not available as a measurement then the decision making capabilities provided by neural networks can be utilized to establish the mode of flex. This function will be performed by a different network whose outputs are applied to the controller in place of measured values. The information required to determine the mode of each member consists partly of information specific to the dynamics of the members but also on the sequence of the links positions. This network should be able to be trained from knowledge of the configuration (links position) before implementation.
As a result of this a large complicated network has been replaced by a number of functional networks, which are smaller, some of which can be trained independently, possibly by simulation prior to system assembly and then integrated and optimised. This has the advantage that the stability of the solution will be much easier to ensure since it will be a function of the smaller blocks. The decision as to the type and size of the neural network to be used for each functional unit of the controller is a problem which needs to be addressed, however using a network which is possibly much larger than required will increase the fault tolerance and flexibility of the resultant controller. The network used to determine the link resonant mode could be implemented with a Kohonen network, possibly with lateral feedback. This type of network is ideal for recognising patterns of inputs, and since some of the inputs will be previous values, this system will be able to recognise sequence patterns. The map network must represent a dynamic system and a multilayer backpropagation network would be suitable. The backpropagation network also provides a relatively straightforward method of training the map network.

The objective of the system is to minimise

$$\|k(z^* - \tilde{z})\| \qquad \forall t \in R_0^+ \tag{4.10}$$

$$z^* = f(x^{r*}, 0) \tag{4.11}$$

where k is a function specifying which of the z_i are more significant. Applying the result of theorem 2.2.2 the rigid controller will ensure that

$$\|x^{r*} - x^r\| < \epsilon \qquad \forall t > T_a \tag{4.12}$$

The function of the neural network is to find a $\widehat{x_r}$ such that $V(\widehat{x^r}, t)$ is minimized w.r.t. $\widehat{x^r}$ where

$$V(\widehat{x^r}, t) = \|kz^* - kf(\widehat{x^r}, x^e)\| \tag{4.13}$$

and $kf(\widehat{x^r}, x^e) \to kz^* \quad \text{as} \quad t \to \infty$.

Now

$$\widehat{x^r_{n+1}} = N(x^{r*}_{n+1}, x^r_n, x^e_n) \tag{4.14}$$

where $N(\cdot)$ is the function performed by the neural network. The dimension of x^e_n is at best uncertain and in this case is assumed to be unknown. However using the fact that $u^e_0 = 0$.

$$\begin{aligned} x^e_{n+1} &= G(x^r_n, u^e_n, u_{n+1}) \\ &= G(x^r_n, G(x^r_{n-1}, u^e_{n-1}, u_n), u_{n+1}) \\ &\vdots \\ &= G(x^r_n, G(x^r_{n-1}, \ldots, u^e_0, \ldots, u_n), u_{n+1}) \\ &= G_n(x^r_n, x^r_{n-1}, \ldots, x^r_0, u_{n+1}, u_n, \ldots, u_1) \end{aligned} \tag{4.15}$$

It is possible using this approach to redefine $N(\cdot)$ so that u^e does not appear.
Another aspect of the problem is how to correct for errors in the neural model. As pointed out in an earlier section, the neural network does require some feedback information about the position of significant links. Now (4.15) is a simple invariant geometric calculation which can be derived analytically (or may even be available directly.) The way this function is implemented is not important so long as it is invariant, since the training error for the neural network is given by

$$tr^e_n = k(f(x^{r*}, x^e) - Z) \tag{4.16}$$

where Z is the position feedback information about arm or particular link positions (and $Z = f(\hat{x}^r, x^e)$ after training). This will generally be z_n but may also include other z_i. The neural network is initially configured along the lines of (4.9). Since the system is dissipative we have before training :

$$u_{n+1} \to 0 \text{ and } x^e_n \to \text{constant} \tag{4.17}$$

We also have a large training error i.e. $tr(t_0) > M$. We want to achieve $tr^e_n \to 0$ for $t > T_a$ i.e learning in finite time. Mapping (4.14) together with a rigid controller will now ensure convergence in finite time.

References

[1] Andronov, A.A., Vitt, A.A., Khaikin, S.E., *Theory of Oscillators*, Pergamon Press Ltd., 1966.

[2] Cesari, L. *Asymptotic Behaviour and Stability Problems in Ordinary Differential Equations*, Springer-Verlag, Berlin-Göttingen-Heidelberg, 1963.

[3] Chen, V.C., Pao, Y.-H., Learning control with neural networks, In *Proceedings of IEEE Conference on Robotics and Automation*, Phoenix, Arizona, 1989

[4] Fillipov, A.F., Classical solutions of differential equations with multivalued right-hand-side, *SIAM J. Control*, Vol. 5, No. 4, 1967, pp. 609-621.

[5] Flashner, H., Skowroński, J.M., Model tracking of hamiltonian systems , In *Trans. ASME, J. Dyn. Syst. Meas. Control, Special Issue, Workshop on Control Mechanics*, 1989.

[6] Hahn, W., *Stability of motion*, Springer-Verlag, Berlin-Heidelberg-New York, 1967.

[7] Hsu, C.S., *Cell-to-Cell Mapping. A Method of Global Analysis for Nonlinear System*, Springer Verlag, New York 1987.

[8] Kawato, M., Uno, Y., Isobe, M., Suzuki, R., Hierarchical neural network model for voluntary movement with application to robotics, *IEEE Control System Magazine*, April 1988.

[9] Krassovskii, N.N., *Control Theory*, Nauka, Moscow, 1968.

[10] Oziraner, A.S. On certain theorems of Liapunov Second Method, *PMM*, Vol. 36, No. 3, 1972, pp. 396-404.

[11] - , On stability of motion relative to a part of variables under constantly acting perturbations, *PMM*, Vol. 45, No. 3, 1981, pp. 419-427.

[12] Oziraner, A.S., Rumiantsev, V.V., The method of Liapunov functions in the stability problems for motion with respect to a part of the variables, *PMM*, Vol. 36, No. 2, 1972, pp. 364-384.

[13] Psaltis, D., Sideris, A., Yamamura A.A., A multilayered neural network controller, *IEEE Control System Magazine*, April 1988

[14] Pszczel, M.B.,*Adaptive Liapunov Controllers for Nonlinear Robotic Dynamics*, PhD thesis, University of Queensland, 1988.

[15] Skowronski, J.M., *Nonlinear Liapunov Dynamics*, World Sci. Publ., New Jersey-Singapore, 1990.

[16] - , Adaptive identification and model reference tracking by a flexible spacecraft, In *Proc. AIAA Aerospace Science Conf.*, Reno, Nev., Jan. 1989.

[17] -, *Control of Nonlinear Mechanical Systems*, Plenum, N.Y. 1991

[18] Truckenbrodt, A., Effects of elesticity on the performance of industrial robots, *Proc. 2nd IAESTED Danos Int. Symp. Robotics*, Switzerland, 1982, pp. 52-56.

[19] Yoshizawa, T., *Stability Theory by Liapunov Second Method*, Math. Soc. Japan Publ., 1966.

[20] Zeman, V., Patel, R.V., Khorasani K., A neural network based control strategy for flexible-joint manipulators. In *Proceedings of 28th Conference on Decision and Control*, Tampa, Florida, Dec. 1989

[21] Zubov, V.I., *Methods of A.M. Liapunov and Their Application*, P. Noordhoff Ltd., Groningen, Netherlands, 1964.

TRACKING AND FORCE CONTROL FOR A CLASS OF ROBOTIC MANIPULATORS

E. Reithmeier, Berkeley [1]
G. Leitmann, Berkeley [2]

Abstract

The increasing utilization of robotic manipulators in industrial tasks, such as assembly, forming or shaping of surfaces, as well as the handling of hazardous materials depends greatly on available hybrid force and position control schemes. Since the robot as well as its environment are often subject to parameter uncertainties which can not be neglected, it is necessary to design controllers which are robust with respect to these uncertainties. In addition, the dynamics of the robot are nonlinear requiring consideration of nonlinear control concepts. Another aspect to be taken into account is the relative stiffness of the robot, the force sensor and the manipulated surface. That is, the behavior of the system normal to the surface is relatively stiff while that tangential to the surface is relatively free. Separation of the controller for these two directions is therefore indicated. We propose a controller design which accounts for this point of view and demonstrate its efficacy with respect to robustness and accuracy of position and force tracking by means of numerical simulations. The design is based on the control concept of [CORLESS, LEITMANN 1989]. The example, considered is a Manutecr3 robot with three degrees of freedom. In addition, we account for the dynamics of the actuator which also possesses three degrees of freedom. The considered parameter uncertainties are friction moments in the links, friction between the end effector and the manipulated object, as well as nonlinear dynamics which are difficult to characterize.

1 Introduction

Industrial processes, such as welding, forming, assembly and so on, are are increasingly being carried out by robots. In general, complex tasks require not only position control but also control of the normal contact force along a prescribed path if the end effector is in contact with an object to be operated on. Usually, the force sensor on the end effector as well as the manipulated object are very stiff. As a result, small deformations at the contact point result in large forces normal to the surface of the object, whereas motion tangential to the surface leads to small contributions to the contact forces. Thus, it seems appropriate to design controllers which take separate account of motions normal and tangential to the surface.

Another important consideration are parameter uncertainties which are always present in practice in systems like industrial robots. Among these uncertainties are especially unknown or not readily determined friction moments and forces in the links and gears, difficult to characterize dynamics in actuators and sensors, friction forces at the point of contact between the end effector and the surface of the operated on object., as well as the uncertainties

[1]Alexander v. Humboldt Research Fellow, Dept. of Mech. Eng., UC Berkeley
[2]Professor, Dept. of Mech. Eng., UC Berkeley

in the position of the manipulated object. In order to assure satisfactory quality of the product, these and similar uncertainties must not substantially affect the prescribed track to be followed and force to be exerted. In other words, it is essential to design hybrid position and force controllers which are robust with respect to parameter uncertainties.

In most existing position and force control schemes, system parameter uncertainties are of secondary concern; e.g. see [SALISBURG 1980], [RAIBERT, CRAIG 1981], [KHATIB, BURDIK1986], [AN, HOLLERBACH 1987], [EPPINGER, SEERING 1986], [WHITNEY 1987], [GOLDENBERG 1988]. In addition, there are proposals for robust hybrid control [DESA, ROTH 1985], [DJAFERIS, MURAH 1987], [KUO, WANG 1990], which are based on linear feedback in terms of end effector coordinates and which, in general, cannot assure asymptotic stability. The subsequent proposal is based on nonlinear feedback and guarantees essentially asymptotic stability despite the presence of parameter uncertainties. Furthermore, a transformation of the equations of motion in surface oriented coordinates permits a separation into a stiff part (force control) and a non–stiff one (position control).

2 Modelling

In this section we describe the dynamical model of a class of industrial robots. The class of robots in question is subject to the the commonly accepted considerations summarized in following subsection.

2.1 Assumptions and Simplifications

Figure 1 shows a typical example of the class of robots considered here. We assume:

1. The robot consists of $2n, (n \in \mathbb{N})$, rigid bodies, namely, n rigid rotors and n rigid arms. The rotors are parts of the actuators which drive the arms.
2. The robot possesses tree structure with n degrees of freedom.
3. Each arm (i) possesses one degree of freedom with respect to arm (i-1) which is influenced by just one actuator (i). The angle between arm (i-1) and arm (i) is denoted by q_i.
4. The possibly occuring gyroscopic moments of the rotors, due to the motion of the arms, are not included in the model dynamics.
5. Rotor (i) is connected to arm (i) by an elastic gear (stiffnes coefficient c_i, damping coefficient d_i); or else, the connection is inelastic (e.g., "direct drive").

2.2 Coordinate Systems

To describe our problem appropriately it is useful to introduce various coordinate systems. We suppose, that each arm (i) is endowed with a fixed orthonormal basis B_i. The origin of basis B_i is at the mass center S_i of arm (i). The axes of B_i are not necessarily the principal axes of inertia of arm (i). The inertia matrix of arm (i) relative to B_i is denoted by $\mathbf{I}_i \in \mathbb{R}^{3,3}$ The basis B_0 is an inertial frame of reference fixed in the base of the robot. The Cartesian coordinates of mass center S_i of arm (i) relative to B_0 are denoted by $\mathbf{x}^{(i)} := (x_1^{(i)}, x_2^{(i)}, x_3^{(i)})^T \in \mathbb{R}^3$ (cf. [PFEIFFER, REITHMEIER 1987]).

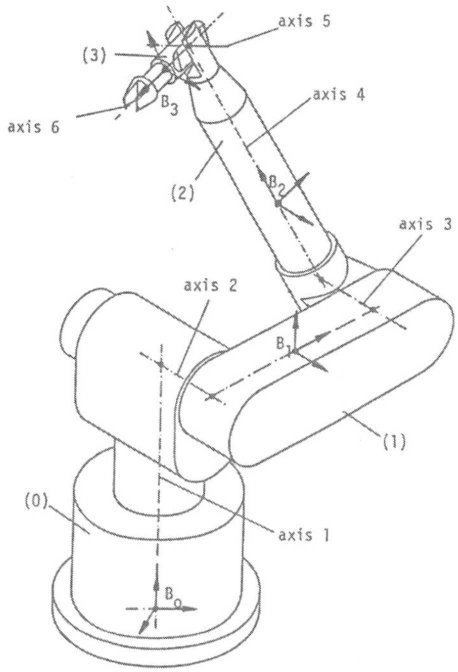

Fig. 1: Manutecr3 robot [TÜRK 1988]

We suppose throughout that the end effector remains in contact with the surface of the operated on object. That is, the robot end effector follows a path on a smooth two dimensional surface. If we denote the Cartesian working space coordinates (relative to B_0) by $\mathbf{x}$, then we can define the surface by means of two parameters z_1 and z_2 ("surface–oriented coordinates") and a C^1–function $\boldsymbol{\chi} : \mathbb{R}^3 \rightarrow \mathbb{R}^3$ via

$$\mathbf{x} = \begin{bmatrix} x_1 \\ x_2 \\ x_3 \end{bmatrix} := \boldsymbol{\chi}(z_1, z_2, z_3 = 0) = \begin{bmatrix} \chi_1(z_1, z_2, z_3 = 0) \\ \chi_2(z_1, z_2, z_3 = 0) \\ \chi_3(z_1, z_2, z_3 = 0) \end{bmatrix} . \tag{2.1}$$

Parameter z_3 is the coordinate along the normal to the tangent plane of the surface at point (z_1, z_2) of the surface. We assume that function χ is bijective. Thus every point of the working space can also be given in $\mathbf{z}$ coordinates. Furthermore, corresponding to each angular position $\mathbf{q} := (q_1, ..., q_n)^T$ there is a point in the robot's working space, e.g., a point on the end effector, say between the tip of the robot and the force sensor. Thus, there exists a C^1–transformation $\boldsymbol{\zeta} : \mathbb{R}^n \rightarrow \mathbb{R}^3$ of $\mathbf{q}$ into $\mathbf{z}$ (e.g., see Fig. 2):

$$\mathbf{z} = \begin{bmatrix} z_1 \\ z_2 \\ z_3 \end{bmatrix} := \boldsymbol{\zeta}(\mathbf{q}) = \begin{bmatrix} \zeta_1(q_1, ..., q_n) \\ \zeta_2(q_1, ..., q_n) \\ \zeta_3(q_1, ..., q_n) \end{bmatrix} \tag{2.2}$$

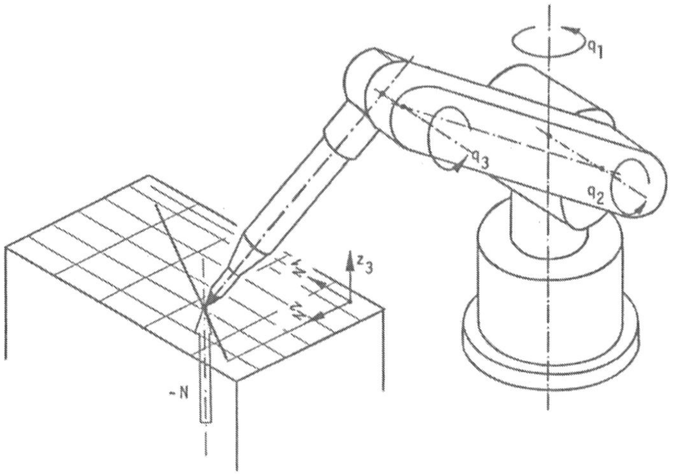

Fig. 2: Plane as an example for the surface of an object

Since, according to (2.1), $\mathbf{x}$ is a function of $\mathbf{z}$, one can employ (2.2) to express $\mathbf{x}$ (of points in the working space) in terms of $\mathbf{q}$:

$$\mathbf{x} = (\chi \circ \zeta)(\mathbf{q}). \tag{2.3}$$

If the robot possesses only three degrees of freedom ($n = 3$), in general, for obtainable angles between robot arms, there exists an inverse transformation $(\chi \circ \zeta)^{-1} : \mathbf{x} \mapsto \mathbf{q}$. If the robot possesses more than three degrees of freedom ($n > 3$), the transformation $\chi \circ \zeta$ is no longer bijective. In other words, following a given path on the surface of an object and exerting a desired normal force might be accomplished with arbitrarily many angular positions $\mathbf{q}$. That is, in case of more than three degrees of freedom, in addition to specifying the desired path and normal contact force history, additional constraints are required in order to obtain a unique angular position history.

2.3 Angular and Mass Center Velocities

In order to formulate the robot equations of motion in an efficient manner, it is useful to determine the velocities of the centers of mass and the angular velocities of each arm. In formulating these velocities there arise functional matrices of translation and rotation; these occur seperately in the formulation of the equations of motion. Usually it is convenient to express the center of mass velocities in components along the axes of inertial basis B_0, but the angular velocity of body (i) in components along the axes of body–fixed basis B_i for $i = 1, ..., n$. In view of the assumed tree structure of the robot, these velocities can be computed recursively; e.g., see [PFEIFFER, REITHMEIER 1987]. For the centers of mass velocity of body (i), expressed in B_0 components, we have :

$$\dot{\mathbf{x}}^{(i)} = \mathbf{J}_T^{(i)}(\mathbf{q}) \cdot \dot{\mathbf{q}} \ , \tag{2.4}$$

where $\mathbf{J}_T^{(i)}$ is the functional matrix or JACOBI matrix of translation. Similarily, for the angular velocity of the mass center of body (i), in B_i components, we have :

$$\omega^{(i)} = \mathbf{J}_R^{(i)}(\mathbf{q}) \cdot \dot{\mathbf{q}} \ , \tag{2.5}$$

where $\mathbf{J}_R^{(i)}(\mathbf{q})$ is the JACOBI matrix of rotation. On employing

$$\mathbf{J}_T(\mathbf{q}) := \begin{bmatrix} \mathbf{J}_T^{(1)}(\mathbf{q}) \\ \vdots \\ \mathbf{J}_T^{(n)}(\mathbf{q}) \end{bmatrix} \in \mathbb{R}^{3n,n} \ ; \ \mathbf{J}_R(\mathbf{q}) := \begin{bmatrix} \mathbf{J}_R^{(1)}(\mathbf{q}) \\ \vdots \\ \mathbf{J}_R^{(n)}(\mathbf{q}) \end{bmatrix} \in \mathbb{R}^{3n,n} \ . \tag{2.6}$$

and

$$\dot{\mathbf{x}} := \begin{bmatrix} \dot{\mathbf{x}}^{(1)} \\ \vdots \\ \dot{\mathbf{x}}^{(n)} \end{bmatrix} \in \mathbb{R}^{3n} \ ; \ \omega := \begin{bmatrix} \omega^{(1)} \\ \vdots \\ \omega^{(n)} \end{bmatrix} \in \mathbb{R}^{3n} \ . \tag{2.7}$$

we obtain :

$$\dot{\mathbf{x}} = \mathbf{J}_T(\mathbf{q}) \cdot \dot{\mathbf{q}} \quad ; \quad \omega = \mathbf{J}_R(\mathbf{q}) \cdot \dot{\mathbf{q}} \ . \tag{2.8}$$

2.4 Link Model

Usually there is one actuator for a link between to successive robot arms which moves the following arm by means of an elastic gear; e.g., see Fig. 3.

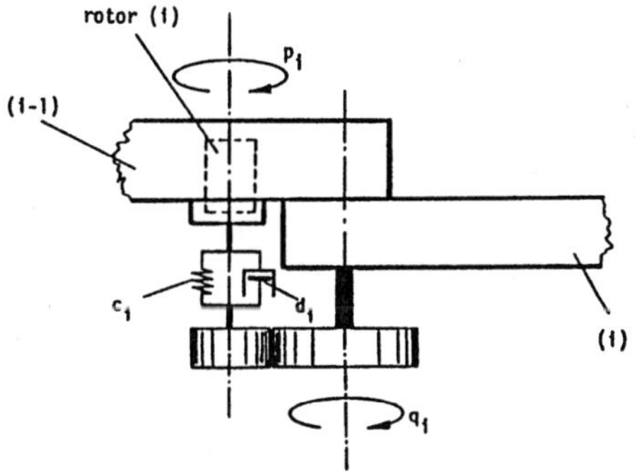

Fig. 3: Mechanical model of an elastic link

In this case, the rotor for arm (i) lies in the stator which is attached to arm (i-1). Then u_i denotes the torque exerted by the stator on rotor (i), and T_i is the torque exerted by the gear. As stated earlier, q_i is the angle between arm (i-1) and arm (i); p_i is the rotor angle of actuator (i), and $p_i - l_i q_i$ denotes the angle of deformation of the gear of the i–th link (between arm (i-1) and arm (i)). The dynamics of the links (including the elastic gears) are usually approximated (see assumption 4 of subsection 2.1) by linear differential equations of the form

$$\Theta\ddot{\mathbf{p}} + \mathbf{D}(\dot{\mathbf{p}} - \mathbf{L}\dot{\mathbf{q}}) + \mathbf{C}(\mathbf{p} - \mathbf{L}\mathbf{q}) + \mathbf{u} = \mathbf{0}, \tag{2.9}$$

respectively

$$\boldsymbol{\Theta}\ddot{\mathbf{p}} - \mathbf{L}^{-1}\mathbf{T} + \mathbf{u} = \mathbf{0} \tag{2.10}$$

where

$\mathbf{T} = (T_1, ..., T_n)^T$	is the vector of link torques exerted by the gears on the following arm,
$\boldsymbol{\Theta} = diag(\Theta_1, ..., \Theta_n)$	is the diagonal matrix of the moments of inertia of the rotors about their axes,
$\mathbf{C} = diag(c_1, ..., c_n)$	is the diagonal matrix of the stiffness coefficients of the elastic gears,
$\mathbf{D} = diag(d_1, ..., d_n)$	is the diagonal matrix of the damping coefficients of the elastic gears,
$\mathbf{L} = diag(l_1, ..., l_n)$	is the diagonal matrix of the gear ratios.

In some situation, e.g. for stiff gears, from a practical point of view (cf. section 3) it is appropriate to neglect the gear elasticity. The simplest way to express this mathematically is by way of an algebraic relation of the form

$$\mathbf{Rq} + \mathbf{Sp} = \mathbf{0} \tag{2.11}$$

where $\mathbf{R}$ and $\mathbf{S}$ are constant matrices of dimension $ne \times n$, and $ne < n$ is the number of rotors which are inelastically coupled to the affected arm.

2.5 Parameter Uncertainties

Based on numerous experiments, it is well known that friction in the links plays a role that can not be ignored in the design of controllers. In the case of hybrid position and force control there is the additional effect of friction between the end effector or gripper and the operated on object. Since it is very difficult or costly to determine these friction moments and forces, it is convenient and useful to interpret their magnitudes as uncertain parameters. Furthermore, it is appropriate to interpret difficult to characterize or unmodelled dynamical effects, such as nonlinearities or neglected sensors and other dynamics, as uncertain parameters. Then the controller is to be designed so as to be robust relative to the parameter uncertainties. The mathematical formulation of the uncertainties considered here is discussed in the following subsection.

2.6 Equation of Motion in Link Coordinates

For a robot with n degrees of freedom, its equations of motion in terms of the link coordinates q_i, introduced in subsection 2.2, are

$$\mathbf{M}(\mathbf{q})\ddot{\mathbf{q}} + \mathbf{g}(\mathbf{q},\dot{\mathbf{q}},\boldsymbol{\mu}) + \boldsymbol{\Delta}\cdot\mathbf{T} + \mathbf{B}(\mathbf{q})\mathbf{u} = \mathbf{0} \tag{2.12}$$

where

$\mathbf{M}(\mathbf{q}) :=$	$\mathbf{J}_T^T(\mathbf{q})\cdot\mathbf{m}\cdot\mathbf{J}_T(\mathbf{q}) + \mathbf{J}_R^T(\mathbf{q})\cdot\mathbf{I}\cdot\mathbf{J}_R(\mathbf{q}) \in \mathbb{R}^{n,n}$ is the mass matrix. The diagonal matrix $\mathbf{m} \in \mathbb{R}^{3n,3n}$ contains the masses of the individual rigid parts (arms) of the robot. The block diagonal matrix $\mathbf{I} \in \mathbb{R}^{3n,3n}$ contains the inertia matrices $\mathbf{I}_i \in \mathbb{R}^{3,3}$ of the arms; the moments and products of inertia are relative to body-fixed basis B_i.
$\mathbf{B}(\mathbf{q})$	$\in \mathbb{R}^{n,n}$ is the input matrix.
$\boldsymbol{\Delta}$	$\in \mathbb{R}^{n,n}$ is a constant matrix which describes the influence of the torques, transmitted by the gears.
$\mathbf{u}$	$\in \mathbb{R}^{n}$ is the control vector.
$\mathbf{g}(\mathbf{q},\dot{\mathbf{q}},\boldsymbol{\mu}) :=$	$(\mathbf{g}_Z + \mathbf{g}_C + \mathbf{g}_G + \mathbf{g}_R)(\mathbf{q},\dot{\mathbf{q}},\boldsymbol{\mu}) \in \mathbb{R}^{n}$ is the vector of the generalized centrifugal-, Coriolis-, weight- and friction forces; it is a continous function of its arguments.
$\boldsymbol{\mu}$	$\in \boldsymbol{\Omega}$ is the vector of parameter uncertainties. $\boldsymbol{\Omega} \subset \mathbb{R}^{m}, m \in \mathbb{N}$, is the compact set of possible parameter values.

We suppose that the following are unknown or uncertain:

- the magnitude $\| \mathbf{F}_R \|$ of the sliding friction force $\mathbf{F}_R$ between robot and the object. The direction of $\mathbf{F}_R$ is tangent to and opposite to the motion (velocity) along the actual path of the contact point on the object surface. That is, if $\mathbf{x}_C$ denotes the position vector of the contact point C (e.g., see Fig. 4) measured from the origin of B_0 and expressed in coordinates of B_0, then

$$\mathbf{F}_R = \frac{\dot{\mathbf{x}}_C}{\| \dot{\mathbf{x}}_C \|}\| \mathbf{F}_R \| \quad , \quad \dot{\mathbf{x}}_C \neq \mathbf{0}. \tag{2.13}$$

- the sliding friction moments $\mathbf{M}_R \in \mathbb{R}^n$ in the links which are mainly due to dry friction.
- difficult to characterize nonlinearities such as Coriolis and centrifugal forces.

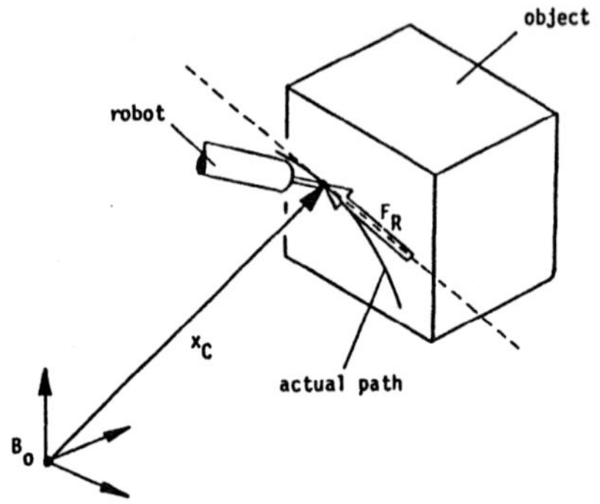

Fig. 4: direction of the friction force between sensor and object

We suppose further that the point of contact between the normal force sensor and the object surface is always in a state of sliding friction. For purposes of numerical simulation we assume the following form of the generalized forces:

$$\mathbf{g}_Z(\mathbf{q},\dot{\mathbf{q}}) := -\mathbf{J}_R^T(\mathbf{q}) \cdot \tilde{\omega}(\mathbf{q},\dot{\mathbf{q}}) \cdot \mathbf{I} \cdot \omega(\mathbf{q},\dot{\mathbf{q}}) \tag{2.14}$$

with

$$\tilde{\omega} := diag(\tilde{\omega}^{(1)}, ..., \tilde{\omega}^{(n)}) \quad \in \mathrm{IR}^{3n,3n} \ , \tag{2.15}$$

$$\mathbf{g}_C(\mathbf{q}) := \mathbf{J}_T^T(\mathbf{q}) \cdot \mathbf{m} \cdot \dot{\mathbf{J}}_T(\mathbf{q},\dot{\mathbf{q}}) + \mathbf{J}_R^T(\mathbf{q}) \cdot \mathbf{I} \cdot \dot{\mathbf{J}}_R(\mathbf{q},\dot{\mathbf{q}}) \ , \tag{2.16}$$

$$\mathbf{g}_G(\mathbf{q}) := \mathbf{J}_T^T(\mathbf{q})\mathbf{F}_G \tag{2.17}$$

with

$$\mathbf{F}_G := \begin{bmatrix} \mathbf{F}_G^{(1)} \\ \vdots \\ \mathbf{F}_G^{(n)} \end{bmatrix} \in \mathrm{IR}^{3n} \ . \tag{2.18}$$

$\mathbf{F}_G^i \in \mathrm{IR}^3$ is the weight of the i–th arm expressed in components along the axes of inertial basis B_0.

$$\mathbf{g}_R(\mathbf{q},\dot{\mathbf{q}},\mu) := \mathbf{M}_R + \left(\frac{\partial \dot{\mathbf{x}}_C}{\partial \dot{\mathbf{q}}}(\mathbf{q},\dot{\mathbf{q}})\right)^T [\mathbf{F}_R(\mathbf{q},\dot{\mathbf{q}}) + \mathbf{F}_N(\mathbf{q},\dot{\mathbf{q}})] \tag{2.19}$$

$\mathbf{M}_R := (M_1, ..., M_n)^T \in \mathrm{IR}^n$ is the vector of link friction torques. The generalized force due to the magnitude of the normal contact force N is

$$\mathbf{F}_N(\mathbf{q},\dot{\mathbf{q}}) := \frac{\frac{\partial \chi}{\partial z_3}(\zeta(\mathbf{q}))}{\| \frac{\partial \chi}{\partial z_3}(\zeta(\mathbf{q})) \|} \cdot N(\mathbf{q},\dot{\mathbf{q}}) \in \mathrm{IR}^n \ . \tag{2.20}$$

where ζ and χ are the coordinate transformations according to (2.1) and (2.2). In the case of elastic gears, recalling (2.9) and (2.10), we have the link torques

$$\mathbf{T} = -\mathbf{L}[\mathbf{D}(\dot{\mathbf{p}} - \mathbf{L}\dot{\mathbf{q}}) + \mathbf{C}(\mathbf{p} - \mathbf{L}\mathbf{q})] \quad . \tag{2.21}$$

The sensors used in practice to measure the normal force exerted at the contact point are elastic elements. Otherwise, relatively small disturbances of the object surface might result in loss of contact ($N = 0$) or, in view of the inertia of the robot arms, to impulsive forces ($N \to \infty$). Usually, both the force sensor and the object are very stiff so that deformations normal to the surface are relatively small. Consequently, it suffices to employ a linear expression for the normal contact force:

$$N := \begin{cases} -c(z_3 - h) - d\dot{z}_3 & if \quad z_3 < h \\ 0 & if \quad z_3 \geq h \end{cases} \tag{2.22}$$

Here c and d are the combined stiffness and (material) damping coefficients, respectively, of the sensor and object, and h is the thickness of the undistorted sensor.

2.7 Equation of Motion in Surface Oriented Coordinates

Without loss of generality, we shall suppose that the robot possesses three degrees of freedom, for instance, as shown in Fig. 3. In the case of more than three degrees of freedom, one must account for the orientation of the gripper at the contact point (cf. subsection 2.2). Since motion normal to the object surface, z_3, is relatively very small, it seems useful to scale this coordinate and to use coordinates

$$\mathbf{y} := \begin{bmatrix} z_1 \\ z_2 \\ -cz_3 \end{bmatrix} = \begin{bmatrix} 1 & 0 & 0 \\ 0 & 1 & 0 \\ 0 & 0 & -c \end{bmatrix} \mathbf{z} =: \mathbf{K}_s \mathbf{z} \; \in \mathbb{R}^3 \; . \tag{2.23}$$

If one neglects damping ($d \to 0$), y_3 represents directly the normal force magnitude N. In order to express the equations of motion in terms of coordinates $\mathbf{y}$, we require the transformation after (2.1) as well as its functional matrix; that is,

$$\mathbf{y} = \mathbf{K}_s \zeta(\mathbf{q}) \quad , \quad \dot{\mathbf{y}} = \mathbf{K}_s D\zeta(\mathbf{q})\dot{\mathbf{q}} \; . \tag{2.24}$$

As a consequence of the implicit function theorem, this transformation is locally invertible, if there exists an open neigborhood of $\mathbf{q}$ within which the inverse of $D\zeta(\mathbf{q})$ exists. In practical situations one may assume that this is the case. Otherwise the robot can reach a given point on the surface in a non-unique manner.

Taking into account the second time derivative

$$\ddot{\mathbf{q}} = D\zeta(\mathbf{q})^{-1} \left[\mathbf{K}_s^{-1}\ddot{\mathbf{y}} - \sum_{i=1}^{n} \left(\frac{\partial(D\zeta(\mathbf{q}))}{\partial q_i} \cdot \dot{\mathbf{q}} \cdot \dot{q}_i \right) \right] , \tag{2.25}$$

equation (2.11) leads to the transformed equation of motion

$$\mathbf{M}^*(\mathbf{y})\ddot{\mathbf{y}} + \mathbf{g}^*(\mathbf{y},\dot{\mathbf{y}},\boldsymbol{\mu}) + \boldsymbol{\Delta}\cdot\mathbf{T} + \mathbf{B}(\mathbf{q}(\mathbf{y}))\mathbf{u} = \mathbf{0} \tag{2.26}$$

with

$$\mathbf{M}^*(\mathbf{y}) := \mathbf{M}(\mathbf{q}(\mathbf{y}))[\mathbf{K}_s D\zeta(\mathbf{q}(\mathbf{y}))]^{-1} \tag{2.27}$$

and

$$\mathbf{g}^*(\mathbf{y},\dot{\mathbf{y}},\boldsymbol{\mu}) := \mathbf{g}(\mathbf{q}(\mathbf{y}),\dot{\mathbf{q}}(\mathbf{y}),\boldsymbol{\mu}) - [D\zeta(\mathbf{q}(\mathbf{y}))]^{-1}\cdot\sum_{i=1}^{n}\left[\frac{\partial(D\zeta(\mathbf{q}(\mathbf{y})))}{\partial q_i}\dot{\mathbf{q}}(\mathbf{y},\dot{\mathbf{y}})\dot{q}_i(\mathbf{y},\dot{\mathbf{y}})\right] \tag{2.28}$$

The reason for the coordinate transformation $\mathbf{y} \mapsto \mathbf{q}$ is the utility of basing the controller design on coordinates which represent directly position on the surface and normal force magnitude.

3 Controller Design

It is the aim of the controller design to assure as accurately as possible that the end effector follow a prescribed C^2–path

$$t \mapsto \begin{bmatrix} \bar{z}_1(t) \\ \bar{z}_2(t) \end{bmatrix}, \qquad t \in [t_0, t_1] \tag{3.1}$$

on the surface of operated upon object, while exerting a prescribed C^2–normal force

$$t \mapsto \bar{N}(t), \qquad t \in [t_0, t_1] \tag{3.2}$$

in the presence of bounded uncertainties. That is, it is desired that the controller be robust vis–a–vis parameter variations (see subsection 2.5).

Our design of a nonlinear feedback is based on the constructive use of LJAPUNOV stability theory, (e.g., see [GUTMAN, LEITMANN 1976] or [CORLESS, LEITMANN 1989]). Although not a necessary condition, (e.g., see [CHEN, LEITMANN 1987]), the controller design is greatly simplified, if a so called "matching condition" is imposed on system (2.10),(2.12). [CHEN, PANDEY 1990] use this approach in conjunction with a linearizing control to obtain a hybrid controller.

If link gears are stiff (inelastic) there is a direct relation between link coordinates $\mathbf{q}$ and rotor coordinates $\mathbf{p}$ according to (2.11). Then, with (2.10), one arrives at a system of inhomogenous equations for the determination of link moments $\mathbf{T}$ depnding on the control $\mathbf{u}$ and the second time derivative of link coordinates $\mathbf{q}$:

$$\left(\mathbf{S}\boldsymbol{\Theta}^{-1}\mathbf{L}^{-1}\right)\mathbf{T} = \left(\mathbf{S}\boldsymbol{\Theta}^{-1}\right)\mathbf{u} + \mathbf{R}\ddot{\mathbf{q}} \tag{3.3}$$

The solution for $\mathbf{T}$ consists of a homogenous and a particular part:

$$\mathbf{T} = \mathbf{H}\boldsymbol{\lambda} + (\mathbf{S}\Theta^{-1}\mathbf{L}^{-1})^{+}\left[\mathbf{S}\Theta^{-1}\mathbf{u} + \mathbf{R}\ddot{\mathbf{q}}\right] \tag{3.4}$$

Here $(\mathbf{S}\Theta^{-1}\mathbf{L}^{-1})^{+} \in \mathbb{R}^{n,ne}$ is the pseudoinverse of $\mathbf{S}\Theta^{-1}\mathbf{L}^{-1}$. $\mathbf{H} \in \mathbb{R}^{n,n-ne}$ is an arbitrary solution of the homogenous system of rank $n - ne$. $\boldsymbol{\lambda} \in \mathbb{R}^{n-ne}$ is a yet to be determined vector which is determined from the differential equations of the remaining $n - ne$ elastic links (see equation (2.11)) as a function of $\mathbf{q}, \dot{\mathbf{q}}, \mathbf{p}, \dot{\mathbf{p}}$ and $\boldsymbol{\mu}$. On substitution of (3.6) in equation (2.13), it follows that

$$\tilde{\mathbf{M}}(\mathbf{q})\ddot{\mathbf{q}} + \mathbf{g}(\mathbf{q}, \dot{\mathbf{q}}, \boldsymbol{\mu}) + \Delta \cdot \mathbf{H} \cdot \boldsymbol{\lambda} + \tilde{\mathbf{B}}(\mathbf{q})\mathbf{u} = \mathbf{0} \tag{3.5}$$

where

$$\tilde{\mathbf{M}}(\mathbf{q}) := \mathbf{M}(\mathbf{q}) - \Delta(\mathbf{S}\Theta^{-1}\mathbf{L}^{-1})^{+} \cdot \mathbf{R} \tag{3.6}$$

and

$$\tilde{\mathbf{B}}(\mathbf{q}) := \mathbf{B}(\mathbf{q}) + \Delta\left[(\mathbf{S}\Theta^{-1}\mathbf{L}^{-1})^{+}(\mathbf{S}\Theta^{-1})\right]. \tag{3.7}$$

Equation (2.12) and (3.5) have the same structure. The transformation of link coordinates to surface oriented ones leads to an equation of motion of the same form as (2.26) as described in subsection 2.7. The matching condition is assured provided $\tilde{\mathbf{B}}(\mathbf{q})$ is invertible. The design of the controller is then based on (2.26) since we are interested in stabilizing $\mathbf{y}$ rather than the link and rotor coordinates. Upon employing (3.5), equation (2.26) becomes

$$\mathbf{M}^{*}(\mathbf{y})\ddot{\mathbf{y}} + \mathbf{g}^{*}(\mathbf{y}, \dot{\mathbf{y}}, \boldsymbol{\mu}) + \Delta \cdot \mathbf{H} \cdot \boldsymbol{\lambda} + \mathbf{B}^{*}(\mathbf{y})\mathbf{u} = \mathbf{0} \tag{3.8}$$

with

$$\mathbf{M}^{*}(\mathbf{y}) := \tilde{\mathbf{M}}(\mathbf{q}(\mathbf{y}))\left[\mathbf{K}_{s} D\zeta(\mathbf{q}(\mathbf{y}))\right]^{-1} \tag{3.9}$$

and

$$\mathbf{B}^{*}(\mathbf{y}) := \tilde{\mathbf{B}}(\mathbf{q}(\mathbf{y})). \tag{3.10}$$

$\mathbf{g}^{*}(\mathbf{y}, \dot{\mathbf{y}}, \boldsymbol{\mu})$ follows directly from (2.28). If we neglect uncertainties in the component weights of the robot and introduce the transformed control vector

$$\mathbf{u}^{*} := \mathbf{M}^{*}(\mathbf{y})^{-1}\left[\mathbf{B}^{*}(\mathbf{y})\mathbf{u} + \mathbf{g}^{*}{}_{G}(\mathbf{y})\right] - \ddot{\bar{\mathbf{y}}} \tag{3.11}$$

and the difference vector between actual state $(\mathbf{y}^{T}, \dot{\mathbf{y}}^{T})$ and desired state $\bar{\mathbf{y}}^{T}, \dot{\bar{\mathbf{y}}}^{T}$:

$$\mathbf{Y} := \begin{bmatrix} \mathbf{y} - \bar{\mathbf{y}} \\ \dot{\mathbf{y}} - \dot{\bar{\mathbf{y}}} \end{bmatrix}, \tag{3.12}$$

we arrive at the standard form

$$\dot{\mathbf{Y}} = \mathbf{A} \cdot \mathbf{Y} + \bar{\mathbf{B}}[\mathbf{u}^* + \mathbf{e}^*(\mathbf{Y}, \boldsymbol{\mu})] \tag{3.13}$$

with

$$\mathbf{A} := \begin{bmatrix} 0 & \mathbf{I} \\ 0 & 0 \end{bmatrix} \in \mathbb{R}^{2n,2n} \quad , \quad \bar{\mathbf{B}} := \begin{bmatrix} 0 \\ \mathbf{E} \end{bmatrix} \in \mathbb{R}^{2n,n} \ , \tag{3.14}$$

and

$$\mathbf{e}^*(\mathbf{Y}, \mathbf{p}, \dot{\mathbf{p}}, t, \boldsymbol{\mu}) := \mathbf{M}^*(\mathbf{y})^{-1} \left[(\mathbf{g}_C + \mathbf{g}_Z + \mathbf{g}_R)(\mathbf{y}, \dot{\mathbf{y}}, \boldsymbol{\mu}) + \boldsymbol{\Delta} \cdot \mathbf{H} \cdot \boldsymbol{\lambda}\right] \ . \tag{3.15}$$

Utilizing the the design theory of [CORLESS,LEITMANN 1989] we can readily arrive at a stabilizing feedback for $\mathbf{Y}$. Upon transforming $(\mathbf{q}, \dot{\mathbf{q}}) \mapsto \mathbf{y}$ we arrive at a control $\mathbf{u}$ in (2.12) which assures nearly asymptotic stability ("uniform ultimate boundedness") of $\mathbf{Y}$, and hence of position and normal force. This controller is given by

$$\mathbf{u} = (\mathbf{M}(\mathbf{q}) - \boldsymbol{\Delta}(\mathbf{S}\boldsymbol{\Theta}^{-1}\mathbf{L}^{-1})^{+}\mathbf{R}) \left[\mathbf{K}_s D\zeta(\mathbf{q})\right]^{-1} \left[\mathbf{u}^* + \dddot{\mathbf{y}}\right] - \mathbf{g}_G(\mathbf{q}) \tag{3.16}$$

with

$$\mathbf{u}^* = -\mathbf{K}\mathbf{Y} - \frac{\bar{\mathbf{B}}^T\mathbf{P}\mathbf{Y}}{\|\bar{\mathbf{B}}^T\mathbf{P}\mathbf{Y}\| + \varepsilon\rho(\mathbf{Y}, \mathbf{p}, \dot{\mathbf{p}}, t)^{-1}} \cdot \rho(\mathbf{Y}, \mathbf{p}, \dot{\mathbf{p}}, t), \tag{3.17}$$

where $\mathbf{Y}$ is a function of $\mathbf{q}, \dot{\mathbf{q}}$ and t. Since $\mathbf{e}^*$ is continous with respect to $\mathbf{Y}, \mathbf{p}, \dot{\mathbf{p}}$ and, in particular, $\boldsymbol{\mu}$, the gain $\rho(\mathbf{Y}, \mathbf{p}, \dot{\mathbf{p}}, t)$ is given by

$$\rho(\mathbf{Y}, \mathbf{p}, \dot{\mathbf{p}}, t) := \max\{\|\mathbf{e}^*(\mathbf{Y}, \mathbf{p}, \dot{\mathbf{p}}, t, \boldsymbol{\mu})\| \mid \boldsymbol{\mu} \in \boldsymbol{\Omega}\}. \tag{3.18}$$

Note that $\mathbf{q}, \dot{\mathbf{q}}$ and only the components of $\mathbf{p}, \dot{\mathbf{p}}$ corresponding to the elastic links need to be determined to employ control (3.17). Furthermore, in practice, instead of determining Y_3 as a function of $\mathbf{q}$ and $\dot{\mathbf{q}}$, it may be better and easier to employ the direct measurement of N and $\dot{N}$.

4 Simulations

As an ilustrative example we consider the Manutecr3 robot shown in Fig. 1. The numerical data for this robot may be found in [TÜRK 1988]. We suppose, that axes 4, 5 and 6 are not moveable (blocked). Axes 1, 2 and 3 are moveable and controlled (angles q_1, q_2 und q_3). Each of these axes is moved indepenently by means of an actuator through a gear. For purpose of controller design we assume that links 1 and 2 are not elastic. In that case, the matching condition is satisfied. All uncertainties discussed in subsection 2.5 are supposed to be present. The friction force between sensor and object is proportional to the normal force; that is,

$$\|\mathbf{F}_R\| = \mu_R N. \tag{4.19}$$

The sliding friction coefficent μ_r depends on a number of variables which are difficult to quantify, such as local roughness, temperature, speed, etc. Here we suppose that this uncertain parameter $\mu_R \in [0, 0.5]$. The friction moments $M_i, i = 1, 2, 3$ are taken as

$$M_i = \mu_i M_0, \tag{4.20}$$

with $M_0 = 10[Nm]$. Again, the uncertain coefficients are assumed to lie in $[0, 0.5]$. In addition, the Coriolis and centrifugal force are included among the uncertainties; they are approximated in accordance with equs. (3.15) and (3.18).

The operated upon object is a horizontal planar table. We consider two cases 1) The end effector is to move at constant speed along straight line while exerting a constant normal force, and 2) the end effector is to move at a circular path at constant speed while the desired normal force is again constant. For the sake of brevity only case 2) is discussed here; details and case 1) can be found in [REITHMEIER, LEITMANN 1990] (to appear).

Fig. 5a is a sketch of the plane (surface of the object). The desired path, shown as a dashed circle, starts at $t_0 = 0[s]$ at

$$\begin{bmatrix} \bar{z}_1(t_0) \\ \bar{z}_2(t_0) \end{bmatrix} := \begin{bmatrix} 0.8[m] \\ 0.5[m] \end{bmatrix} \tag{4.21}$$

and ends at $t_1 = 2[s]$ at the same point. The radius of the circle is $r := 0.25[m]$.

The simulated path is shown as a solid curve. The desired normal force is given by $\bar{N}(t) = 10[N], t \in [t_0, t_1]$. Fig. 5b shows both the desired (dashed) and simulated (solid) normal force history.

The initial values of the simulated path and force history $(z_i(0), \dot{z}_i(t_0); i = 1, 2; N(t_0), \dot{N}(t_0))$ are taken to be different from the desired ones, since there is always uncertainty in position and normal force at the outset (cf. Fig.6b).

Figures 6a,b show the time histories of the deviations $\Delta z_i(t) := z_i(t) - \bar{z_i}(t)$. This process is not aperiodically damped; the reason is the fact that link damping is very small and the sensor is very stiff and possesses very little damping. Furthermore, the controller parameters were not optimized nor tuned with respect to the process.

The time histories of the actuator moments u_1, u_2 and u_3 are shown in Figs. 7a,b and c. According to (3.17), each moment u_i consists of a linear and a nonlinear part. The linear parts are shown by solid curves, the nonlinear ones by dashed curves. It is noteworthy that, in general, the linear and nonlinear parts are in opposite directions.

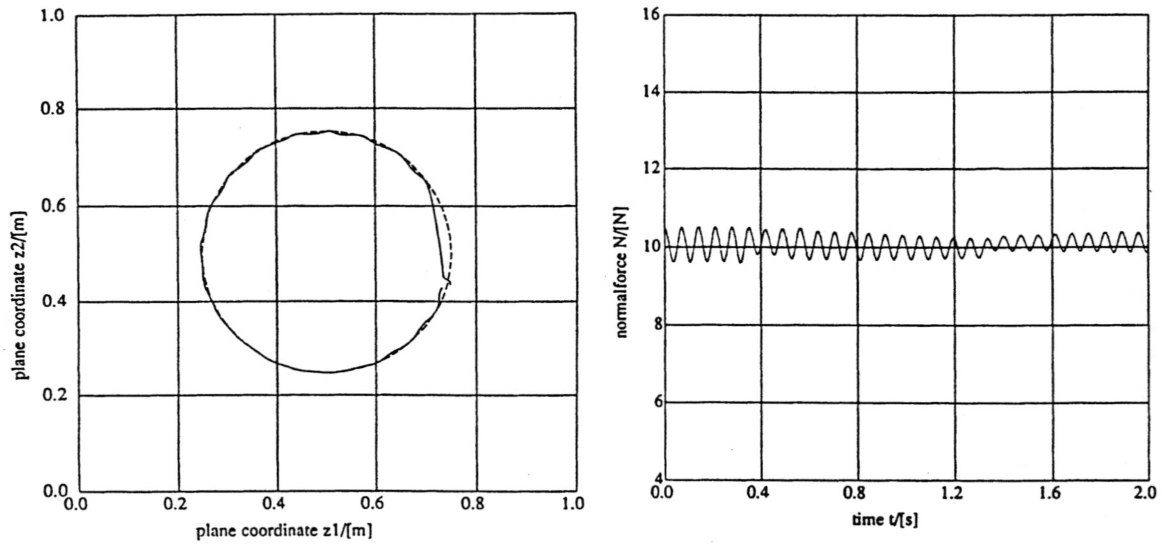

Fig. 5a,b: simulated (—) and desired (- -) path and normal force history

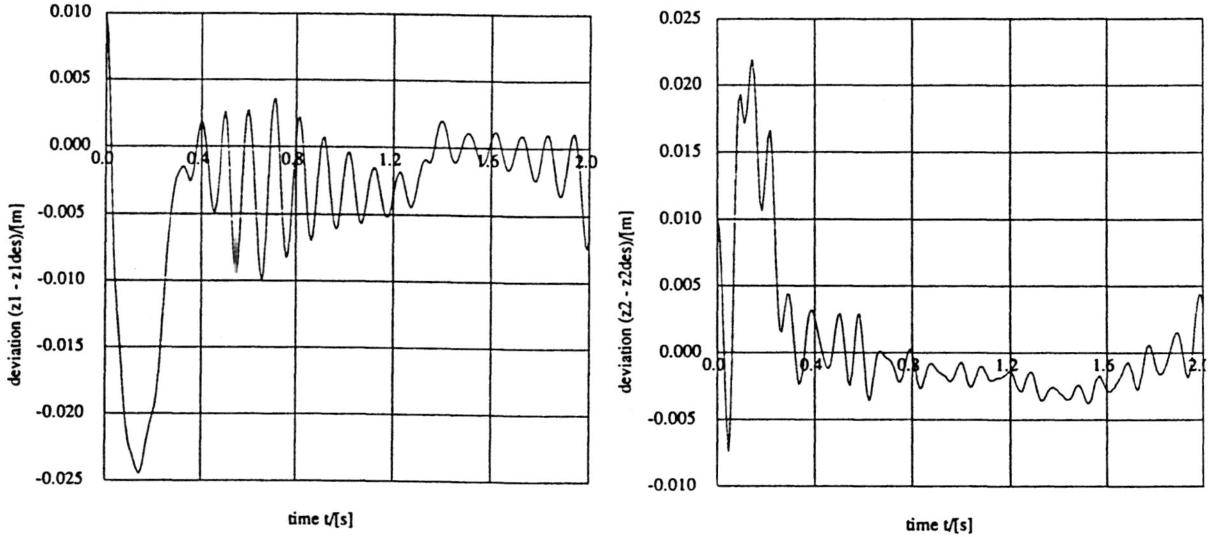

Fig. 6a,b: Deviation Δz_1 and Δz_2

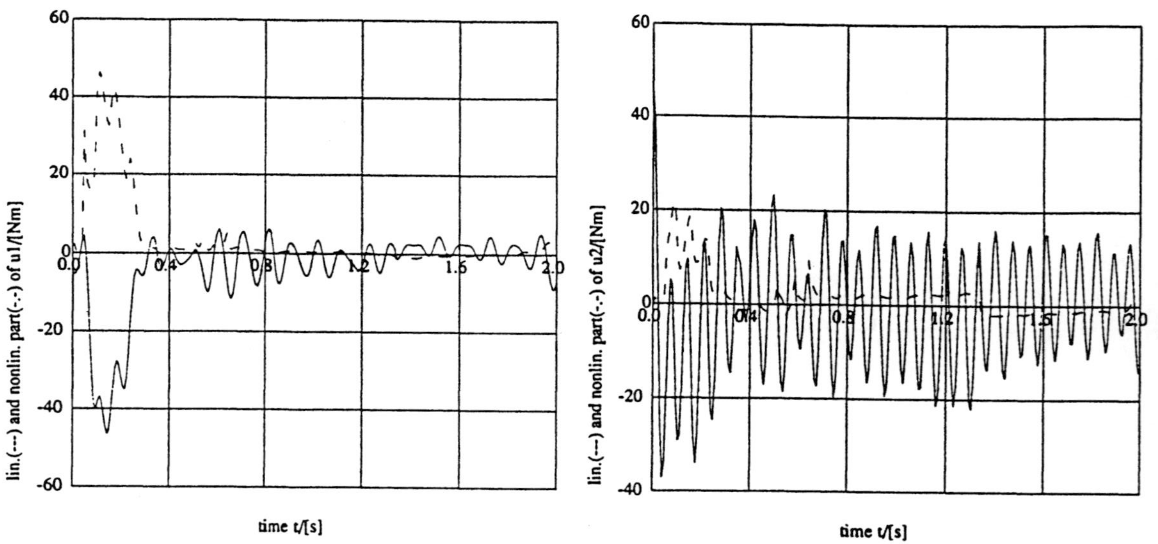

Fig. 7a,b: Linear (—) and nonlinear (- -) part of the actuator torques u_1, u_2

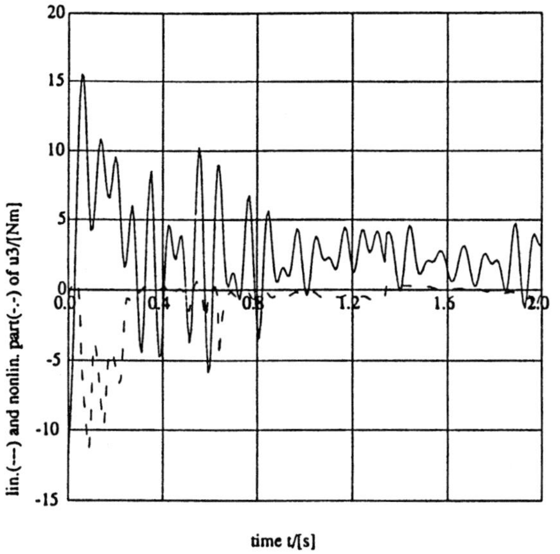

Fig. 7c: Linear (—) and nonlinear (- -) part of the actuator torque u_3

5 References

[AN, HOLLERBACH 1987]
An, C.H.; Hollerbach, J.M.: Dynamic Stability Issues in Force Control of Manipulators. Proceedings of American Control Conference, pp 821 - 827 (1987)

[ASADA, SLOTINE 1986]
Asada, H.; Slotine, E.: Robot Analysis and Control. John Wiley and Sons. Interscience Publication (1986)

[CORLESS, LEITMANN 1981]
Corless, M.; Leitmann, G.: Continous State Feedback Guaranteeing Uniform Ultimate Boundedness for Uncertain Dynamic Systems. IEEE Trans. Automat. Contr. 26, pp 1139 - 1144 (1981)

[CHEN, PANDEY 1990]
Chen, Y.; Pandey, S.: Uncertainty Bound-Based Hybrid Control for Robot Manipulators. IEEE Transactions on Robotics and Automation. Vol. 6, no. 3, pp 303 - 311 (1990)

[CORLESS, LEITMANN 1989]
Corless, M.; Leitmann, G.: Deterministic Control of Uncertain Systems. In Modern Optimal Control, edited by E.O. Roxin. Marcel Dekker, Inc., New York, Basel (1989)

[DESA, ROTH 1985]
Desa, S.; Roth, B.: Synthesis of Control Systems for Manipulator using Multivariable Robust Servo Mechanism Theory. Int. J. of Robotic Research 4 no. 3, pp 18 - 34 (1985)

[DJAFERIS, MURAH 1987]
Djaferis, T.E.; Mrah, B.; Franklin, J.: Compliant Control Using Robust Multivariable Feedback Methods. Proc. American Contr. Conf., pp 2021 - 2026 (1987)

[EPPINGER, SEERING 1986]
Eppinger, S.D.; Seering, W.P.: On Dynamic Models of Robot Force Control. Proceedings of IEEE Int. Conf. on Robotics and Automation, pp 29-34 (1986)

[GOLDENBERG 1988]
Goldenberg, A.A.: Implementation of Force and Impedance Control in Robot Manipulators. Proceedings of IEEE Int. Conf. on Robotics and Automation, pp 1626 - 1632 (1988)

[GUTMAN, LEITMANN 1976]
Gutman, S.; Leitmann, G.: Stabilizing Feedback Control for Dynamical Systems with Bounded Uncertainty. Proc. IEEE Conf. Decision Control (1976)

[KANG, LEITMANN 1989]
Kang, C.G.; Leitmann, G.; Horowitz, R.: Robust Deterministic Controller Design of a Two Degree of Freedom SCARA Manipulator. In Proc. American Contr. Conf., pp 1457 - 1462 (1989)

[KATHIB, BURDICK 1986]
Kathib, O.; Burdick, J.: Motion and Force Control of Robot Manipulators. Proc. of IEEE Int. Conf. on Robotics and Automation, pp 1381 - 1386 (1986)

[KUO, WANG 1990]
Kuo, C.Y., Wang, S.: Nonlinear Robust Control of Robotic Manipulators. Trans. of the ASME 48 no. 112, pp 48 - 54 (1990)

[LEITMANN, RYAN 1986]
Leitmann, G.; Ryan, E.P.; Steinberg, A.: Feedback Control of Uncertain Systems: Robustness with Respect to Neglected Actuator and Sensor Dynamics. Int. J. Contr., 43, pp 1243 - 1256 (1986)

[LEITMANN, RYAN 1987]
Leitmann, G.; Ryan, E.P.: Output Feedback Control of a Class of Singularly Perturbed Uncertain Dynamical Systems. In Proc. American Contr. Conf., pp 1590 - 1594 (1987)

[PFEIFFER, REITHMEIER 1987]
Pfeiffer, F.; Reithmeier, E.: Roboterdynamik. Teubner Studienbuecher Stuttgart, FRG (1987)

[RAIBERT, CRAIG 1981]
Raibert, M.H.; Craig, J.J.: Hybrid Position/Force Control of Manipulators. ASME J. of Dyn. Syst., Measurement and Control, 103 no. 2, pp 126 - 133 (1981)

[REITHMEIER, LEITMANN 1990]
Reithmeier, E.; Leitmann, G.: Robuste Positions- und Kraftregelung für Industrieroboter mit Parameterunsicherheiten. To appear.

[SALISBURY 1980]
Salisbury, J. K.: Active Stiffness Control of a Manipulator in Cartesian Coordinates. Proc of 19th IEEE Conf. on Decision and Control, pp 95 - 100 (1980)

[TÜRK 1988]
Türk, S.: Dynamische Robotermodelle am Beispiel des Manutecr3. DFVLR Mitteilungen 88–16 (1988)

[WHITNEY 1987]
Whitney, D.E.: Historical Perspective and State of the Art in Robot Force Control. Int. J. of Robotic Research 6 no. 1, pp 3 - 14 (1987)

MUTUAL REFERENCE ADAPTIVE CONTROL OF NONLINEAR UNCERTAIN SYSTEMS

by

J.M. Skowronski
Mechanical Engineering
University of Southern California
Los Angeles, CA 90089-1453

Abstract

The popular model reference adaptive control (MRAC) technique is extended to the case of two nonlinear dynamical systems mutually tracking and/or avoiding each other in a specified sense, subject to bounded uncertainty of their parameters (MURAC). The error equation approach classical in MRAC is replaced by a different approach which may be called the product-state-space method. Signal adaptive feedback controllers and adaptive laws are proposed together with sufficient conditions for the required tracking and avoidance. Application to a mechanical system illustrates the results.

1. Introduction

There is a number of technical situations where the tracking control has to be mutual, i.e. with both systems actively involved in tracking and/or avoiding each other, but still controlled by a single agency. As typical cases we may quote a two-arm robotic manipulator reaching for an object while avoiding collision between the arms, or some spacecraft assembly problems like rendezvous, docking, formation keeping, etc. With uncertainties present, the model reference adaptive control (MRAC) technique seems very appropriate but must be extended to allow for mutual references of two or more plants in tracking, and particularly in avoiding collision. We shall call this extension the Mutual Reference Adaptive Control (MURAC).

In mutual tracking we may have two approaches: either the objective lies in mutual tracking only, i.e. it does not matter where the converging systems go, as long as they converge, or we aim at mutual tracking to a stipulated target set in the state space. Obviously the second case is more practical. The target may be fixed or moving. If moving, then the motion may be given by a time-parametrized curve or more generally by a dynamic model. In the latter case the problem becomes a double MRAC with reference to a single model. We design the reference model which is relatively simple but compatible with the plants. The design may include choosing a controller to secure a desired behavior. Then, in particular, choosing suitable initial conditions we may obtain a trajectory of the model that coincides with the planned target-curve for both plants to follow, or two trajectories to coincide with a pair of target-curves, each to be followed by a plant. The latter objective may additionally secure avoidance of collision between the plants concerned. However, even if the reference trajectories to be followed avoid each other, the collision of plants may still occur due to the uncertainty included in the plant parameters and external loads. It thus seems reasonable to secure the avoidance of collision between the two plants independently of the reference model, by a suitable interfacing or switching-off of the tracking controllers on some safety zone.

In passing from MRAC to MURAC there is also another need of adjustment. In most of the modern technical applications like robotics, spacecraft dynamics, ..., etc, the plants concerned are systems which must be considered highly nonlinear and nonlinearly coupled, while the classical MRAC technique is linear, see Ref. 1. Various modernizations have been proposed to allow for a nonlinear plant while the reference model was to be designed as simple as possible, preferably linear, see Ref. 2, 3, 4. For computational purposes such a requirement seems reasonable, as long as the model is compatible with the nonlinear plant – for instance, in the case of mechanical systems, as long as it has the same number of equilibria, see Ref. 5. However with several plants tracking the same model the compatibility with linearization seems doubtful, and if we have mutual reference only – say in avoidance – there is no room for linearization at all. We shall thus need the nonlinear (a nonlinearized) MRAC introduced in Ref. 5, 6 as the basis for an extension to MURAC. It means that we abandon the traditional technique aimed at asymptotic stability of zero-error in the classical error equation obtained by subtracting the plant from the model state equation. Instead we attempt convergence to diagonal set in the Cartesian product of the state spaces of the systems concerned. In our MURAC approach we generalize the technique proposed for particular cases in Ref's 7 and 8.

2. The Plants and the Reference Model

For a review of the linear MRAC technique the reader is referred to Ref. 1. Its nonlinear extension, see Ref. 5, is described in Ref's 12 and 13. We shall use the latter in our case of two plants tracking the same reference model in state space, in real time, with stipulated accuracy, while avoiding collision within a stipulated safety zone.

Consider two plants in general state format

$$\dot{x}^i = f^i(x^i, u^i, \lambda^i, w^i),\ i = 1, 2 \tag{1}$$

where $x^i(t) \in \Delta \in \mathbb{R}^N$ are the state vectors ranging in a given bounded set Δ for $t \geq t_0 = 0$. Here and for the remainder of this paper we omit the notation of the transpose vectors when confusion is not likely to occur. The vectors $\lambda^i(t) = (\lambda_1^i(t), ..., \lambda_l^i(t)) \subset \Lambda \subset \mathbb{R}^l$ represent adjustable parameters with values in a given compact set Λ, the adjustment being governed by adaptive laws

$$\dot{\lambda}^i = f_a^i(x^i, \lambda^i, t),\ i = 1, 2 \tag{2}$$

later specified precisely. Further, $w^i(t) \in W_i \subset \mathbb{R}^s$ are unknown vectors ranging within known compact sets W_i. The control vectors $u^i(t) \in U \subset \mathbb{R}^r$, with U a given compact set, are to be selected by feedback signal adaptive programs, in general set-valued, $P^i(.) : \Delta \times \Lambda \times \mathbb{R} \to$ subsets of U, to accommodate discontinuities in the controls. The programs are at this stage defined in the general format of

$$u^i(t) \in P^i(x(t), \lambda^i(t), t) \tag{3}$$

later specified more precisely. In consequence of the set-valuedness of the programs and the uncertainty of $w^i(t)$ as well as the possible chattering of $\lambda^i(t)$, the equation (1) should be written in the contingent form

$$\dot{x}^i \in \{f^i(x^i, u^i, \lambda^i, w^i) | u^i \in P^i(x^i, \lambda^i), \lambda^i \in \Lambda, w^i \in W_i\}. \tag{4}$$

The known functions $f_j^i(x^i, u^i, \lambda^i, w^i)$, $j = 1, ..., N$, $f^i = (f_1^i, ..., f_N^i)$, of (1) and (4) are assumed to be such that, together with (3) they secure through each $x^{io} = x^i(0)$ at least one solution $k^i(x^{io}, .) : \mathbb{R} \to \Delta$ to (4), and such that for all $w^i(.)$ there exist $u^i(.)$ for which $k^i(x^{io}, t)$ satisfies (1) for almost all $t \geq 0$, see Ref. 9. We denote the class of such solutions by $\mathcal{K}(x^{io})$.

The reference model is taken generally as a prescribed nonlinear system

$$\dot{x}_m = f_m(x_m, u_m, \lambda_m) \tag{5}$$

with $x_m(t) \in \Delta \subset \mathbb{R}^N$, $u_m(t) \in U \subset \mathbb{R}^r$, the vector $\lambda_m = (\lambda_{m1}, ..., \lambda_{ml}) = \text{const} \in \Lambda$, of given parameters, and the vector $f_m = (f_{m1}, ..., f_{mN})$ consisting of specified functions which, given u_m, λ_m, allow unique solutions $k_m(x_m^o, .) : \mathbb{R} \to \Delta$, $x_m^o = x_m(0)$, for almost all $t \geq 0$. The control $u_m(t)$ is selected such that the corresponding trajectories $k_m(x_m^o, t)$, $t \geq 0$ in Δ achieve a desired goal. In particular we may require that one or several trajectories (for various initial conditions) coincide with specific paths to be tracked. In our fundamental discussion, to focus attention on something definite but generally applicable, let us concentrate on a fairly common feature demanded for (5), namely uniform boundedness.

There would not be much use for tracking, if the trajectories of (5) were not uniformly equibounded in some, possibly large set $\Delta_0 \subset \Delta$, where we want the tracking to occur. More precisely we assume that there is a constant $\beta > 0$ such that $x_m^o \in \Delta_o$ implies $\| k_m(x_m^0, t) \| < \beta$, $t \geq 0$. Let $\Delta_L \subset \Delta_0$ be a set enclosing all equilibria of (5) in Δ_o and let $C\Delta_L \triangleq \Delta_o - \Delta_L$, called the region in-the-large. Using the same argument as in Yoshizawa, Ref. 10, it is readily seen that the above uniform boundedness implies the existence of an at least piece-wise C^1-function $V_m(.) : \Delta_o \to \mathbb{R}$ such that for all $x_m^{\cdot} \in C\Delta_L$,

$$\gamma^1(\| x_m \|) \leq V_m(x_m) \leq \gamma^2(\| x_m \|), \tag{6}$$

$$\dot{V}_m(x_m) < 0, \tag{7}$$

where $\gamma^1(.)$ is a continuous non-negative increasing function while $\gamma^2(.)$ is a continuous positive increasing function, $\| \cdot \|$ is any norm in $\mathbb{R}^N$ and $\dot{V}_m(x_m)$ is the time derivative of $V_m(.)$, along arcs of the trajectories of (5).

The pair of functions $f_m(x_m, \lambda_m)$, $V_m(x_m)$ specify our reference model for the purpose of this work. In other cases, particularly in case studies, more conditions could be imposed on the model.

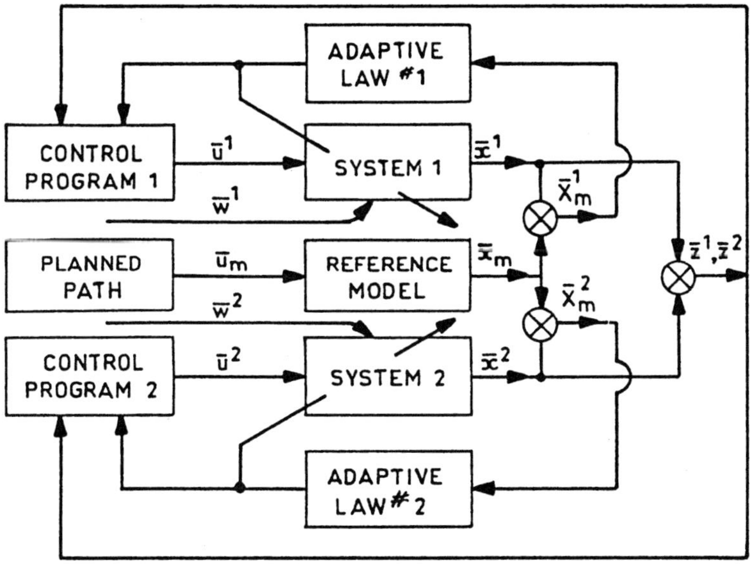

Fig. 1

3. Mutual Reference Adaptive Control

The system scheme is represented in Fig. 1. We introduce two $2N$-vectors $X^i \triangleq (x^i, x_m) \in \Delta^2 \subset \mathbb{R}^{2N}$, $\Delta^2 \triangleq \Delta \times \Delta$, and two ℓ-vectors $\alpha^i \triangleq \lambda^i - \lambda_m$, $i = 1,2$, as well as the vector $F^i(X^i, \alpha^i, u^i, w^i) \triangleq (f^i, f_m, f_a^i)$ comprising the right hand sides of (1), (2), (5). Then the product system in $\Delta^2 \times \Lambda \subset \mathbb{R}^{2N+\ell}$ can be written as

$$(\dot{X}^i, \dot{\alpha}^i) \in \{F^i(X^i, \alpha^i, u^i, w^i) | u^i \in P^i(X^i, \lambda^i),\ \alpha^i \in \Lambda,\ w^i \in W_i\} \tag{8}$$

with solutions $K^i(X^{io}, \alpha^{io}, t)$, $t \geq 0$. Regarding the latter, note that since λ_m = const implies $\dot{\lambda}^i(t) = \dot{\alpha}^i(t)$, the parameter errors $\alpha^i(t)$ are solutions of (2). Let us denote the family of $K^i(X^{io}, \alpha^{io}, .)$ by $\mathcal{K}(X^{io}, \alpha^{io})$.

Then we define the "diagonals" in $\Delta^2 \times \Lambda$

$$M^i \triangleq \{(X^i, \alpha^i) \in \Delta^2 \times \Lambda | x^i = x_m,\ \alpha^i = 0\},\ i = 1,2$$

and their neighborhoods

$$M_\eta^i \triangleq \{(X^i, \alpha^i) \in \Delta^2 \times \Lambda |\ \| x^i - x_m \| < \eta^i, \| \alpha^i \| < \eta^i\},\ i = 1,2$$

where $\| \cdot \|$ is any norm in $\mathbb{R}^{2N+\ell}$ and $\eta^i > 0$ is a stipulated constant, a precision estimate for tracking. Let us recall that $\Delta_o \subset \Delta$ was the desired set for tracking. We define $\Delta_o^2 \triangleq \Delta_o \times \Delta_o$, and formalize our **first objective** in the following two definitions.

Definition 3.1. *The set Δ_o is positively invariant under control, iff there is a pair of Programs $p^i(.)$ and functions $f_a^i(.)$ such that $(X^{io}, \alpha^{io}) \in \Delta_o^2 \times \Lambda$, $K^i(X^{io}, \alpha^{io}, .) \in \mathcal{K}(X^{io}, \alpha^{io})$, imply $K(X^{io}, \alpha^{io}, t) \in \Delta_o^2 \times \Lambda$, $t \geq 0$; see Fig. 2.*

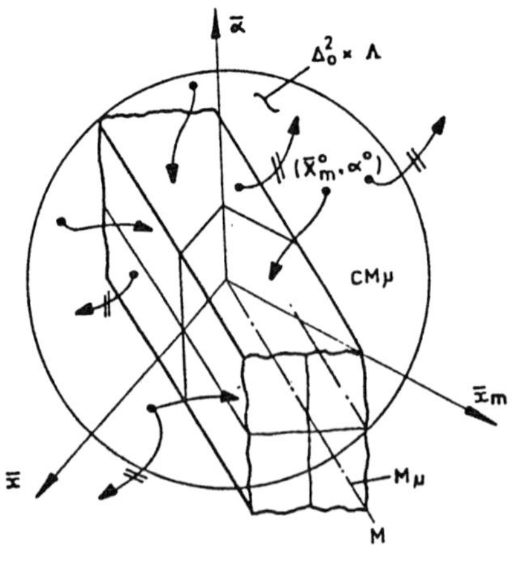

Fig. 2

Definition 3.2. *The plants (4) are mutually η-tracking the reference model (5) on $\Delta_o \subset \Delta$ iff Δ_o is positively invariant under control with the Programs $p^i(.)$ and the adaptive laws $f_a^i(.)$ such that there is some $t_\eta > 0$, possibly stipulated, for which every $K^i(X^{io}, \alpha^{io}, .) \in \mathcal{K}(X^{io}, \alpha^{io})$, $(X^{io}, \alpha^{io}) \in \Delta_o^2 \times \Lambda$, generates $K(X^{io}, \alpha^{io}, t) \in M_\eta^i$, $\forall t \geq t_\eta$. Then we define the anti-target*

$$M_A \triangleq \{(x^1, x^2) \in \Delta_o^2 | x^1 = x^2\}$$

as well as, given a stipulated constant $d > 0$, the enveloping avoidance set (see Ref. 11):

$$\mathcal{A} \triangleq \{(x^1, x^2) \in \Delta_o^2 | \quad | \parallel x^1 \parallel - \parallel x^2 \parallel | \leq d\}$$

enclosing M_A. Moreover, we denote $C\mathcal{A} \triangleq \Delta_o^2 = \mathcal{A}$, choose the $\mathbb{R}^N$-norm $\parallel . \parallel$ in terms of the scalar function $V_m(.)$, and introduce the safety zone about $\mathcal{A}$ (see Ref. 9) in $C\mathcal{A}$:

$$\Delta_{\mathcal{A}} \triangleq \{(x^1, x^2) \in \Delta_o^2 | d < |V_m(x^1) - V_m(x^2)| \leq \epsilon\}.$$

Here $\epsilon > d$ is some suitable constant. Our **second objective** is then expressed by

Definition 3.3. *The mutually tracking plants (4) of Definition 3.2 avoid collision iff there is an $\epsilon > 0$ such that $x^{io} \in \Delta$, $k^i(x^{io}, .) \in \mathcal{K}(x^{io})$, $i = 1, 2$ imply*

$$(k^1(x^{io}, t), k^2(x^{io}, t)) \in C\mathcal{A}, \forall T \geq 0. \tag{9}$$

4. Sufficient Conditions

Let $N[\partial(\Delta_o^2 \times \Lambda)]$ be a neighborhood of the boundary $\partial(\Delta_o^2 \times \Lambda)$ of the region $\Delta_o^2 \times \Lambda$ and define the semi-neighborhood $N_S \triangleq N[\partial(\Delta_o^2 \times \Lambda)] \cap \overline{\Delta_o \times \Lambda}$ and the relative complement $CM_\eta^i \triangleq (\Delta_o^2 \times \Lambda) - M_\eta^i$. Moreover introduce S^i (open) $\supset CM_\eta^i$, such that $S^i \cap M^i = \emptyset$. Furthermore let $V_s^i(.) : \overline{N}_s \to \mathbb{R}$ and $V_\eta^i(.) : S^i \to \mathbb{R}, i = 1, 2$ be C^1-functions with the positive constants

$$v_s^i = V_s^i(X^i, \alpha^i), \forall (X^i, \alpha^i) \in \partial(\Delta_o^2 \times \Lambda);$$
$$v_\eta^{i-} = \inf V_\eta^i(X^i, \alpha^i) | (X^i, \alpha^i) \in \partial M_\eta \cap \overline{CM}_\eta^i;$$
$$v_\eta^{i+} = \sup V_\eta^i(X^i, \alpha^i) | (X^i, \alpha^i) \in \partial(\Delta_o^2 \times \Lambda) \cap \overline{CM}_\eta^i.$$

The first relation obviously requires forming $V_s^i(.)$ from suitable $\partial(\Delta_o^2 \times \Lambda)$ taken as a level curve of this function, or forming Δ_o, Λ from level curves of suitable $V_s^i(.)$. In the latter case smaller Δ_o, Λ than these really desired will be the secure choice. The following theorem can be proved by the same argument as its single system correspondents in Ref's 5, 12, 13.

Theorem 4.1. *The plants (4) are mutually tracking the reference model (5) according to Definition 3.2 if, given Δ_o, Λ, η, there are admissible Programs $p^i(.)$ and test functions $V_s^i(.)$, $V_\eta^i(.)$, $i = 1,2$ such that for all $(X^i, \alpha^i) \in \Delta_o^2 \times \Lambda$, $i = 1,2$ we have*

(i) $V_s^i(X^i, \alpha^i) \leq v_s^i$, $\forall (X^i, \alpha_i) \in N_s$, $i = 1,2$;

(ii) *for each $u^i \in P^i(x^i, \lambda^i)$,*

$$\nabla V_s^i(X^i, \alpha). F^i(X^i, \alpha^i, u^i, w^i) < 0, \tag{10}$$

for all $w^i \in W_i$, $i = 1,2$;

(iii) $0 \leq V_\eta^i(X^i, \alpha^i) < v_\eta^{i+}$, $v_\eta^{i+} > 0$, $\forall (X^i, \alpha^i) \in \overline{CM}_\eta^i$, $i = 1,2$;

(iv) *for each $u^i \in P^i(x^i, \lambda^i)$, there is $c_i(r), r > 0$, monotone increasing, such that*

$$\nabla V_\eta^i(X^i, \alpha^i). F(X^i, \alpha^i, u^i, w^i) \leq -c_i(\| (X^i, \alpha^i) \|) \tag{11}$$

for all $w^i \in W_i$, $i = 1,2$.

In the above $\| \, . \, \|$ is any norm in $\mathbb{R}^N$. When t_η is stipulated: $t_\eta = \bar{t}_\eta$, we choose c_1, c_2 such that

$$\sup_{\Delta_o^2} c_i(\| (x^i, \alpha^i)^T \|) = c_i = \frac{v_\eta^{i+}}{\bar{t}_{\eta i}}, \; i = 1,2 \tag{12}$$

with $\bar{t}_\eta = \max(\bar{t}_{\eta 1}, \bar{t}_{\eta 2})$ to secure satisfying of Definition 3.2 after the stipulated $\bar{t}_\eta$.

The following theorem is based on the result obtained in Ref. 11.

Theorem 4.2. *Given $\Delta_o, \Lambda, d > 0$, the plants (4) mutually tracking according to Def. 3.2, avoid collision (see Def. 3.3), if the conditions of Theorem 4.1 are satisfied and if there is $\epsilon > 0$ and a C^1-function $V_A(.) : S \to \mathbb{R}$, S (open) $\supset \overline{\Delta}_A$, such that all for all $(x^1, x^2) \in \Delta_A$,*

(vi) $0 < v_a < V_A(x^1, x^2)$, *where $v_a = \sup V_A(x^1, x^2) | (x^1, x^2) \in \partial A$;*

(vii) *for each $u^i \in P^i(X^i, \lambda^i)$,*

$$\nabla V_A(x^1, x^2). [f(x^1, u^1, \lambda^1, w^1), \, f^2(x^2, u^2, \lambda^2, w^2)] \geq 0 \tag{13}$$

for all $w^i \in W_i$, $i = 1,2$.

Proof. Consider the trajectory $(k^1(x^{1o}, t), k^2(x^{2o}, t))$, $t \geq 0$, from $(x^{1o}, x^{2o}) \in \Delta_A$ which intersects ∂A (if not, there is no problem). Then there is $t_1 > 0$ such that $(x^1(t_1), x^2(t_1)) \in \partial A$ for the first time. This, by (vi), implies $V_A(x^1(t_1), \, x^2(t_2)) < V_A(x^{1o}, x^{2o})$, which contradicts (vii).

5. Control Programs and Adaptive Laws

Consider

$$V_s^i(X^i,\alpha^i) = V_m(x^i) + V_m(x_m) + a^i\alpha^i; \tag{14}$$

$$V_\eta^i(X^i,\alpha^i) = \begin{cases} |V_m(x^i) - V_m(x_m)| + a^i\alpha^i, & \text{for } (X^i,\alpha^i) \in CM_\eta^i, \\ a^i.\alpha^i, & \text{for } (X^i,\alpha^i) \in M_\eta^i \cap S^i \end{cases}; \tag{15}$$

$$V_A(x^1,x^2) = \frac{1}{2}[V_m(x^1) - V_m(x^2)], \tag{16}$$

where $a^i = (\text{sign } \alpha_1^i, ..., \text{sign } \alpha_l^i)$, $\alpha^i \neq 0$, and $V_m(x^i)$ specifies a norm $\| x^i \|$ that satisfies (6). Note that for the zero-error, $\alpha^i = 0$, no adaptation is needed.

Choosing Δ_o, Λ so that $\partial(\Delta_o^2 \times \Lambda)$ is defined by a level curve of $V_s^i(.)$ in (14) and choosing $N_s \subset C\Delta_L$, the assumption (6) on $V_m(.)$ implies conditions (i) of Theorem 4.1. From the geometry of $\partial(\Delta_o \times \Lambda)$, we have $v_\eta^{i+} = v_s^i$, and since the sum of functions satisfying (6) is no less than the absolute value of their difference, conditions (iii) holds as well. Conditions (iv), (vi) follow directly from (15) and (16), respectively. To check upon (ii), (v) and (vii) we differentiate (14) – (16) with respect to time:

$$\dot{V}_s^i(X^i,\alpha^i) = \dot{V}_m(x^i) + \dot{V}_m(x_m) + a^i\dot{\alpha}^i; \tag{17}$$

$$\dot{V}_\eta^i(X^i,\alpha^i) = \begin{cases} \dot{V}_m(x^i) - \dot{V}_m(x_m) + a^i\dot{\alpha}^i, & (X^i,\alpha^i) \in C^+M_\eta^i; \\ \dot{V}_m(x_m) - \dot{V}_m(x^i) + a^i\dot{\alpha}, & (X^i,\alpha^i) \in C^-M_\eta^i; \\ a^i\dot{\alpha}, & (X^i,\alpha^i) \in M_\eta^i \cap S^i; \end{cases} \tag{18}$$

$$\dot{V}_A(x^1,x^2) = [V_m(x^1) - V_m(x^2)].[\dot{V}_m(x^1) - \dot{V}_m(x^2)]. \tag{19}$$

In the above $C^\pm M_\eta^i$ are subsets of CM_η^i defined by $V_m(x^i) \geq V_m(x_m)$, $V_m(x^i) < V_m(x_m)$ respectively, and

$$\dot{V}_m(x_m) = \nabla V_m(x_m).f_m(x_m,u_m,\lambda_m), \tag{20}$$

$$\dot{V}_m(x^i) = \nabla V_m(x^i).f^i(x^i,u^i,\lambda^i,w^i). \tag{21}$$

Here (21) specifies the derivative of the norm $V_m(.)$ along trajectories of (1). The following three sets of conditions imply conditions (ii), (v) and (viii) of Theorem 4.1.

(A) TRACKING CONTROL CONDITIONS

$$\min_{u^i} \max_{w^i} \dot{V}_m(x^i) \leq \dot{V}_m(x_m),\ (X^i,\alpha^i) \in C^+M_\eta^i; \tag{22}$$

$$\max_{u^i} \min_{w^i} \dot{V}_m(x^i) \geq \dot{V}_m(x_m),\ (X^i,\alpha^i) \in C^-M_\eta^i; \tag{23}$$

given u^i,

$$\max_{w^i} |\dot{V}_m(x^i)| < 2|\dot{V}_m(x_m)| \tag{24}$$

(B) COORDINATION CONTROL CONDITIONS

$$\max_{u^1} \min_{w^1} \dot{V}_m(x^1) \geq \min_{u^2} \max_{w^2} \dot{V}_m(x^2), \text{ for } V_m(x^1) \geq V_m(x^2), \tag{25}$$

$$\min_{u^1} \max_{w^1} \dot{V}_m(x^1) \leq \max_{u^2} \min_{w^2} \dot{V}_m(x^2), \text{ for } V_m(x^1) < V_m(x^2). \tag{26}$$

(C) ADAPTATION CONDITIONS

$$a^i \dot{\alpha}^i \leq -c_i\eta - |\dot{V}_m(x^i)|, \; i = 1,2 \tag{27}$$

for suitable $c_i = c_i(\| (x^i, \alpha^i) \|) > 0$. We shall now justify the above statement. For simplicity of the exposition it will be assumed that the extremizing $w^i = \tilde{w}^i$ have been already substituted.

Consider (ii). By (7) and (C), $\dot{V}_s^i(X^i, \alpha^i) \leq -c_i\eta +$ negative terms. Substituting (A) and (C) into (18) yields (v).

To check upon (vii) we have to use conditions (A) and (B) combined – with interfaced control functions. Let us investigate the possibilities.

Case $(X^i, \alpha^i) \in C^+M_\eta^i$, $i = 1,2$. *We have two subcases:*

(a) $V_m(x^1) \geq V_m(x^2) \geq V_m(x_m)$. *Here we use (A):* $\dot{V}_m(x^1) \leq \dot{V}_m(x_m)$ *to calculate the control* u^1 *and then substituting it, we use (B) to find* u^2 *such that* $\dot{V}_m(x^2) \leq \dot{V}_m(x^1)$. *Since in turn* $\dot{V}_m(x^1) \leq \dot{V}_m(x_m)$ *then* $\dot{V}_m(x^2) \leq \dot{V}_m(x_m)$ *and* u^2 *also satisfies (A).*

(b) $V_m(x^2) \geq V_m(x^1) \geq V_m(x_m)$. *The roles of* $V_m(x^i)$ *are reversed. Using (A) we find* u^2 *and then using (B) we find* u^1 *such that* $\dot{V}_m(x^1) \leq \dot{V}_m(x^2) \leq \dot{V}_m(x_m)$ *thus also satisfying (A).*

Case $(X^i, \alpha^i) \in C^-M_\eta^i$, $i = 1,2$. *Again there are two subcases:*

(c) $V_m(x^1) \leq V_m(x^2) < V_m(x_m)$. *We use (A):* $\dot{V}_m(x^1) > \dot{V}_m(x_m)$ *to find* u^1, *and substituting it, we use (B) to find* u^2 *such that* $\dot{V}_m(x^2) \geq \dot{V}_m(x^1) > \dot{V}_m(x_m)$, *thus also satisfying (A).*

(d) $V_m(x^2) \leq V_m(x^1) < V_m(x_m)$, *we use (A) to find* u^2 *and (B) to find* u^1 *such that* $\dot{V}_m(x^1) \geq \dot{V}_m(x^2) > \dot{V}_m(x_m)$ *thus also satisfying (A), which closes our discussion.*

Substituting now (20), (21) into the above discussion, the combined conditions (A) and (B) become

$$\min_{u^i} \max_{w^i} \nabla V_m(x^i).f^i(x^i,u^i,\lambda^i,w^i) \le \nabla V_m(x_m).f_m(x_m,\lambda_m) \tag{28}$$

$$\min_{u^k} \max_{w^k} \nabla V_m(x^k).f^k(x^k,u^k,\lambda^k,w^k) \le \max_{\tilde{u}^i} \min_{w^i} \nabla V_m(x^i).f^i(x^i,\tilde{u}^i,\lambda^i,w^i) \tag{29}$$

for all $(X^{i,k},\alpha^{i,k}) \in C^+M_\eta^{i,k}$, $V_m(x^i) \ge V_m(x^k)$, $i,k = 1,2$, $i \ne k$; and

$$\max_{u^i} \min_{w^i} \nabla V_m(x^i).f^i(x^i,u^i,\lambda^i,w^i) \ge \nabla V_m(x_m).f_m(x_m,\lambda_m) \tag{30}$$

$$\max_{u^k} \min_{w^k} \nabla V_m(x^k).f^k(x^k,u^k,\lambda^k,w^k) \ge \min_{\tilde{u}^i} \max_{w^i} \nabla V_m(x^i).f^i(x^i,\tilde{u}^i,\lambda^i,w^i) \tag{31}$$

for all $(X^{i,k}, \alpha^{i,k}) \in C^-M_\eta^{i,k}$, $V_m(x^i) < V_m(x^k)$, $i,k = 1,2$, $i \ne k$. In the above $\tilde{u}^i \in \tilde{P}^i(X^i,\lambda)$, with $\tilde{P}^i(X^i,\lambda^i)$ in each case specified by the first inequality.

Then the tracking control u^i obtained from (28), (30) is substituted into (29), (31), respectively, to obtain the coordinated tracking control of the other plant u^k. Should we have more plants than two to track the model while avoiding themselves (e.g. formation keeping aircraft), the control-coordination technique may remain the same: for each u^i of m plants, we choose coordinated and converging set of u^k, $k = 1,...,m$.

The conditions (28) – (31) reach as far as we can go towards defining the programs $P^i(.)$, $P^k(.)$ in general terms i.e. without specifying $f^i(.)$, $f_m(.)$ and $V_m(.)$. Slightly more detailed format may be obtained if we assume (1), (5) with control additive and linear: $f^i(x^i,u^i,\lambda^i,w^i) \triangleq A^i(x^i,\lambda^i,w^i) + B^i(x^i)u^i$. Then (28), (29) become

$$\min_{u^i} \nabla V_m(x^i).B^i(x^i)u^i \le \nabla V_m(x_m).f_m(x_m,\lambda_m) - \max_{w^i} \nabla V_m(x^i).A^i(x^i,\lambda^i,w^i) \tag{28'}$$

$$\begin{aligned}\min_{u^k} \nabla V_m(x^k).B^k(x^k)u^k \le{} & \max_{u^i} \min_{w^i} \nabla V_m(x^i).(A^i(x^i,\lambda^i,w^i) + B^i(x^i)u^i) \\ & - \min_{w^k} \nabla V_m(x^k).A^k(x^k,\lambda^k,w^k).\end{aligned} \tag{29'}$$

Similarly for (30), (31). For practical purposes we want to calculate all components of the vectors u^i, u^k. The convenient rule of doing so follows from the observation, that both sides of the inequalities in (28) – (31) are inner products. Thus, for instance (28)' is implied by the sequence of conditions imposed upon the components of the product:

$$\min_{u_j^i} \frac{\partial V_m}{\partial x_j^i} B_j^i(x^i)u_j^i \le \frac{\partial V_m}{\partial x_{mj}} f_{mj}(x_m,\lambda_m) - \min_{w^i} \frac{\partial V_m}{\partial x_j^i} A_j^i(x^i,\lambda^i,w^i), \quad j = 1,...,N, \tag{28''}$$

where B_j^i, A_j^i are the j-th columns of B^i, A^i. We can use (28)" to design the controllers $u_j^i(t)$, $j = 1,...,N$. Depending upon the sign of $\frac{\partial V_m}{\partial x_j^i} B_j^i(x^i)$, the control u_j^i is immediately calculable from

(28)". Note, that in all the above cases, for specified $f(.)$, $f_m(.)$ we can obtain algorithms for control programs in closed form, computed immediately from x^i, x_m without any digital search.

In turn, the adaptation conditions (C) is implied by the following adpative laws

$$\dot{\alpha}_j^i = -\alpha_j^i \left(c_i + \frac{1}{\eta}\left|\dot{V}_m(x)\right|\right), \; |\alpha^i| \geq \eta \qquad j = 1, ..., \ell, \; i = 1, 2 \tag{32}$$

where $\dot{V}_m(x)$ is obtained from (21). The condition $|\alpha^i| \geq \eta$ is obvious for otherwise the product trajectories are already in M_η and there is no need for tracking. The said implication of (C) by (32) can be proved by the following argument. From (32) we have

$$(\text{sign}\ \alpha_j^i)\dot{\alpha}_j^i = -|\alpha_j^i|\left(c_i + \frac{1}{\eta}|\dot{V}_m(x^i)|\right), \; j = 1, ..., \ell \qquad i = 1, 2$$

Summing up over j,

$$a\dot{\alpha}^i = -c_i \sum_j |\alpha_j^i| = \frac{1}{\eta}|\dot{V}_m(x^i)| \sum_j |\alpha_j^i|, \tag{33}$$

but

$$\sum_i |\alpha_j^i| \geq |\alpha^i| \geq \eta$$

whence also

$$\frac{1}{\eta}|\dot{V}_m(x^i)|. \sum_j |\alpha_j^i| = |\dot{V}_m(x^i)|. \frac{\sum_j |\alpha_j^i|}{\eta} \geq |\dot{V}_m(x^i)|.$$

Consequently

$$-c_i \sum_j |\alpha_j^i| \leq -c_i\eta$$

$$-\frac{1}{\eta}|\dot{V}_m(x)| \sum_j |\alpha_j^i| \leq -|\dot{V}_m(x^i)|$$

which substituted into (33) gives the condition (C).

Observe that (32) are exactly integrable with solutions $\alpha^i(\alpha^{io}, t)$, $t \geq 0$, representing the $V_m(t)$-flux which becomes positive (influx) or negative (outflux) depending where $\alpha^{io} = \alpha^i(o)$ is located (below or above) with respect to the zero-error surface $\alpha^i = 0$. Thus (32) act as a 0-regulator uniformly outside $|\alpha^i| < \eta$, attracting $\alpha^i(t)$ to M_η. Observe also that $\alpha^i(t)$ does not change sign: sign $\alpha^i(t) \equiv$ sign α^{io}, i.e. the surface $\alpha^i = 0$ is never crossed, which eliminates chattering in adaptation.

6. Example

Consider two point-mass modelled craft

$$\ddot{q}^i + w^i\lambda^i\dot{q}^i + aq^i + b(q^i)^3 = u^i,\ a, b > 0 \tag{34}$$

and design the controllers to "follow the leader"

$$\ddot{q}_m + \lambda_m\dot{q}_m + q_m = 0,\ \lambda_m = \text{ const } > 0 \tag{35}$$

while avoiding collision. We have $w^i \in [1,2]$, $\lambda^i \in [1,2]$, $|u^i| \leq \hat{u} = \text{const}$, $i = 1,2$. The demand on the reference model is to have $(0,0)$ asymptotically stable which commands

$$V_m(q_m, \dot{q}_m) = \frac{1}{2}(q_m^2 + \dot{q}_m^2), \tag{36}$$

$$\dot{V}_m(q_m, \dot{q}_m) = -\lambda_m\dot{q}^2 \leq 0,\ \forall\dot{q}_m \neq 0. \tag{37}$$

Then also

$$V_m(q^i, \dot{q}^i) = \frac{1}{2}[(q^i)^2 + (\dot{q}^i)^2], \tag{38}$$

$$\dot{V}_m(q^i, \dot{q}^i) = u^i\dot{q}^i - w^i\lambda^i(\dot{q}^i)^2. \tag{39}$$

We define $\Delta_0 = \{(q^i, \dot{q}^i) \in \Delta | (q^i)^2 + (\dot{q}^i)^2 \leq 10\}$. The control conditions (28) – (31) require first (28):

$$u^1 \text{ sign } \dot{q}^1 \leq \left[\lambda^1 - \lambda_m \left(\frac{\dot{q}_m}{\dot{q}^1}\right)^2\right] \text{ sign } \dot{q}^1 \tag{40}$$

with the extremizing $\bar{w}^1 = 1$. Then we choose $\bar{u}^1$ and demand (29):

$$\min_{u^2}(u^2\dot{q}^2) \leq \min_{w^2}[w^2\lambda^2(\dot{q}^2)^2] + \min_{w^1}[\bar{u}^1\dot{q}^1 - w^1\lambda^1(\dot{q}^1)^2]$$

or

$$u^2 \text{ sign } \dot{q}^2 \leq \left[\lambda^2\dot{q}^2 + \bar{u}^1\frac{\dot{q}^1}{\dot{q}^2} - 2\lambda^1\frac{(\dot{q}^1)^2}{\dot{q}^2}\right] \text{ sign } \dot{q}^2 \tag{41}$$

with extremizing $\bar{w}^2 = 2$, everywhere on $C^+M_\eta^1$, $V_m(q^1, \dot{q}^1) \geq V_m(q^2, \dot{q}^2)$. Identical technique applies to (30), (31) for $C^-M_\eta^1$, $V_m(q^1, \dot{q}^1) < V_m(q^2, \dot{q}^2)$. However (40), (41) do not determine the programs yet. Observe that when $\dot{q}^1, \dot{q}^2 = 0$, some terms of these inequalities became undefined. This itself represents no trouble as the axis' $\dot{q}^i = 0$, $i = 1,2$ is crossed by the trajectories of (34) instantaneously (perpendicularly) and the controller may be switched off for that instant without harm, as the trajectory is continuous. In fact it will do so automatically, as it must change sign at $\dot{q}^i = 0$, see (40), (41). However when $\dot{q}^i \to 0$, the terms divided by $\dot{q}^i$ will blow up, pulling the control above the admissible boundary of $U : |u^i| \leq \hat{u}$. Then we have to use some suitable constant value $\tilde{u}$ between 0 and $\hat{u}$, which may or may not bring the trajectory over on the other side of $\dot{q}^i = 0$. The latter depends very much on when we switch to this saturated control. The values $|\dot{q}^i| = \beta^i$ for which to do so, may be calculated by letting $|u^{i,k}| = \hat{u}$ with $u^{i,k}$ found from (40), (41). Hence the programs $P^1(.)$, $P^2(.)$ will take the shape of

$$u^1 \text{ sign } \dot{q}^1 \begin{cases} \leq \text{ sign } \dot{q}^1 \left[\lambda^1 - \lambda_m \left(\frac{\dot{q}_m}{\dot{q}_1}\right)^2\right], \forall|\dot{q}^1| \geq \beta^1 \\ = \text{suitable constant}, \forall|\dot{q}^1| < \beta^1 \end{cases} \tag{42}$$

and

$$u^2 \text{ sign } \dot{q}^2 \begin{cases} \leq \left[\lambda^2 \dot{q}^2 + \tilde{u}^1 \frac{\dot{q}^1}{\dot{q}^2} + 2\lambda^1 \frac{(\dot{q}^1)^2}{\dot{q}^2}\right] \text{ sign } \dot{q}^2, \forall|\dot{q}^2| \geq \beta^2; \\ = \text{suitable constant}, \forall|\dot{q}^2| < \beta^2. \end{cases} \tag{43}$$

Note that when $\dot{q}^2$ changes sign, the inequality in (42), (43) changes direction, thus we may avoid chattering by taking equality instead. Similar controls come out of the second part of the program, defined on $C^- M_\eta$, $V_m(q^1, \dot{q}^1) < V_m(q^2, \dot{q}^2)$.

Turning now to the adaptive laws, we have

$$\dot{\alpha}^i = -\alpha^i [c_i + \lambda_m (\dot{q}^i)^2 / \eta], \ |\alpha^{io}| \geq \eta, \ i = 1, 2 \tag{44}$$

where $c_i = c_i(\| (X^i, \alpha^i)^T \|) > 0$, or in case of stipulated t_c, is calculated from $\bar{c}_i = \frac{v_\eta^{i+}}{t_c} = \frac{12}{t_c}$. Note here that $v_\eta^{i+} = v_s^i = \frac{1}{2}(10 + 10) + 2$.

The equations (44) are integrable, with Sign α^{io} = sign $\alpha^i(t)$, $\forall t \geq 0$. Substituting it into (42), (43) we obtain the controls and substituting both into (34), (35) we obtain the trajectories.

References

1. Landau, I.D., *Adaptive Control*, M. Dekker, New York, 1979.
2. Balestrino, A., De Maria, G. and Sciavicco, L., *An adaptive model following control for robotic manipulators*, Trans. ASME, J. Dyn. Syst. Meas. Control **105** (1985), 3.
3. Dubovsky, S. and DesForges, D.I., *The application of model referenced adaptive control to robotic manipulators*, Trans. ASME., J. Dyn. Syst. Meas. Control **101** (1979), 3.
4. Singh, S.N., *Adaptive model following of nonlinear robotic systems*, IEEE Trans. Autom. Control AC-30 (1985), 1.
5. Skowronski, J.M., *Nonlinear model reference adaptive control*, J. Australian Math Soc., Series B, Applied Math. **28** (1986), 147–157.

6. Skowronski, J.M., *Liapunov type playability for adaptive physical systems*, Proc. Nat. System Conf., Combaitore PSG-College of Technology, India **Section Q11** (1977), 1–5.

7. Skowronski, J.M., *Algorithms for adaptive control of two arm flexible manipulator under uncertainty*, IEEE Trans. Aerospace & Elec. Systems **AES 24** (1988), 5.

8. Skowronski, J.M., *Control Theory of Robotic Systems*, World Scientific Publ., New Jersey – Singapore, 1990.

9. Filippov, A.F., *Existence of solutions of generalized differential equations*, Math. Notes **10** (1971), 608–611.

10. Yoshizawa, T., *Stability Theory by Liapunov Second Method*, Publ. Math. Soc., Japan, Tokyo, 1966.

11. Leitmann, G. and Skowronski, J.M., *Avoidance Control*, J. Opt. Th. Applic. **23** (1977), #4, 581–591.

12. Skowronski, J.M., *Control Dynamics of Robot Manipulators*, Academic Press, 1986.

13. Skowronski, J.M., *Control of Nonlinear Mechanical Systems*, Plenum, New York, 1991.

ADAPTIVE IDENTIFICATION AND MODEL TRACKING BY A FLEXIBLE SPACECRAFT

J.M. Skowronski
University of Southern California
Los Angeles, CA 90089–1453

Abstract. A rigid–flexible spacecraft structure subject to bounded uncertainty in structural parameters and payload, with large articulation angles, is modelled by a hybrid multidimensional system with high (untruncated) geometric nonlinearity and Coriolis forces. It is to be controlled adaptively to track a rigid body reference model with desired dynamics. To this aim the system is replaced by a nonlinear adaptive, state and parameter identifier with considerably reduced number of DOF and made exactly integrable, i.e. with solutions in closed form. The technique used is that of nonlinear extension to MRAC introduced by the author a few years ago. The results lie in obtaining feedback signal adaptive controller in analytic form and exactly integrable adaptive laws, both the controller and the laws based on the state information supplied by the identifier, thus robust to uncertainty.

The technique allows for the tracking to occur with stipulated precision obtained in stipulated real time. The reduced dynamics and the exact integrability of the identifier and the adaptive laws make on–line computation of the algorithms simple enough to be made sufficiently fast on a small on–board computer.

I. Introduction

With the present, and even more so, future demands on performance of flexible spacecraft, the precise mathematical modelling becomes essential for effective control.

It is at least debatable whether continuum models are more physically justified than discretization with dense grid, cf. Ref. 1. Practical spacecraft configurations do not generate simple boundary conditions or simple shapes, making the partial differential equation models inaccurate and discretization overwhelmingly accepted. On the other hand we may no longer ignore features leading to high (untruncable) nonlinearity in dynamics, like the need for large angle maneuvers about several equilibria, interface between attitude and structural control, cf. Ref. 2, the latter with obvious geometric nonlinearities, appearance of Coriolis and centrifugal forces, etc. Thus it seems that the motion equations must admit untruncated nonlinearity and large amplitudes in general, though in their elastic part, with sufficiently dense discretization the amplitudes may be considered small. However, even in this case it does not sound justified if we assume the elastic motion a perturbation decoupled from that of the rigid–body, cf. Ref. 3,4.

Moreover, precise modelling may not ignore uncertainty in the spacecraft parameters like inertia, flexural rigidity and trade–off between the two, structural damping, the uncertainty caused by varying loads, on–line disturbances which cannot be foreseen from ground texts, etc. Apart from that, we need to manage spillover destabilization. Robust control against all this needs to be feedback and signal adaptive, which in turn requires on–line information on state and parameters for both the control programs and adaptive laws. We may supply such information via equilibria–compatible adaptive identifiers, see Ref. 5., which better be with reduced dimensions, or even closed–form integrable, so that the computation time on a small on–board computer is within tolerable limits. For the latter reason one may also avoid a digital search for candidates of optimal solutions obtained via necessary conditions, and design controllers and adaptive laws which imply sufficient conditions for the desired control objectives, not necessarily optimal.

If we want to control the spacecraft, say, to track a desired rigid reference model – with the above long list of demands implemented simultaneously, we better think of methods quite different than small modifications or augmentation of the existing linear techniques. It may be worthwhile to adopt at least some of the recent results in Nonlinear Mechanics, so far overlooked by the control designers. Implementing such results, or for that matter any other approach to the problems outlined, requires new mathematical models which should be able to accommodate the full list of demands with minimum level of formal complications. The study of such models must be fundamental and may be long. Some steps have already been made – this work proposes some more. In tracking we shall use the nonlinear version (Ref.'s 6–9) of the model reference adaptive control (MRAC) technique, see Ref. 10, which proved very successful for aircraft flight control, cf. Ref.'s 11–18.

II. Dynamical Model

Suppose the rigid–body configuration of the spacecraft is described by the set of Lagrangian coordinates $q_1,\dots,q_n$ forming the vector $q(t) = (q_1(t),\dots,q_n(t))^T$ varying in the configuration envelope $\Delta_q \subset \mathbb{R}^n$, with the corresponding velocities $\dot{q}(t) \in \Delta_{\dot{q}} \subset \mathbb{R}^n$. Then considering the assembly of elastic links we introduce the deformation coordinates for the i–th link, $i = 1,\dots,m$, as follows: $r_i(y_i,t)$ along the axis of the link positions at y_i, $0 \le y_i \le \ell_i$, where ℓ_i is the length of the link, $v_i(y_i,t)$ and $w_i(y_i,t)$ perpendicularly to the link axis in horizontal and vertical direction, respectively.

Using the Ritz–Kantorovitch series expansion we have

$$r_i(y_i,t) = \sum_{\nu=1}^{p} r_i^{\nu}(y_i) r_i^{\nu}(t) = r_i(y_i) r_i(t) \tag{1}$$

and for $v_i(y_i,t)$, $w_i(y_i,t)$ analogously, with the exact deformation expected for $p \to \infty$. We take p large enough so that the Kantorovitch linearization is physically justified. The technical way about it is to subdivide stepwise the distances between grid for so many times (obtaining many modes) until the difference between the obtained steps of successive approximation becomes acceptably small. Having agreed to this technique we consider the first step only (first mode), with convenience assumption that $p \cdot m = n$. Having specified (1) we form the vector $\eta(t) \triangleq (\eta_1(t),\dots,\eta_n(t))^T \in \Delta_\eta$, where $\eta_i(t) \triangleq (r_i(t),v_i(t),w_i(t))^T$ and following Ref. 19, write the hybrid system as

$$\begin{bmatrix} M & M_c \\ M_c^T & M_\eta \end{bmatrix} \begin{bmatrix} \ddot{q} \\ \ddot{\eta} \end{bmatrix} + \begin{bmatrix} 0 & D_c \\ 0 & D \end{bmatrix} \begin{bmatrix} \dot{q} \\ \dot{\eta} \end{bmatrix} + \begin{bmatrix} 0 & P_c \\ 0 & P \end{bmatrix} \begin{bmatrix} q \\ \eta \end{bmatrix} \tag{2}$$

$$+ \begin{bmatrix} C \\ C_\eta \end{bmatrix} + \begin{bmatrix} K \\ K_\eta \end{bmatrix} = \begin{bmatrix} F \\ 0 \end{bmatrix}$$

where $M(q,\lambda,s)$ is the inertia $n \times n$ matrix, $C(q,\dot{q},\lambda,s)$ is the vector of nonpotential forces (Coriolis, gyro, damping) and $K(q,\lambda,s)$ is the vector of the nonlinear elastic forces, all three for the rigid–body part of the structure. Then $M_\eta(\eta,\lambda,s)$, $C_\eta(\eta,\dot{\eta},\lambda,s)$ and $K_\eta(\eta,\lambda,s)$ are their elastic correspondents, while $M_c(q,\eta)$, $D_c(q,\dot{q},\eta,\dot{\eta})$, $P_c(q,\eta)$ and the internal damping $D(q,\dot{q},\eta,\dot{\eta})$ as well as the hybrid restoring coefficients $P(q,\eta)$ are coupling matrices. These matrices are formed by integrals over the shape functions concerned, see Ref. 19. Finally $F(q,\dot{q},u)$ is the actuator force vector. Within the above functions, $\lambda(t) = (\lambda_1(t),...,\lambda_\ell(t)) \in \Lambda \subset \mathbb{R}^\ell$ is the vector of kinetic and structural parameters to be adjusted, with values within a given bounded band Λ and with the structural parameters subvector $\xi = (\xi_1,...,\xi_p)$, ξ_i = const > 0, to be designed as part of λ. The adaptation will be governed by a later designed adaptive laws, which can be generally written as

$$\dot{\lambda} = f_a(q,\dot{q},\eta,\dot{\eta},\lambda) \tag{3}$$

Then $s(t) = (s_1(t),...,s_k(t))^T \in S \subset \mathbb{R}^k$ is the uncertainty vector, unknown except for the known bounded envelope S. Finally $u(t) = (u_1(t),...,u_n(t))^T$ is the rigid body control vector with values constrained to the bounded and closed set $U \subset \mathbb{R}^n$, to be specified by a later designed control program $\mathcal{P}(\cdot)$: $\Delta_q \times \Delta_{\dot{q}} \to$ subsets of U, generally set valued to accommodate discontinuities of the function $u(\cdot)$. The reader must have noticed that only the rigid body is controlled. It is indeed reasonable to assume that not all the elastic modes are controlled, leaving some of them as so called residual (uncontrolled part of the system). As the technique of control after discretization does not differ between the rigid and elastic subsystems, we do not narrow generality taking the whole n elastic modes as residual.

Introducing the hybrid inertia matrix

$$\mathcal{M}(q,\lambda,s) = \begin{bmatrix} M & M_c \\ M_c^T & M_\eta \end{bmatrix}$$

which is nonsingular positive definite, and defining $\mathcal{D} = \mathcal{M}^{-1}(D_c\dot{\eta}+C, D\dot{\eta}+C_\eta)^T$, $\mathcal{P} = \mathcal{M}^{-1}(P_c\eta+K, P\eta+K_\eta)^T$, $\mathcal{F} = \mathcal{M}^{-1}(F,0)^T$, we may write (2) in the decoupled format

$$(\ddot{q},\ddot{\eta})^T + \mathcal{D}(q,\dot{q},\eta,\dot{\eta},\lambda,s) + \mathcal{P}(q,\eta,\lambda,s) = \mathcal{F}(q,\dot{q},u). \tag{4}$$

The reference model to be followed is a rigid body system corresponding to (4), thus with 2n dimensional displacement vectors $q_m \in \Delta_q \times \Delta_\eta$, and velocities $\dot{q}_m \in \Delta_{\dot{q}} \times \Delta_{\dot{\eta}}$, and the motion equations

$$\ddot{q}_m + \mathcal{D}_m(q_m,\dot{q}_m) + \mathcal{P}_m(q_m,\lambda_m) = 0 \tag{5}$$

where $\lambda_m = (\lambda_{m1},...,\lambda_{m\ell})$ = const $\in \Lambda$ is the parameter vector to which $\lambda(t)$ will be adjusted and $\mathcal{D}_m$, $\mathcal{P}_m$ represent successively non–potential and potential model forces such that the equilibria q^e, η^e of (2) and (5) coincide:

$$\mathcal{D}_m(q_m,0) = 0 \, , \; \forall q_m \tag{6}$$

$$\mathcal{P}_m(q^e,\eta^e,\lambda_m) = 0 \, , \; \forall \lambda_m \tag{7}$$

on the surface $\dot{q} = 0 \, , \; \dot{\eta} = 0$.

The total energy of the model is

$$E_m(q_m,\dot{q}_m) = \frac{1}{2}\,\dot{q}_m^T\dot{q}_m + \int \mathcal{P}_m(q_m,\lambda_m)\, dq_m \tag{8}$$

and its time derivative along trajectories of (5) is

$$\dot{E}_m((q_m,\dot{q}_m) = -\mathcal{D}_m(q_m,\dot{q}_m)\,\dot{q}_m \tag{9}$$

for all $\dot{q}_m \neq 0$. The model is selected such as to demonstrate a stipulated behavior in the state space. Let $\Delta = \Delta_q \times \Delta_{\dot{q}} = \Delta_\eta \times \Delta_{\dot{\eta}}$ and $\Delta^2 = \Delta \times \Delta$. Then suppose that we want to track the model on some tracking region Δ_o^2 in Δ^2 . It is then reasonable to assume that the trajectories of (5) will be ultimately bounded in Δ_o^2 , which may be considered the "flying envelope" for (5). This assumption requires that for all points in $C\Delta_o^2 = \Delta^2 - \Delta_o^2$, we have

$$\nabla E_m(q_m,\dot{q}_m)^T q_m \geq 0, \tag{10}$$

while the power

$$\dot{E}_m(q_m,\dot{q}_m) < 0 \, . \tag{11}$$

Indeed, if any trajectory of (5) was to leave Δ_o^2 it would make (10), (11) contradict, see Ref. 6.

III. Nonlinear Tracking

We want now to specify a control program $\mathcal{P}(\cdot)$ and an adaptive law (3) such that the trajectories and parameters of (2) will come close to these of (5) in real, possibly stipulated time $T_\mu < \infty$ and stay there within a stipulated precision. The block diagram of the system is shown in the figure below. More technically, we introduce the objective of the nonlinear MRAC in the following way. Suppose for the time being that $q(t),\dot{q}(t),\eta(t),\dot{\eta}(t),\lambda(t)$ are all well defined, and let us form the state vectors: $x(t) = (q,\eta,\dot{q},\dot{\eta})^T \in \Delta^2$, $x_m(t) = (q_m,\dot{q}_m) \in \Delta^2$ and the product vector $X(t) = (x,x_m)^T \in \mathbb{R}^{8n}$, as well as $\alpha(t) = \lambda(t) - \lambda_m \in \Lambda \subset \mathbb{R}^{\ell}$. Then we aggregate (4), (5) and (3) in the joint system in state format by introducing the vectors $f = [(\dot{q},\dot{\eta})^T,(\mathcal{F}-\mathcal{D}-\mathcal{P})]^T$, $f_m = [\dot{q}_m,(-\mathcal{D}_m-\mathcal{P}_m)]^T$, and $f = (f,f_m,f_a)^T$, which allows us to write the system as

$$(\dot{X},\dot{\alpha})^T = f(X,\alpha,s,u)\ . \tag{12}$$

In consequence of the bounded uncertainty s and setvaluedness of the program for u , (12) should be formally written in the contingent format

$$(\dot{X},\dot{\alpha})^T \in \{f(X,\alpha,s,u)\,|\,s \in S,\ u \in \mathcal{P}(X,\alpha)\} \tag{13}$$

with $f(\cdot)$'s of (12) called the selectors of the orientor field defined by the right hand side of (13). For suitable $f(\cdot)$, $\lambda(\cdot)$, $\mathcal{P}(\cdot)$, see Ref. 6, given $u(\cdot)$, $s(\cdot)$, through each X^o,α^o there is at least one solution $k(X^o,\alpha^o,t)$, $t \geq t_o$ to (13) which is an absolutely continuous curve in $\Delta^4 \subset \mathbb{R}^{8n}$, and conversely there are $u(\cdot)$, $s(\cdot)$ that make the corresponding $k(X^o,\alpha^o,t)$ to satisfy (12). We denote the family of such solutions by $\mathcal{K}(X^o,\alpha^o)$. Let us now define the "diagonal" set

$$M = \{(X,\alpha)\in\Delta_o^4\times\Lambda \mid x=x_m,\ \alpha=0\}$$

and its $\mu > 0$ neighborhood:

$$M_\mu = \{(X,\alpha)\in\Delta_o^4\times\Lambda \mid \|x-x_m\|<\mu,\ \|\alpha\|<\mu\}\ ,$$

with $\|\cdot\|$ any norm in $\mathbb{R}^{8n}$ and Δ_o^4 being the $\mathbb{R}^{8n}$ counterpart of Δ_o^2 .

<u>MRAC OBJECTIVE</u>: The structure (4) tracks the reference model (5) on Δ_o^2 iff there is a controller $\mathcal{P}(\cdot)$ such that for each $k(\cdot)\in\mathcal{K}(X^o,\alpha^o)$ the set $\Delta_o^4\times\Lambda$ is positively invariant i.e. $(X^o,\alpha^o) \in \Delta_o^4\times\Lambda$ implies $k(X^o,\alpha^o,t) \in \Delta_o^4\times\Lambda$, $\forall t \geq t_o$, and that there is $t_\mu < \infty$, possibly stipulated, for which $k(X^o,\alpha^o,t) \in M_\mu$, $t \geq t_\mu$.

Let $N[\partial(\Delta_o^4\times\Lambda)]$ be a neighborhood of the boundary $\partial(\Delta_o^4\times\Lambda)$ and define $N_s \triangleq N[\partial(\Delta_o^4\times\Lambda)] \cap \overline{\Delta_o^4\times\Lambda}$, and the relative complement $CM_\mu \triangleq (\Delta_o^4\times\Lambda) - M_\mu$. Moreover, let $D(\text{open}) \supset \overline{CM}_\mu$ such that $D\cap M = \phi$ and introduce two C^1–functions $V_s: N_s\to\mathbb{R}$, $V_\mu: D\to\mathbb{R}$ with positive constants

$$v_s \triangleq V_s(X,\alpha),\ \forall(X,\alpha) \in \partial(\Delta_o^4\times\Lambda)\ ,$$

$$v_\mu^- \triangleq \inf V_\mu(X,\alpha)\,|\,(X,\alpha) \in \partial M_\mu \cap \overline{CM}_\mu\ ,$$

$$v_\mu^+ \triangleq \sup V_\mu(X,\alpha)\,|\,(X,\alpha) \in \partial(\Delta_o^4\times\Lambda) \cap \overline{CM}_\mu\ .$$

Here we choose Δ_o as a V_s–level or choose V_s with $\partial\Delta_o$ as a level. The following theorem has been proved in Ref. 6.

Theorem 1. The MRAC Objective is attained if there are $\mathcal{P}(\cdot)$, $V_s(\cdot)$, $V_\mu(\cdot)$ such that for all $(X,\alpha) \in \Delta_0^4 \times \Lambda$ we have

(i) $V_s(X,\alpha) \le v_s$, $\forall (X,\alpha) \in N_s$;

(ii) $\forall u \in \mathcal{P}(X,\alpha)$,

$$\nabla V_s(X,\alpha)^T f(X,\alpha,s,u) \le 0, \ \forall s \in S \ ;$$

(iii) $0 \le V_\mu(X,\alpha) < v_\mu^+$, $v_\mu^+ > 0$, $\forall (X,\alpha) \in \overline{CM}_\mu$;

(iv) $V_\mu(X,\alpha) \le v_\mu^-$, $\forall (X,\alpha) \in D \cap M_\mu$;

(v) $\forall u \in \mathcal{P}(X,\alpha)$, there is $c = \text{const} > 0$ such that

$$\nabla V_\mu(X,\alpha)^T f(X,\alpha,s,u) \le -c, \ \forall s \in S \ .$$

When $t_\mu < \infty$ is stipulated, we impose

$$c = \frac{v_\mu^+}{t_\mu} . \tag{14}$$

We now select

$$V_s = E_m(x) + E_m(x_m) + a\alpha \ , \tag{15}$$

$$V_\mu = \begin{cases} |E_m(x) - E_m(x_m)| + a\alpha, & (X,\alpha) \in CM_\mu \\ a\alpha \ , & (X,\alpha) \in M_\mu \cap D \ , \end{cases} , \tag{16}$$

where $E_m(x)$ is the model energy function $E_m(\cdot)$ with x_m replaced by x , i.e. along the trajectories of (4), moreover $a = (\text{sign}\ \alpha_1, \dots, \text{sign}\ \alpha_\ell)$, with $|\alpha_i(t)| \not\geq \mu$ for all $t \ge t$, as otherwise adaptation is redundant. By the assumed character of $E_m(\cdot)$, conditions (i), (iii), (iv) hold. To check (ii), (v) define $\delta E_m = E_m(x) - E_m(x_m)$ and differentiate:

$$\dot V_s = \dot E_m(x) + \dot E_m(x_m) + a\dot\alpha \ ,$$

$$\dot V_\mu = \begin{cases} \dot E_m(x) - \dot E_m(x_m) + a\dot\alpha, & (X,\alpha) \in CM_\mu, \ \delta E_m \ge 0, \\ \dot E_m(x_m) - \dot E_m(x) + a\dot\alpha, & (X,\alpha) \in CM_\mu, \ \delta E_m < 0, \\ a\dot\alpha, & (X,\alpha) \in M_\mu \cap D, \end{cases}$$

where

$$\begin{aligned} \dot E_m(x_m) &= \nabla E_m(x_m)^T \cdot f_m(x_m, \lambda_m), \\ \dot E_m(x) &= \nabla E_m(x)^T f(x,u,\lambda,s) \ . \end{aligned} \tag{17}$$

Little calculation shows that the following two sets of conditions imply (ii), (v):

<u>CONTROL CONDITIONS</u>

$$\min_u \max_s \nabla E_m(x)^T f(x,u,\lambda,s) \le \dot E_m(x_m), \text{ for } \delta E_m \ge 0 \tag{18}$$

$$\min_u \max_s \nabla E_m(x)^T f(x,u,\lambda,s) \le \dot E_m(x_m), \text{ for } \delta E_m \ge 0 \tag{19}$$

ADAPTATION CONDITIONS

$$a\dot{\alpha} \leq - |\alpha| \, [c+|\dot{E}_m(x)|] \, . \tag{20}$$

Substituting (4), (5) with the extremizing values s* of s we obtain that the following implies (18):

$$\begin{aligned}
\min \mathcal{F}_i(q,\dot{q},u) & \\
\leq & - \mathcal{D}_{mi}(q_m,\dot{q}_m)\dot{q}_m - \mathcal{D}_{m(n+1)}(q_m,\dot{q}_m)\dot{q}_{m(n+1)} \\
& + \mathcal{D}_i(q,\dot{q},\eta,\dot{\eta},\lambda,s^*)\dot{q}_i + \mathcal{D}_{n+i}(\dot{q},q,\dot{\eta},\eta,\lambda,s^*)\dot{q}_{n+i} \\
& + [\mathcal{P}_i(q,\eta,\lambda,s^*) - \mathcal{P}_{mi}(q,\eta,\lambda_m)]\dot{q}_i \\
& + [\mathcal{P}_{n+i}(q,\eta,\lambda,s^*) - \mathcal{P}_{m(n+i)}(q,\eta,\lambda_m)]\dot{q}_{n+i} \, ,
\end{aligned}$$

$$i=1,\ldots,n$$

for $\delta E_m \geq 0$. Here $\mathcal{F}_i,\ldots\mathcal{P}_{m(n+i)}$ are the indicated components of the vectors $\mathcal{F},\ldots,\mathcal{P}_m$. The $\min_u \mathcal{F}_i$ is replaced by $\max_u \mathcal{F}_i$ with the inequality inverted and s* adjusted to imply (19) for $\delta E_m < 0$. Specifying the functions $\mathcal{F}(\cdot),\ldots,\mathcal{P}_m(\cdot)$ one determines the control program which implies the above two control conditions. The condition (20) is implied by the adaptive laws

$$\dot{\alpha}_i = -\alpha_i[c + \mathcal{D}_m(q,\dot{q},\eta,\dot{\eta})^T(\dot{q},\dot{\eta})], \tag{21}$$

$$i=1,\ldots,\ell .$$

Note here that since $\lambda_m = \text{const}$, $\dot{\alpha} = \dot{\lambda}$.

More detailed derivation of the control program and adaptive laws together with case studies and simulation results may be found in Ref. 20.

IV State and Parameter Identifier

The control programs and adaptive laws of last section require the state information x, available with all the difficulties described in the Introduction, not the least difficulty being the considerable dimensionality of the product system (12). We avoid these problems by applying the model reference identification technique, see Ref. 9, 6. One assumes for a time being that x, s are known and proceeds to design an identifier which is a dynamical system with significantly reduced dimensionality, say $k<2n$, whose trajectories converge in stipulated real time smaller than t_μ to x and whose parameters approach s in the same way.

A few comments are needed before we choose the identifier. Consider (4) and let $y = g(x)$ be a flight monitored output, with $g(\cdot)$ known vector function but without univalued inverse. Then let us write the identifier dynamics in the general format, later specified,

$$\dot{z} = f_z(z,y,\lambda_p) \tag{22}$$

where $z(t) \in \Delta_z \subset \mathbb{R}^k$, $k < 4n, \Delta_z$ bounded set embedded in Δ^2, and $\lambda_p(t) \in \Lambda \cap \mathbb{R}^\ell$ is a vector of adjustable parameters. From within the unknown vector $x = (x_1,\ldots,x_{4n})^T$ there would be $x_1,\ldots,x_k$ variable to be identified, let us call them "identifiable", the remainder $x_{k+1},\ldots,x_{4n}$ representing the residual or neglected dynamics. The selection of the $x_1,\ldots,x_k$ is arbitrary in this general setting, and a matter of convenience in case studies. We have taken $k=2n$ at the beginning for reasons explained, but this has been by no means necessary. If some state variables can be monitored and thus directly form $y(t)$, or part of it, we let them enter the identifiable subvector to reduce the dimensions of (22) even further. The spillover dynamics must be kept reasonably bounded, while the identifiable x and s are predicted. To this aim we introduce the vectors $Z = (Z_1,\ldots,Z_{4n})^T = (z_1,\ldots,z_k,0,\ldots,0)^T$ and $f_Z = (f_{zi},\ldots,f_{zk},0,\ldots,0)$. Then we define $X_p = (x,Z)^T$, $\beta = \lambda_p(t)-s(t) \in \Lambda$, $f_p = (f,f_z,f_b)^T$, with

$$\dot{\beta} = f_b(x,y,\lambda_p)\,. \tag{23}$$

Note, that in general s is not a constant, whence $\dot{\beta} \neq \dot{\lambda}_p$ and we must know or estimate $\dot{s} = f_s(t)$ for the use in (23), see for details Ref. 9. We may now form the product system

$$(\dot{X}_p,\dot{\beta})^T \in \{f_p(X_p,\beta,y,u,\rho)\,|\,u\in\mathcal{P}(X,\alpha)\}$$

with u given by $\mathcal{P}(\cdot)$ specified subject to (18), (19) and the product trajectories $k(X_p^o,\beta^o,t)$, $t\geq t_o$, formed similarly to those of (13). The increased dimension has only a formal meaning here and is used for the sake of general description. In case investigation it collapses immediately to the space of $k\times(4n-k)$ dimensions, which in turn reduces to the k dimensions of (22) when the convergence is implemented. It is natural that the tracking region Δ_o^2 is the subset of Δ^2 on which we want the identification. It is also natural that the design of (22) is such that it's trajectories must be ultimately bounded in such $\Delta_o^2 \cap \Delta_z^2$. Define

$$M_p = \{(X_p,\beta)\in\Delta_o^4\times\Lambda \mid \|x_r - Z_\sigma\|<\mu_\sigma,\ \sigma=1,\ldots,2n,\ \|\beta\|<\mu\}$$

where $\mu_1,\dots,\mu_k > 0$ represent the precision estimates for identification, while $\mu_{k+1},\dots,\mu_{4n}$ estimate stabilization of the residual (neglected) dynamics, and μ stabilizes the parameter error.

IDENTIFICATION OBJECTIVE: Each equation of (22) is a stabilizing identifier of (4) on Δ_o after $t_p \le t_\mu$ with precision to $\mu_1,\dots,\mu_{2n},\mu$, iff given $f(\cdot)$, $\mathcal{P}(\cdot)$ attaining the tracking of (5), there is $f_z(\cdot)$ of (22) such that for each $k(X_p^o,\beta^o,\cdot) \in \mathcal{K}(X_p^o,\beta^o)$, the set $\Delta_o^4 \times \Lambda$ is made positively invariant, and $(X_p^o,\beta^o) \in \Delta_o^4 \times \Lambda$ implies $k(X_p^o,\beta^o,t) \in M_p$, $\forall t \ge t_p$.

Recall N_s and $V_s(\cdot)$ from Theorem 1, then denote

$CM_p = \Delta_o^4 - M_p$, $D_p(\text{open}) \supset \overline{CM}_p$ with $D_p \cap \{(X_p,\beta) \mid x=Z,\ \beta=0\} = \phi$, and introduce a C^1–function $V_p(\cdot)\colon D_p \to \mathbb{R}$ with

$$v_p^+ = \sup V_p(X_p,\beta) \mid (X_p,\beta) \in \partial(\Delta_o^4 \times \Lambda) \cap \overline{CM}_p$$

$$v_p^- = \inf V_p(X_p,\beta) \mid (X_p,\beta) \in \partial(\Delta_o^4 \times \Lambda) \cap \overline{CM}_p .$$

Theorem 2. The Identification Objective is attained, if there are $f_z(\cdot)$, $V_s(\cdot)$ and $V_p(\cdot)$ such that for all $(X_p,\beta) \in \Delta_o^4 \times \Lambda$,

(i) $V_s(X_p,\beta) \le N_s,\ \forall (X_p,\beta) \in N_s$;

(ii) given $u \in \mathcal{P}(X,\alpha), y$,

$$\nabla V_s(X_p,\beta)^T f_p(X_p,\beta,y,u,s) \le 0,\ \forall s \in S;$$

(iii) $0 \le V_p(X_p,\beta) \le v_p^+,\ v_p^+ > 0,\ \forall (X_p,\beta) \in CM_p$;

(iv) $V_p(X_p,\beta) \le v_p^-,\ \forall (X_p,\beta) \in D_p \cap M_p$;

(v) given $u \in \mathcal{P}(X_m,\alpha), y$,

$$\max_s \nabla V_p(X_p,\beta)^T \cdot f_p(X_p,\beta,y,u,s) \le -\frac{v_p^+}{t_p}.$$

The proof follows by the same argument as for Theorem 1. Let us choose

$$V_s = E_m(x) + E_m(z) + b\beta , \tag{24}$$

$$V_p = \begin{cases} |E_m(x) - E_m(z)| + b\beta, & (X_p,\beta) \in CM_p \\ b\beta, & (X_p,\beta) \in M_p \cap D_p, \end{cases} \tag{25}$$

with $b = (\text{sign}\,\beta_1,\dots,\text{sign}\,\beta_\ell), |\beta| \ge \mu$ for the same reason as α . Conditions (i), (iii), (iv) of Theorem 2 hold by the same argument as for Theorem 1. To check (ii), (v) we denote

$\delta E_m' = E_m(x) - E_m(z)$ and differentiate

$$\dot{V}_s = \dot{E}_m(x) + \dot{E}_m(z) + b\dot{\beta}\,, \tag{26}$$

$$\dot{V}_p = \begin{cases} \dot{E}_m(x) - \dot{E}_m(z) + b\dot{\beta}, & (X_p,\beta) \in CM_p,\ \delta E_m' \geq 0, \\ \dot{E}_m(z) - \dot{E}_m(x) + b\dot{\beta}, & (X_p,\beta) \in CM_p,\ \delta E_m' < 0, \\ b\dot{\beta}, & (X_p,\beta) \in M_p \cap D_p, \end{cases} \tag{27}$$

Similarly like for Theorem 1, we may show that

$$\max_s \dot{E}_m(x) \leq \dot{E}_m(z),\ \delta E_m' \geq 0\,, \tag{28}$$

$$\min_s \dot{E}_m(x) \geq \dot{E}_m(z),\ \delta E_m' < 0\,, \tag{29}$$

$$b\dot{\beta} \leq -\left[\frac{v_p^+}{t_p} + \max_s |E_m(x)|\right], \tag{30}$$

imply (ii), (v). Indeed, by well known Yoshizawa's theorem, see Ref. 6, ultimate boundedness of $z(t)$ in Δ_o^2 implies the existence of a C^1 function of the $E_m(\cdot)$ type for which, for all $z \in C\Delta_o^2$, $\dot{E}_m(z) < 0$. Then (28), (30) substituted into (26) gives $\dot{V}_s < 0$ for $\delta E_m' \geq 0$. For $\delta E_m' \geq 0$, we substitute (29), (30), obtaining $\dot{V}_s \leq -\frac{v_p^+}{t_p}$, thus yielding (ii) satisfied. Condition (v) follows immediately upon substitution of (28)–(30) into (27). Observe now that the tracking controller has been designed to satisfy (18), (19), in view of which the following conditions imply (28), (29), respectively.

<u>DESIGN CONDITIONS FOR f_z:</u>

$$\begin{aligned} \dot{E}_m(x_m) &\leq \nabla E_m(z)^T f_z(z,y,\lambda_p),\ \delta E_m' \geq 0, \\ \dot{E}_m(x_m) &\geq \nabla E_m(z)^T f_z(z,y,\lambda_p),\ \delta E_m' < 0. \end{aligned} \tag{31}$$

On the other hand (30) is implied by

<u>ADAPTATION CONDITION:</u>

$$b\dot{\beta} \leq -|\beta| \left[\frac{v_p^+}{t_p} + |\dot{E}_m(z)|\right]. \tag{32}$$

Indeed, by (28), (29), $|\dot{E}_m(z)| \geq \dot{E}_m|(x,s^*)|$, where s^* is the extremizing value of s, which in turn implies $|\beta| \left[\frac{v_p^+}{t_p} + |\dot{E}_m(z)|\right] \geq \left[\frac{v_p^+}{t_p} + |\dot{E}_m(x,s^*)|\right]$, proving the point. Hence f_z must satisfy (31) and the following adaptation laws

$$\dot{\beta}_i = -\beta_i \left[\frac{v_p^+}{t_p} + |\nabla E_m(z)^T f_z(z,y,\beta)| \right], \qquad i=1,\dots,\ell. \tag{33}$$

which imply (32) by the same argument as for (21).

We may now design our identifier (22). Since $k = 2n$ of the rigid body, we propose

$$\ddot{q}_p + \mathcal{D}_p(y) + \mathcal{P}_p(q_p,\lambda_p) = 0 \tag{34}$$

with $q_p(t) \in \Delta_o \subset \mathbb{R}^n$, $\dot{q}_p \equiv y$, $\mathcal{D}_{pi} = d_i y_i |y_i|$, $i=1,\dots,n$, where $d_i>0$ suitable constants, and $\mathcal{P}_{pi} \equiv \mathcal{P}_{mi}(q_p,\lambda_p)$. Then (31) holds if

$$\left.\begin{aligned} \mathcal{D}_p(y)^T y &\le \mathcal{D}_m(q_m,\dot{q}_m)^T \dot{q}_m, \ \delta E'_m \ge 0 \\ \mathcal{D}_p(y)^T y &\ge \mathcal{D}_m(q_m,\dot{q}_m)^T \dot{q}_m, \ \delta E'_m < 0 \end{aligned}\right\} \tag{35}$$

with $\delta E_m = \frac{1}{2}\dot{q}_m^T \dot{q}_m - \frac{1}{2} y^T y + \oint \mathcal{P}_m(q_m,\lambda_m) dq_m$

$$- \oint \mathcal{P}_m(q_p,\lambda_p) dq_p$$

In turn (35) holds if

$$\left.\begin{aligned} d_i y_i^2 |y_i| &\le \mathcal{D}_{mi}(q_m,\dot{q}_m)\dot{q}_{mi}, \ \delta E'_m \ge 0 \\ d_i y_i^2 |y_i| &\ge \mathcal{D}_{mi}(q_m,\dot{q}_m)\dot{q}_{mi}, \ \delta E'_m < 0 \end{aligned}\right\} \tag{36}$$

which gives the conditions for determining d_i's . Then also the adaptive laws (33) become

$$\dot{\beta}_i = -\beta_i \left[\frac{v_p^+}{t_p} + \sum_{j=1}^{n} y_j^2 |y_j| \right], \qquad i=1,\dots,\ell. \tag{37}$$

We may show now that the identifier so designed is exactly integrable. Indeed (34) may be rewritten as

$$\frac{dy_i}{dq_{pi}} = \frac{-d_i y_i |y_i| - \mathcal{P}_{pi}}{y_i}$$

As $|y_i| = y_i \operatorname{sign} y_i$, the above becomes

$$y_i \frac{dy_i}{dq_{pi}} + d_i y_i^2 \operatorname{sign} y_i = -\mathcal{P}_{pi}, \quad i=1,\dots,n$$

or

$$\frac{dy_i^2}{dq_{pi}} + 2d_i y_i^2 \operatorname{sign} y_i = -2\mathcal{P}_{pi}, \ i=1,\dots,n$$

which is a system of linear equations, integrable with an exact algorithm, when all the variables under each $\mathcal{P}_{pi}$ are fixed except q_{pi} , so that the integration is made stepwise.

References

1. Hughes, P.C., Space structure vibration modes: how many exist? Which ones are important?, Workshop on Space Telerobotics, NASA–JPL, Pasedema, 1987, pp.13–47.

2. Hale, A.L., Lisowski, R.J., Dahl, W.E., Optimal simultaneous structural and control design of maneuvering flexible spacecraft, J. Guidance, Vol. 8, 1985, pp.86–93.

3. Ho, J.Y.L., Gluck, R., Inductive methods for generating the dynamic equations of motion for multibodies flexible systems, Pt.2, Perturbation approach, Synthesis of Vibrating Systems, ASME, N.Y. 1971.

4. Meirovitch, L., Quinn, R.D., Equations of motion for maneuvering flexible spacecraft, J. Guidance, Vol. 10, 1987, pp. 453–464.

5. Balas, M.T., Active control of flexible systems, Proc. Symp. Dynamics & Control of Large Flexible Spacecraft, Blacksburg, VA, 1977.

6. Skowronski, J.M. Control Dynamic of Robotic Manipulators, Academic Press, 1986.

7. Skowronski, J.M. Model Reference adaptive control under uncertainty of nonlinear flexible manipulators, Proc. 1986 AIAA Guidance Navigation and Control Conf., Paper 86–1976CP.

8. Skowronski, J.M. Nonlinear model reference adaptive control, J. Aust. Math. Soc. Ser. Appl. Math., Vol. 28, 1986, 28–36.

9. Skowronski, J.M. Parameter and state identification in nonlinearizable uncertain systems, Int. J. Nonlinear Mechanics, Vol. 19, 1984, 421–429.

10. Landau, I.D. Adaptive Control: The Model Reference Approach, M. Dekker, N.Y., 1979.

11. Landau, I.D., Courtiol, B. Adaptive model following systems for flight control and simulation, J. Aircraft, Vol. 9, pp. 688–674, 1972.

12. Farrington, F.D., Goodson, R.E. Simulated flight tests of a digitally autopiloted STOL–craft on a curved approach, J. Dyn. Syst. Meas. Control, pp. 55–63, March, 1973.

13. Kaufmaan, H., Berry, P. Adaptive flight control using optimal linear regulator techniques, Automatica, Vol. 12, pp. 565–576, 1976.

14. Alag, G., Kaufman, H. An implementable digital adaptive flight controller designed using stabilized single stage algorithms, IEEE Trans. Autom. Control, Vol. AC–22, pp. 780–788, October, 1977.

15. Stein, G., Hartmann, G.L., Hendrick, R.C. Adaptive control laws for F–9 flight test, IEEE Trans. Autom. Control, Vol. AC–22, pp. 758–780, October, 1977.

16. Ridgely, D.B., Banda, S.S., D'Azzo, J.J. Decoupling of high gain multivariable tracking systems, J. Guidance, Vol. 8, pp. 44–49, 1982.

17. Kanai, K., Uchikado, S., Nikiforuk, P., Hori, N. Application of a New Multivariable Model – Following Method to decoupled flight control, j. Guidance, Vol. 8, pp. 637–643, Oct., 1985.

18. Vassar, R.H., Sherwood, R.B. Formation keeping for a pair of satellites in a circular orbit, J. Guidance, Vol. 8, pp. 235–242, April, 1985.

19. Truckenbrodt, A. Effects of Elasticity on the Performance of Industrial Robots, Proc. 2nd IASTED Danos International Symposium on Robotics, Switzerland, 1982, pp. 52–56.

20. Skowronski, J.M., Adaptive nonlinear model following, IEEE Trans. Vol. CAS35, 1988, Sept.

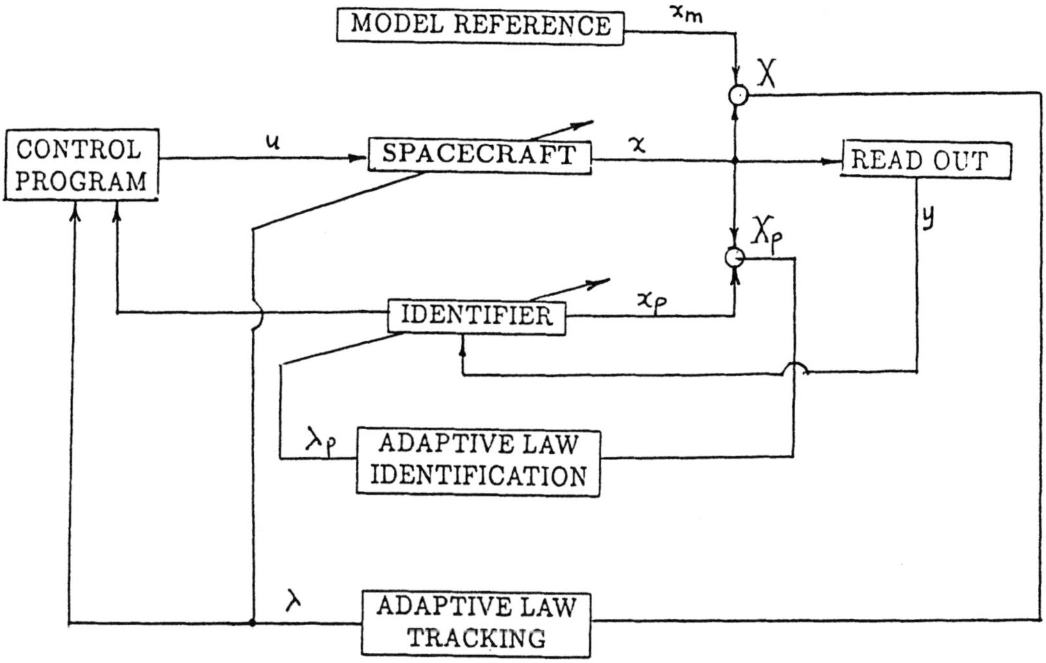

AN INVESTIGATION ON ANTISEISMIC BASE-ISOLATION AND CONTROL

Argiris Gerasimos Soldatos

Vibration and Sound Research Institute
1625 The Alameda, suite 416
San Jose, CA 95126
and
Department of Mechanical Engineering
University of California, Berkeley
Berkeley, CA 94720

ABSTRACT: *Earthquake protection of buildings is achieved by means of isolation and control at the base of the structure. The desired control objective is to keep the building arbitrarily close to its initial, undisturbed configuration. The controllers developed require a control force on the first floor.*

INTRODUCTION

In active control, one of the main research avenues relies upon statistical considerations (Astrom 1970). Loosely speaking, the uncertain quantities involved are assumed to behave according to some probabilistic patterns and the desired behavior is given by a similar description. Another stream of research uses a deterministic approach, (Leitmann 1980) where the system description is not in terms of random variables. Accordingly, the models used stipulate that the conditions under which a process takes place determine its outcome.

Typical system representations used for control purposes contain uncertain elements. These result from imprecise knowledge of inputs, unknown parameters pertinent to the system description and unmodelled dynamics. These "disturbances", internal or external to the system, may adversely affect its performance. Hence, the task is to control the system so as to achieve the desired objective in the presence of these uncertainties.

We use a deterministic approach for base-isolated buildings whose representation involves an "uncertain" earthquake input. Base isolation can be efficacious in greatly reducing disturbances transmitted from the ground to the base floor. Indeed, essentially complete decoupling can be achieved in principle, say, by supporting the structure on ball bearings. Clearly, such a scheme is not practical, since even small disturbances would result in motion of the structure; it would simply slide off the foundation. This, then, is where active control enters the picture, for it can be designed to employ information about the motion of the structure to activate forces to counteract this motion. The only information needed about the earthquake are estimates of the maximum amplitude of the ground velocity and displacement. One may use, for example, the maximum values observed during the worst earthquake on record.

Control Objective

The physical system under consideration is a base-isolated, N-story building viewed as an interconnection of physical components. For the present analysis and design, the physical system is idealized by a linear model. In particular, it is considered as a "lumped parameter" model, i.e., an assemblage of masses, each representing a combined floor mass, and linear springs and dampers for the inter-story connections. Motion is assumed to take place in one horizontal direction. Torsional and external damping effects, although present generally, are not considered in this study. This model is the system under investigation.

Let $y_0(t)$ be the displacement of the ground and $y_i(t)$ be the displacement of the ith floor at time t relative to an inertial frame of reference; see Fig. 1. Let m_i be the mass of the ith floor, c_{i-1} be the damping coefficient and k_{i-1} be the spring constant in the connection of the ith floor to the floor, or ground, below it. Also, let a control force, $u_1(t)$, be applied to the base (first) floor. Subsequently, the time argument, t, is omitted when no confusion is likely to arise.

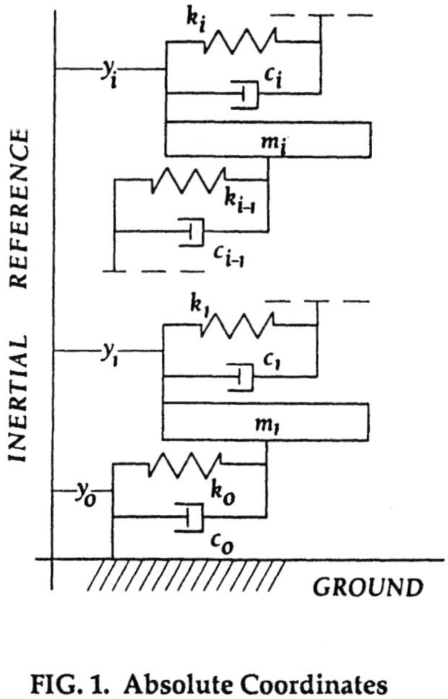

FIG. 1. Absolute Coordinates

Then,

$$m_1\ddot{y}_1 = -c_0(\dot{y}_1-\dot{y}_0)-k_0(y_1-y_0)+c_1(\dot{y}_2-\dot{y}_1)+k_1(y_2-y_1)+u_1,$$

$$\cdot \quad \cdot \quad \cdot \quad \cdot \quad \cdot \quad \cdot$$

$$m_i\ddot{y}_i = -c_{i-1}(\dot{y}_i-\dot{y}_{i-1})-k_{i-1}(y_i-y_{i-1})+c_i(\dot{y}_{i+1}-\dot{y}_i)+k_i(y_{i+1}-y_i), \tag{1}$$

$$\cdot \quad \cdot \quad \cdot \quad \cdot \quad \cdot \quad \cdot$$

$$m_N\ddot{y}_N = -c_{N-1}(\dot{y}_N-\dot{y}_{N-1})-k_{N-1}(y_N-y_{N-1}),$$

$$y_i(t_0) = y_i^0, \quad \dot{y}_i(t_0) = \dot{y}_i^0,$$

where t_0 is some initial time.

We consider our system divided into two subsystems, I and II. Subsystem I contains the first floor, and II the remaining floors of the building. In state space formalism we have,

subsystem I:

$$\dot{z}_1 = A_1 z_1 + B_1 u_1 + D_1 z_0 + C_1 z_2, \tag{2a}$$

$$z_1(t_0) = z_1^0, \tag{2b}$$

and subsystem II:

$$\dot{z}_2 = A_2 z_2 + C_2 z_1, \tag{3a}$$

$$z_2(t_0) = z_2^0, \tag{3b}$$

where

$$z_1 = (\dot{y}_1, y_1)^T, \quad z_0 = (\dot{y}_0, y_0)^T,$$

$$z_2 = (\dot{y}_2,\ldots,\dot{y}_N, y_2,\ldots,y_N)^T,$$

$$A_1 = \begin{bmatrix} (-c_0-c_1)/m_1 & (-k_0-k_1)/m_1 \\ & \\ 1 & 0 \end{bmatrix},$$

$$B_1 = [\quad 1/m_1 \quad 0 \quad]^T,$$

$$D_1 = \begin{bmatrix} c_0/m_1 & k_0/m_1 \\ 0 & 0 \end{bmatrix},$$

$$C_1 = \left[\begin{array}{c|c|c|c} c_1/m_1 & & k_1/m_1 & \\ & O_{2x(N-2)} & & O_{2x(N-2)} \\ 0 & & 0 & \end{array}\right],$$

$$A_2 = \left[\begin{array}{cccc|cccc} (-c_1-c_2)/m_2 & c_2/m_2 & & & (-k_1-k_2)/m_2 & k_2/m_2 & & \\ & c_{i-1}/m_i & (-c_{i-1}-c_i)/m_i & c_i/m_i & & k_{i-1}/m_i & (-k_{i-1}-k_i)/m_i & k_i/m_i \\ & & c_{N-1}/m_N & -c_{N-1}/m_N & & & k_{N-1}/m_N & -k_{N-1}/m_N \\ \hline & I_{(N-1)\,x\,(N-1)} & & & & O_{(N-1)\,x\,(N-1)} & & \end{array}\right],$$

$$C_2 = \left[\begin{array}{c|c} c_1/m_2 & \\ & O_{2x(2N-3)} \\ k_1/m_2 & \end{array}\right]^T.$$

The state of each of the two subsystems affects the dynamic behavior of the other. In particular, the inputs to I are induced from the ground motion and the state of II. The only input to II comes from the state of the first floor.

It is desired to use a control input, u_1, such that the state, $(z_1, z_2)^T$, of the structure is restricted to a sufficiently small neighborhood of its initial (prior the earthquake) value.

PROBLEM STATEMENT

The desired performance of the controlled system is best described by using the concepts of uniform boundedness and uniform ultimate boundedness.

Let $z(t) = (z_1(t), z_2(t))^T$.

Definition 1. A solution $z(\cdot): [t_0, t_1) \longrightarrow R^{2N}, z(t_0) = z^0$ of Eqs. 2a, 2b, 3a and 3b is uniformly bounded, if there exists a $\gamma(z^0) > 0$, such that $\| z(t) \| \leq \gamma(z^0)$ for all $t \in [t_0, t_1)$.

Definition 2. A solution $z(\cdot): [t_0, \infty) \longrightarrow R^{2N}, z(t_0) = z^0$ of Eqs. 2a, 2b, 3a and 3b is uniformly ultimately bounded with respect to a set S if there exists a non-negative bounded $T(z^0, S)$ such that $z(t) \in S$ for all $t \geq t_0 + T(z^0, S)$.

If our system complies with Definition 1 and Definition 2 it is called practically stable. Hence, the task is to design a control, u_1, which assures that, no matter what z_0 is, every solution of Eqs. 2a, 2b, 3a, and 3b is uniformly bounded and uniformly ultimately bounded with respect to a given set S.

Previous activity has resulted in a control scheme which makes use of the entire state of the structure. Here, we use a control input which depends on z_1 alone, i.e. the state of I. This may be considered to be a partial state feedback or an output feedback since only z_1 needs to be measured during an earthquake.

CONTROLLER DESIGN

The uncertain input in Eq. 2a, z_0, is due to the ground motion. We assume that $z_0(t) \in Z$, a known compact set. This assumption is reasonable, since maximum values of ground displacement and velocity are known for the worst earthquake on record.

Theorem. Given a system described by Eqs. 2a, 2b, 3a, and 3b, if $K \in R^{1\times2}$ is such that

a. $A_1 + B_1 K$ is stable,

b. $(A_1 + B_1 K)^T P_1 + P_1 (A_1 + B_1 K) = -Q_1$,

and

c. $\lambda_{min}(Q_1)\,\lambda_{min}(Q_2) - (\| P_1 C_1 \| + \|P_2 C_2 \|)^2 > 0,$

where,

Q_1, P_1, Q_2, P_2 are positive definite matrices of appropriate dimensions, with

$$A_2^T P_2 + P_2 A_2 = -Q_2,$$

and λ_{min} is the minimum eigenvalue, then the system is practically stabilizable by a control input, u_1, which depends only on z_1.

Proof. The pair (A_1, B_1) is controllable, i.e. there exists K such that

$$\bar{A}_1 = A_1 + B_1 K$$

is stable. Also,

$$D_1 = B_1 E_1,$$

where

$$E_1 = (c_0 \quad k_0).$$

Take

$$u_1 = Kz_1 + p_1(z_1), \tag{4}$$

such that for $\varepsilon > 0$,

$$p_1(z_1) = \begin{cases} \dfrac{-B_1^T P_1 z_1}{|B_1^T P_1 z_1|}\rho_1, & \text{if } |B_1^T P_1 z_1| \geq \varepsilon, \\[2ex] \dfrac{-B_1^T P_1 z_1}{\varepsilon}\rho_1, & \text{if } |B_1^T P_1 z_1| < \varepsilon, \end{cases}$$

where P_1 solves

$$\bar{A}_1^T P_1 + P_1 \bar{A}_1 = -Q_1$$

for Q_1 positive definite and

$$\rho_1 = \max_{z_0} | E_1 z_0 | \geq e_1 ,$$

where

$$e_1 = E_1 z_0 .$$

Then, Eq. 2a becomes,

$$\dot{z}_1 = \bar{A}_1 z_1 + B_1 p_1 (z_1) + B_1 e_1 + C_1 z_2 .$$

Let P_2 be the unique positive definite solution of

$$A_2^T P_2 + P_2 A_2 = - Q_2 ,$$

where Q_2 is positive definite and A_2 is stable. Consider the Lyapunov function candidates for the subsystems I and II,

$$V_1 = z_1^T P_1 z_1 ,$$
$$V_2 = z_2^T P_2 z_2 .$$

Upon differentiation of these functions along the solutions of 2a, 2b, 3a, and 3b, one obtains

$$\dot{V}_1 = - z_1^T Q_1 z_1 + 2 B_1^T P_1 z_1 (p_1 (z_1) + e_1) + 2 z_1^T P_1 C_1 z_2$$

$$\leq - \lambda_{min} (Q_1) \| z_1 \|^2 + (\varepsilon/2) \rho_1 + 2 \| z_1 \| \| z_2 \| \| P_1 C_1 \| ,$$

and

$$\dot{V}_2 = - z_2^T Q_2 z_2 + 2 z_2^T P_2 C_2 z_1$$

$$\leq - \lambda_{min} (Q_2) \| z_2 \|^2 + 2 \| z_1 \| \| z_2 \| \| P_2 C_2 \| .$$

Take

$$W = V_1 + V_2$$

as a Lyapunov function candidate for the entire system. Then,

$$\dot{W} = \dot{V}_1 + \dot{V}_2$$

$$\leq - \lambda_{min} (Q_1) \| z_1 \|^2 - \lambda_{min} (Q_2) \| z_2 \|^2 + (\varepsilon/2) \rho_1 +$$

$$+ 2 (\| P_1 C_1 \| + \| P_2 C_2 \|) \| z_1 \| \| z_2 \| ,$$

i.e.

$$\dot{W} \leqslant -\langle \psi, \Pi \psi \rangle + (\varepsilon/2)\, \rho_1 ,$$

where

$$\psi = (\, \| z_1 \| \quad \| z_2 \| \,)^T$$

and

$$\Pi = \begin{bmatrix} \lambda_{min}(Q_1) & -(\| P_1 C_1 \| + \| P_2 C_2 \|) \\ -(\| P_1 C_1 \| + \| P_2 C_2 \|) & \lambda_{min}(Q_2) \end{bmatrix}.$$

Consequently, if

$$\lambda_{min}(Q_1)\, \lambda_{min}(Q_2) - (\| P_1 C_1 \| + \| P_2 C_2 \|)^2 > 0 ,$$

Π is positive definite and $\dot{W}$ is negative definite outside an arbitrarily small spherical region (by choosing ε sufficiently small) containing the origin. Hence, the proposed control input renders the system practically stable (Kelly, Leitmann, and Soldatos 1987b). □

The control input, Eq. 4, is composed of a linear and a nonlinear part. The conditions of the theorem are sufficient for the induction of appropriate "fast dynamics" on the first floor.

Lemma. Given a system described by Eqs. 2a, 2b, 3a, and 3b, there exists $\mathbf{K} \in R^{1\times2}$ such that the control input of Eq. 4 yields the system practically stable.

The proof of the Lemma (Soldatos 1991) suggests a constructive way of deriving the linear state feedback gain matrix **K**.

Open Loop Control

The fact that the ground acceleration is directly measurable has not been used yet. If the ground velocity and displacement are available concurrently with the acceleration, i.e. errors in data processing, including integration, are neglected, then the "uncertain" inputs of Eqs. 1 or 2a are known. Assuming that this is the case and that c_0 and k_0 are also known, take

$$u_1 = -c_0\dot{y}_0 - k_0 y_0 \qquad (6)$$

instead of Eq. 4. Then, the whole structure becomes an unforced stable dynamical system. The input given by Eq. 6 does not require knowledge of the state and is termed open loop or feedforward.

In contradistinction with the feedback scheme, where the corrective action starts only after the state has been affected, here one needs to know c_0 and k_0 exactly for the cancellation to work. Also, the common way of getting velocity and displacement from acceleration through integration is not very accurate in practice. This difficulty is especially true for signals of relatively high frequency content like $\ddot{y}_0$ needed here, but not as severe for $\ddot{y}_1$ used earlier. These points, coupled with the errors due to measuring and the incurred delays render the open loop control unattractive. However, had the feedforward scheme been easy to implement, its use would have yielded the system asymptotically stable. It should be emphasized that since the feedforward control is open loop, its success depends on the constancy and the accuracy of the parameter values involved in the cancellation. Any drift in these values will result in imperfect compensation.

Fortunately, one may combine the merits of both techniques. In other words, it is possible to approximately compensate for the known disturbances before they "materialize" and at the same time take care of the possible errors. Hence, the control inputs given by Eqs. 4 and 6 may be used simultaneously.

The influence of the main disturbance source, i.e. the ground motion, can be greatly reduced by the open loop compensation and the base isolation without requiring a considerably high gain. The effect of other "disturbance" sources such as inaccuracies of the open loop, moderate delays, etc. can be taken care of by the closed loop scheme. Hence, the undesired actions from all causes can be reduced without requiring a relatively large loop gain.

EXAMPLE

A two story experimental structure for housing sensitive equipment is considered. The representation used is Eqs. 1 with N = 2. The subsystems are defined by

$$\mathbf{A}_1 = \begin{bmatrix} -0.51 & -5.1 \\ 1 & 0 \end{bmatrix},$$

$$\mathbf{B}_1 = \begin{bmatrix} 5 \times 10^{-5} & 0 \end{bmatrix}^T,$$

$$D_1 = \begin{bmatrix} 0.01 & 0.1 \\ 0 & 0 \end{bmatrix},$$

$$C_1 = \begin{bmatrix} 0.5 & 5 \\ 0 & 0 \end{bmatrix},$$

$$A_2 = \begin{bmatrix} -1 & -10 \\ 1 & 0 \end{bmatrix},$$

$$C_2 = \begin{bmatrix} 1 & 10 \\ 0 & 0 \end{bmatrix}.$$

The earthquake record used is one component of the 1940 El Centro quake. Plots of the digitized ground velocity and displacement are given in Fig. 2. From these data, the assumed maximum values of ground velocity and displacement are $\dot{y}_0^{max} = 0.35 m/s$ and $y_0^{max} = 0.11 m$, determining the maximum value of $p_1(z_1)$ in Eq. 4. Q_2 is taken to be the identity, $Q_1 = \mathrm{diag}(161, 161)$ and the parameter $\varepsilon = 0.01$.

Fig. 3 shows the velocity records for the first floor before and after the application of the control input under the assumed earthquake excitation. Fig. 4 does the same for the second floor.

Fig. 5 has the displacement time histories of the first floor with and without control. Those of the second floor follow in Fig. 6.

The control force record is shown in Fig. 7.

Conclusions

An output feedback control scheme is proposed for base isolated structures under earthquake excitation. The only information required during the operation of the controller is the current absolute velocity and displacement of the first floor. These may be obtained from the readings of an accelerometer attached to the first floor. Thus we are led to a one-reading-one-force scheme for the control objective set forth. A combination of open and closed loop schemes can be beneficial.

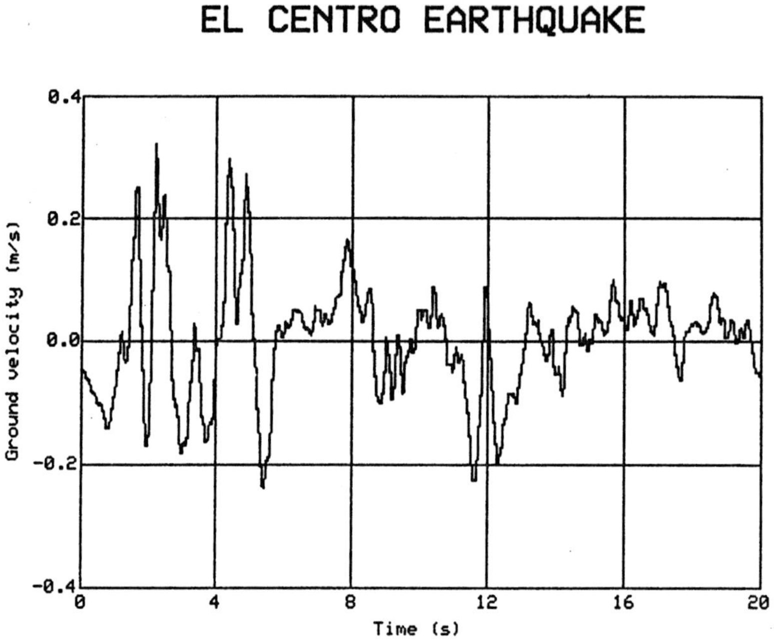

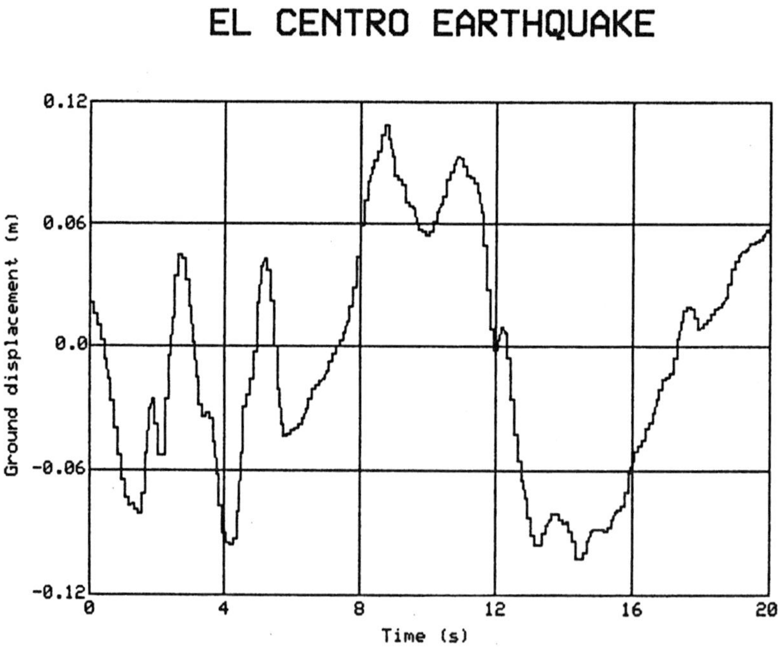

FIG. 2. El Centro Earthquake Inputs

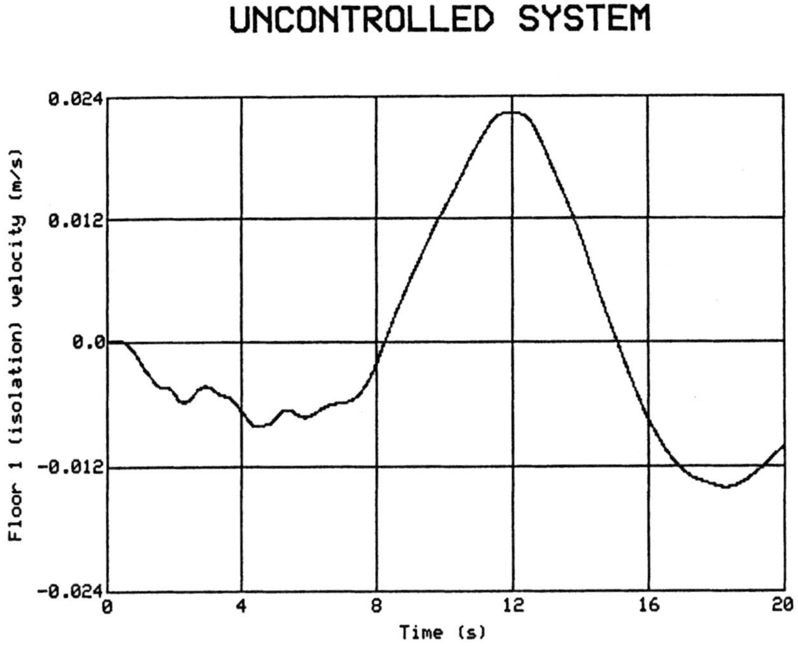

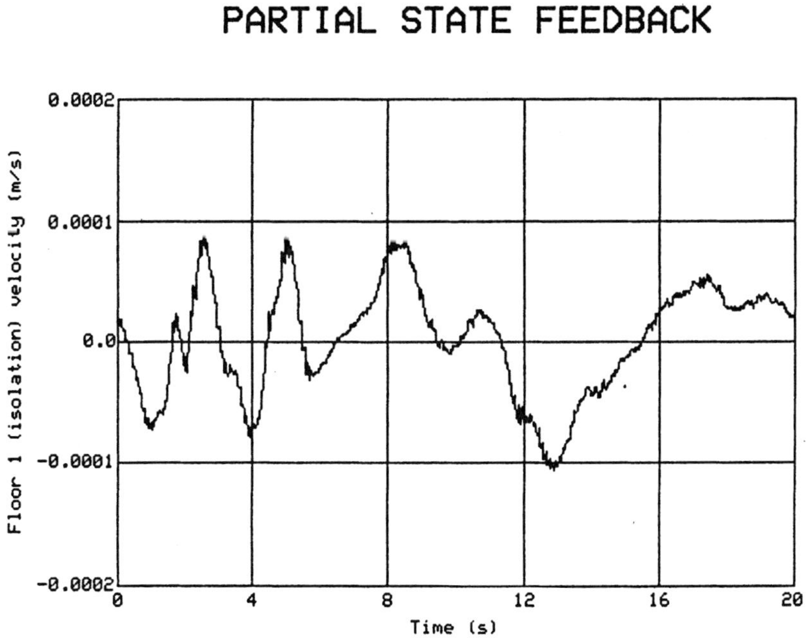

FIG. 3. First Floor Velocity Records

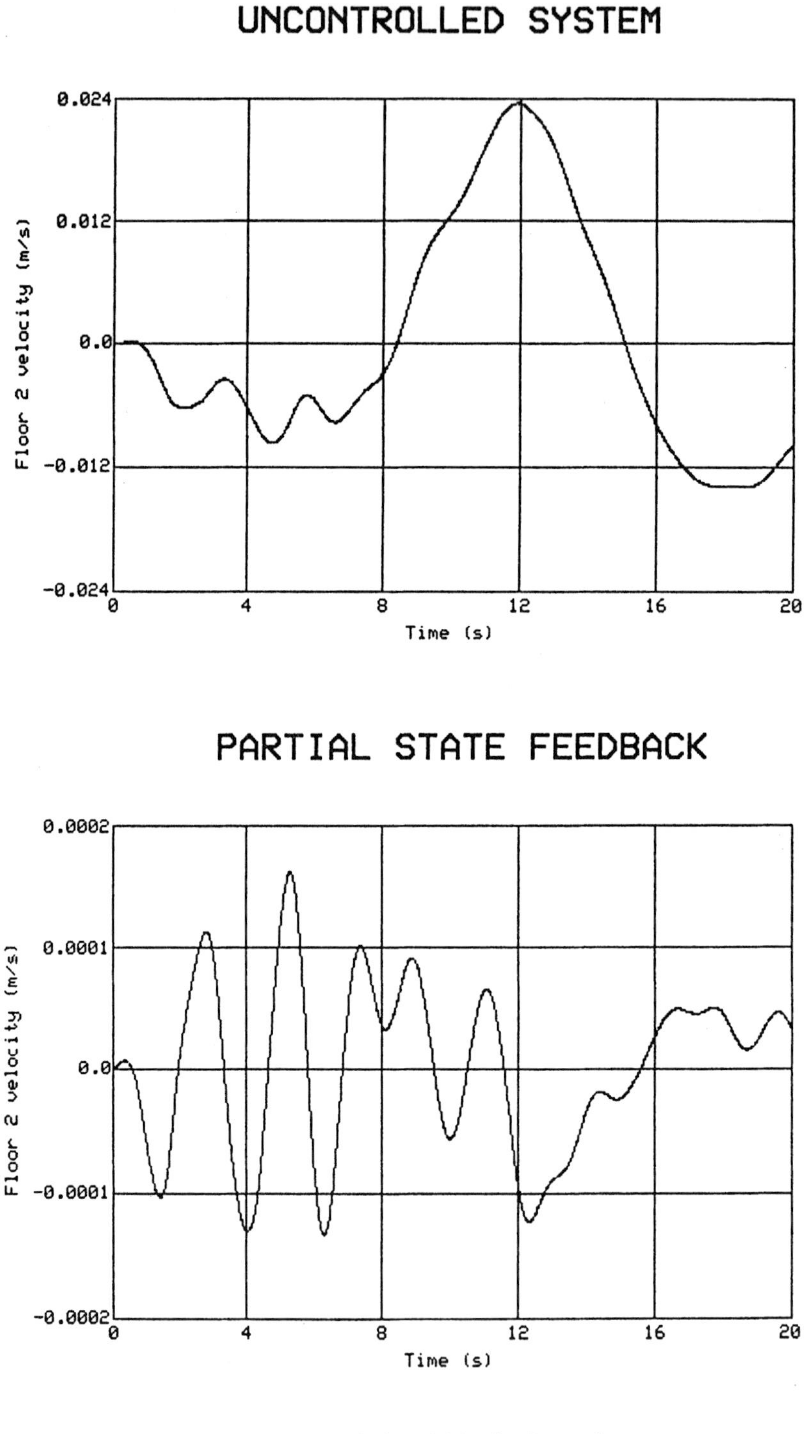

FIG. 4. Second Floor Velocity Records

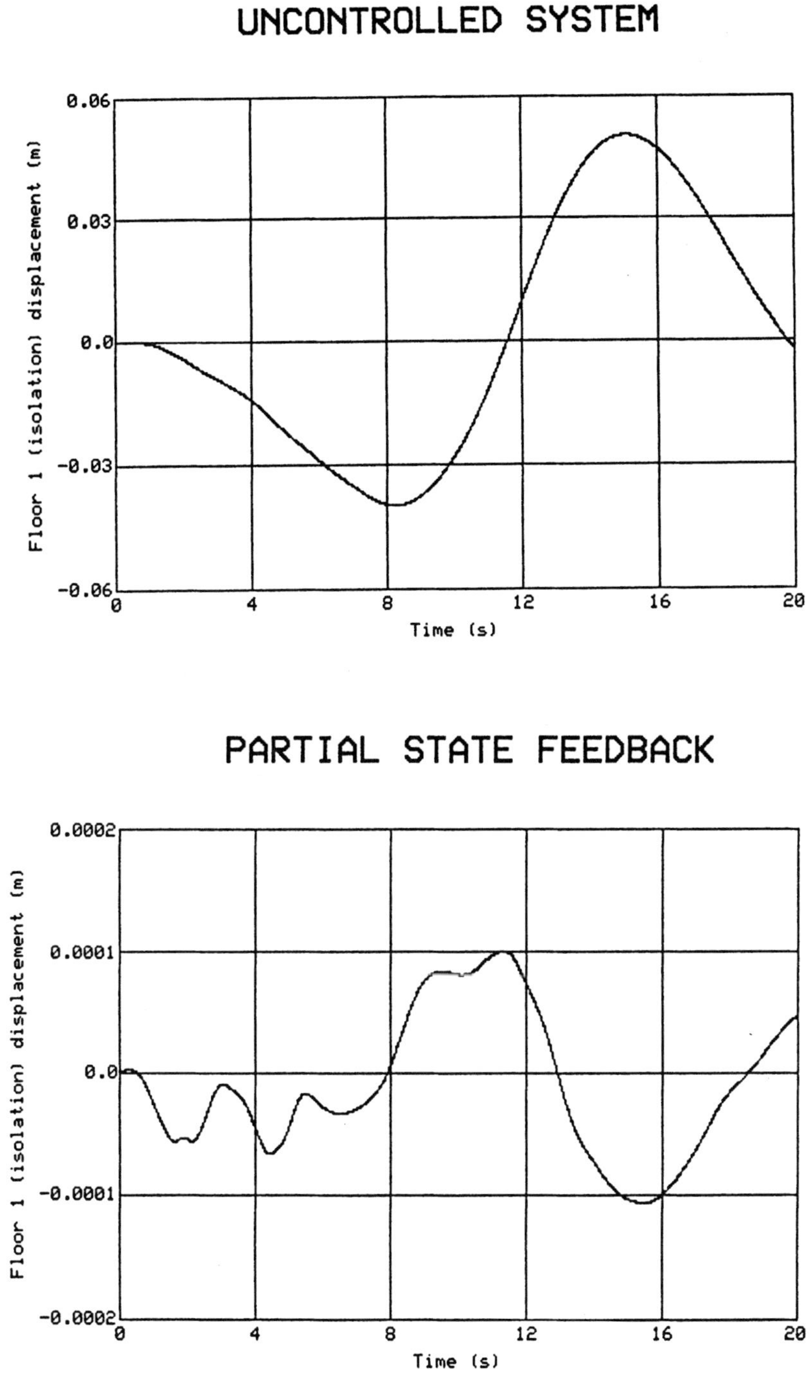

FIG. 5. First Floor Displacement Records

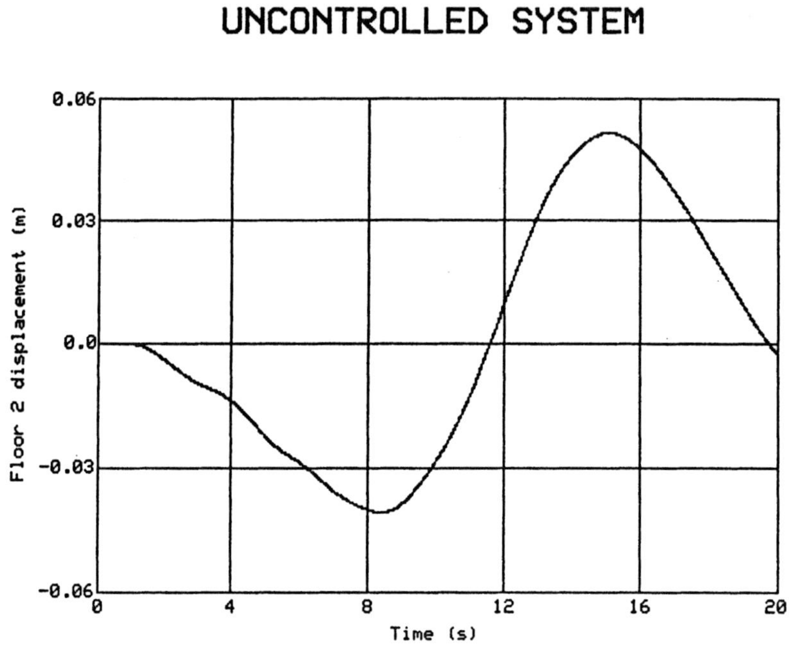

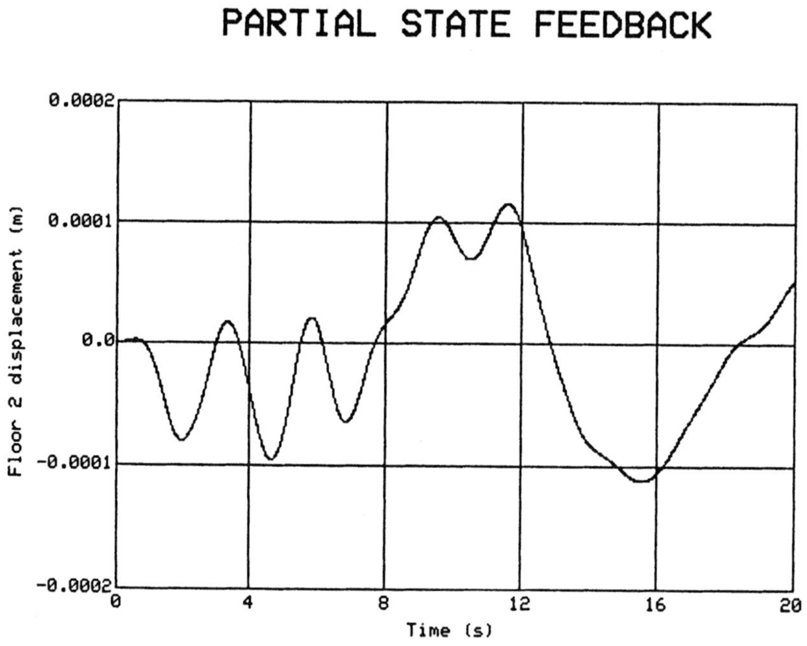

FIG. 6. Second Floor Displacement Records

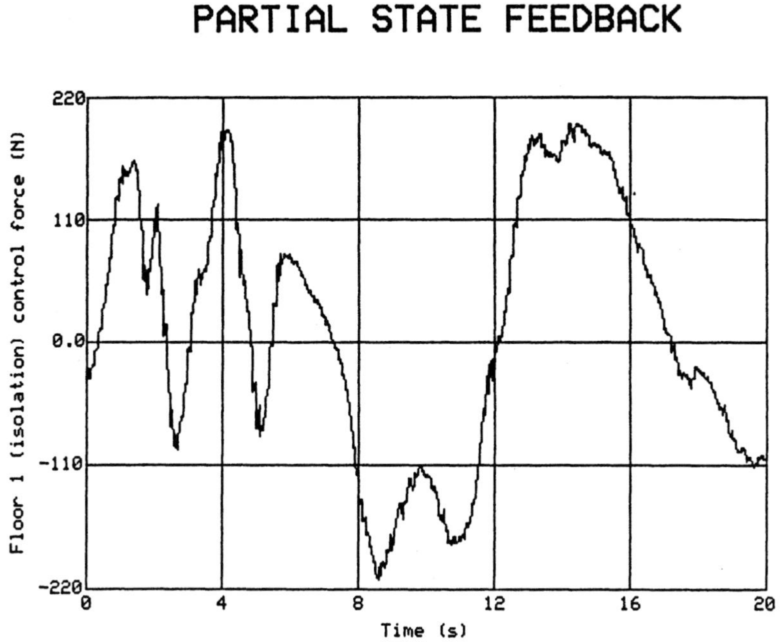

FIG. 7. Control Force Record

REFERENCES

Astrom, K. J. (1970). Introduction to Stochastic Control Theory. Academic Press, New York.

Kelly, J.M. (1986). "Aseismic Base Isolation: Review and Bibliography." International Journal of Soil Dynamics and Earthquake Engineering, 5(3), 202-216.

Kelly, J.M., Leitmann, G., and Soldatos, A.G. (1987a). "Seismic Protection of Structures Using Base Isolation and Active Control." American Control Conference, Minneapolis, Minnesota.

Kelly, J.M., Leitmann, G., and Soldatos, A.G. (1987b) "Robust Control of Base Isolated Structures Under Earthquake Excitation." Journal of Optimization Theory and Applications, 53(2), 159-180.

Leitmann, G. (1978). "Guaranteed Ultimate Boundedness for a Class of Uncertain Linear Dynamical Systems." IEEE Transactions on Automatic Control, AC-23, 1109.

Leitmann, G. (1980). "Deterministic Control of Uncertain Systems." Acta Astronautica, 7, 1457.

Martin, C.R, and Soong, T.T. (1976). "Modal Control of Multistory Structures." J. Engrg. Mech. Div., ASCE, 102 (4), 613-623.

Soldatos, A.G. (1984). "Analysis of Earthquake-Induced Oscillations on Multistory Buildings," MS report submitted to the University of California, Berkeley.

Soldatos, A.G. (1991). "Output Feedback Control of Seismically Excited Structures." under review.

Takahashi, Y., Rabins, M.J., and Auslander, D.M. (1970). Control and Dynamic Systems. Addison Wesley, Reading, Mass.

COLLISION AVOIDANCE
OF TWO PR MANIPULATORS

by

R.J. STONIER

Department of Mathematics & Computing
University College of Central Queensland
Rockhampton Queensland Australia 4702

1. INTRODUCTION

In this paper Liapunov techniques for collision-avoidance introduced in [1] are extended to the collision-avoidance problem of two PR manipulators. Motion of the manipulators occurs in the horizontal plane with a defined common workspace.

2. ROBOT DYNAMICS

We consider two PR manipulators, each described as shown in Figure 2.1. The arm has length ℓ and its mass per unit length is a constant m_A/ℓ. The length of the prismatic radial link changes when it slides through the hub. A force opposes the motion of this link. It is modelled as a spring with spring constant k_s, which imposes a zero force at $r = 2\ell/3$.

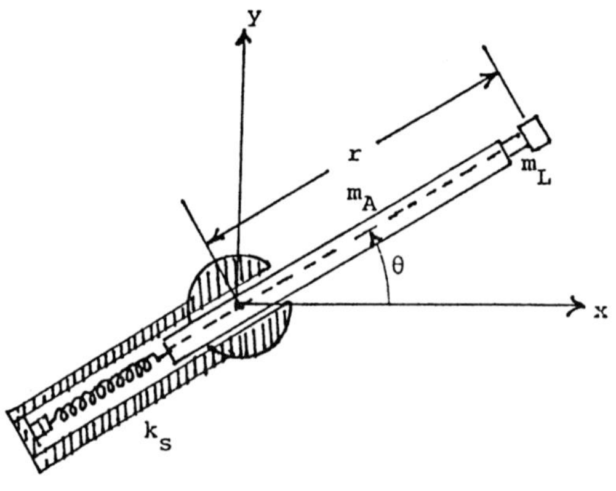

FIGURE 2.1 Cylindrical PR Manipulator

Let F_r be the translational force force acting in the radial direction, τ_θ be the torque causing rotation of the arm, and terms $B_r(\dot{r})$ and $B_\theta(\dot{\theta})$ be frictional forces opposing radial and transverse motion.

Using Lagrangian analysis, see [2], we obtain

$$\ddot{r} = \frac{\bar{m}_\ell(r)(r/2)\dot{\theta}^2}{(m_A + m_L)} - \frac{k_s(r - 2\ell/3)}{(m_A + m_L)} + \frac{(F_r - B_r(\dot{r}))}{(m_A + m_L)} \tag{2.1}$$

$$\ddot{\theta} = \frac{1}{I_t(r)}\left[-[\frac{3m_A}{4\ell}(r^2 - (\ell - r)^2) + 2m_L r]\,\dot{r}\dot{\theta} + \tau_\theta - B_\theta(\dot{\theta})\right] \tag{2.2}$$

where

$\bar{m}_A = m_A + m_L =$ total mass ,

$I_\theta =$ the effective moment of inertia ,

$I_t(r) = [r^3 + (\ell - r)^3]\,m_A/(4\ell) + m_L r^2 + I_\theta$, and

$\bar{m}_\ell(r) = 3[m_A r/\ell - m_A(\ell - r)^2/(r\ell)]/4 + 2m_L$.

For numerical simulation the following numerical values are assumed:

$m_A = 1.0$ kg , $m_L = 0.125$ kg , $I_\theta = 0.1$ kgm^2, $\ell = 1.5$ m , $k_s = 100$ N/m ,

and the coefficients of viscous friction

$B_r = 5.0$ Ns/m , $B_\theta = 0.001$ Nms .

The two PR robots we assumed identical and their configuration in the plane is shown in Figure 2.2. The dynamic equations for the two robots can be written in state variable form setting

$x_1 = \theta_1$, $x_2 = \dot{\theta}_1$, $x_3 = r_1$, $x_4 = \dot{r}_1$,

$y_1 = \theta_2$, $y_2 = \dot{\theta}_2$, $y_3 = r_2$ and $y_4 = \dot{r}_2$.

With appropriate identification of terms we obtain

Robot Arm 1	*Robot Arm 2*	
$\dot{x}_1 = x_2$	$\dot{y}_1 = y_2$	
$\dot{x}_2 = (u_1^1 + f_1^1)/I_1^1$	$\dot{y}_2 = (u_1^2 + f_1^2)/I_1^2$	
$\dot{x}_3 = x_4$	$\dot{y}_3 = y_4$	
$\dot{x}_4 = (u_2^1 + f_2^1)/I_2^1$	$\dot{y}_4 = (u_2^2 + f_2^2)/I_2^2$	(2.3)

where, for example,

$$f_1^1 = -\left[\frac{3m_A}{4\ell}(x_3^2 - (\ell - x_3)^2) + 2m_L x_3\right] x_4 x_2 - B_\theta(x_2) ,$$

$$f_2^1 = \bar{m}_\ell(x_3)(x_3/2)x_2^2 - k_s(x_3 - 2\ell/3) - B_r(x_4) ,$$

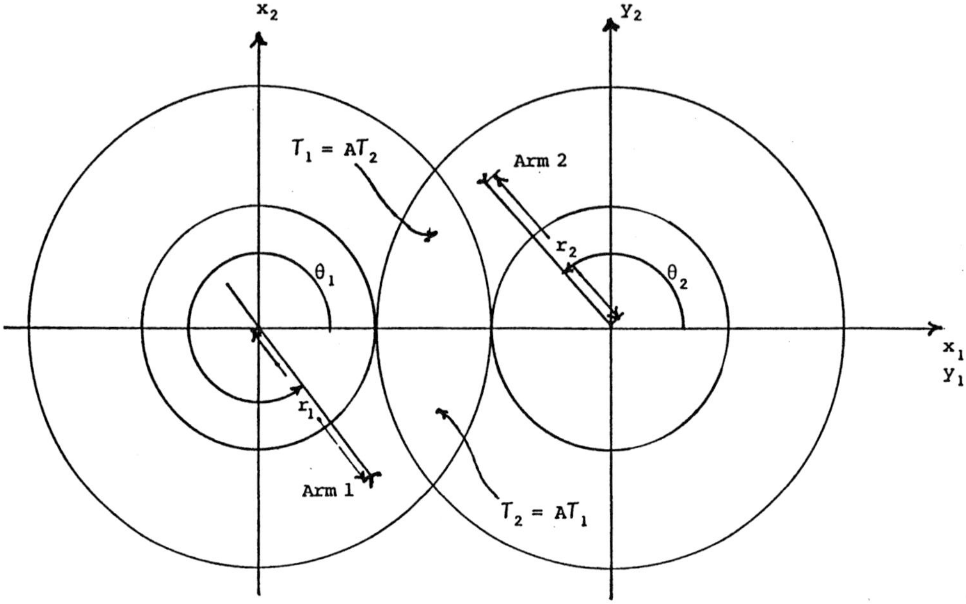

FIGURE 2.2 Two Robot Arm Configuration in the Plane

$$u_1^1 = \tau_\theta , \quad u_2^1 = F_r ,$$

$$I_1^1 = I_t(x_3) \quad \text{and} \quad I_2^1 = m_A + m_L .$$

The equations (2.3) written in vector format are

$$\dot{\underset{\sim}{x}} = \underset{\sim}{F}_1(\underset{\sim}{x},\underset{\sim}{u}^1) \quad \text{and} \quad \dot{\underset{\sim}{y}} = \underset{\sim}{F}_2(\underset{\sim}{y},\underset{\sim}{u}^2) \tag{2.4}$$

where clearly $\underset{\sim}{x} = [x_1 \; x_2 \; x_3 \; x_4]^T$, $\underset{\sim}{u}^1 = [u_1^1 \; u_2^1]^T$, etc.

Constraints:

The constraints on the physical motion for each robot arm are the same. They are described for robot arm 1 below:

$$-1.5 < \dot{\theta}_1 = x_2 < 1.5 \quad \text{(rad/sec)}$$

$$0.5 < r = x_3 < 1.5 \quad \text{(m)}$$

$$-1.5 < \dot{r} = x_4 < 1.5 \quad \text{(m/sec)}$$

$$\max |\dot{x}_2| = \max |\ddot{\theta}_1| = 1$$

$$\max |\dot{x}_4| = \max |\ddot{r}| = 1 .$$

3. COLLISION-AVOIDANCE FORMULATION

Let us consider for the moment the objective of controlled movement of robot arm 1 to a desired target in the presence of obstacles and constraint inequalities. Construction of targets, antitargets and Liapunov functions parallels the construction given in [1].

We wish to position the gripper of robot 1 in target T_1 which is a circle, with center $[p_1c_1, p_1c_2]^T$ and radius rp_1, in the $x_1 - x_3$ plane

$$T_1 = \{\underset{\sim}{x} : (x_3 \cos x_1 - p_1c_1)^2 + (x_3 \sin x_1 - p_1c_2)^2 \le rp_1^2\}$$
$$= \{\underset{\sim}{x} : x_3^2 - 2x_3(p_1c_1 \cos x_1 + p_1c_2 \sin x_1) + p_1c_1^2 + p_1c_2^2 \le rp_1^2\} .$$

The constraint inequalities define six antitargets which must be 'avoided' by robot arm 1 in its motion.

$$AT_1^1 = \{\underset{\sim}{x} : x_2 + 1.5 \le 0\} ,$$
$$AT_2^1 = \{\underset{\sim}{x} : x_2 - 1.5 \ge 0\} ,$$
$$AT_3^1 = \{\underset{\sim}{x} : x_3 - 0.5 \le 0\} ,$$
$$AT_4^1 = \{\underset{\sim}{x} : x_3 - 1.5 \ge 0\} ,$$
$$AT_5^1 = \{\underset{\sim}{x} : x_4 + 1.5 \le 0\} ,$$
$$AT_6^1 = \{\underset{\sim}{x} : x_4 - 1.5 \ge 0\} .$$

The arm is also required to avoid target T_2 of robot 2 (assumed to be set also within the common workspace for computer simulation), and also the arm of robot 2 itself. Considering the former, we define the antitarget:

$$AT_7^1 = \{\underset{\sim}{x} : (x_3 \cos x_1 - p_2c_1)^2 + (x_3 \sin x_1 - p_2c_2)^2 \le rp_2^2\}$$
$$= \{\underset{\sim}{x} : x_3^2 - 2x_3(p_2c_1 \cos x_1 + p_2c_2 \sin x_1) + p_2c_1^2 + p_2c_2^2 \le rp_2^2\}$$
$$= T_2 .$$

(It is assumed that target T_2 for robot 2 is also circular, center (p_2c_1, p_2c_2) and radius rp_2. Coordinates of the center of T_2 are given relative to the $x_1 - x_3$ plane.)

Consider at time t the state $\underset{\sim}{y}(t)$ of robot 2. For a given state $\underset{\sim}{x}$ of robot 1 define a function $D(\underset{\sim}{x},\underset{\sim}{y}(t))$ which measures a distance from robot 1 to robot 2. The definition of D for the specific case of this two PR robot arm configuration is given in the Appendix. The last antitarget for robot 1 is defined by the set

$$AT_8^1 = \{\underset{\sim}{x} : D^2(\underset{\sim}{x},\underset{\sim}{y}(t)) - \varepsilon_1^2 \le 0\}$$

where ε_1 is a prescribed avoidance tolerance set by the user.

Now form the Liapunov function V^1

$$V^1 = V_0^1 + \sum_{i=1}^{8} \beta_1[i]/V_i^1 ,$$

where

$$V_0^1 = x_3^2/2 - x_3(p_1c_1 \cos x_1 + p_1c_2 \sin x_1) + (x_2^2 + x_4^2)/2 \qquad \text{(Positive on } C(T_1)) ,$$

$$V_1^1 = x_2 + 1.5 \qquad \text{(Positive on } C(AT_1^1)) ,$$

$$V_2^1 = 1.5 - x_2 \qquad \text{(Positive on } C(AT_2^1)) ,$$

$$V_3^1 = x_3 - 0.5 \qquad \text{(Positive on } C(AT_3^1)) ,$$

$$V_4^1 = 1.5 - x_3 \qquad \text{(Positive on } C(AT_4^1)) ,$$

$$V_5^1 = x_4 + 1.5 \qquad \text{(Positive on } C(AT_5^1)) ,$$

$$V_6^1 = 1.5 - x_4 \qquad \text{(Positive on } C(AT_6^1)) ,$$

$$V_7^1 = x_3^2/2 - x_3(p_2c_1 \cos x_1 + p_2c_2 \sin x_1) \qquad \text{(Positive on } C(AT_7^1)) ,$$

$$V_8^1 = D(\underset{\sim}{x},\underset{\sim}{y}(t)) - \varepsilon_1^2 \qquad \text{(Positive on } C(AT_8^1)) ,$$

and $\beta_1[i]$ $i = 1,\ldots,8$ are positive constants.

In the formulation of V_0^1 above the last two terms have been included to secure as 'close to' zero speed of the gripper as the target T_1 is approached. Once T_1 is reached, another method of trajectory planning may be used to position the gripper exactly at (p_1c_1, p_1c_2) with zero velocity. With the symmetry given in the problem formulation we can define the objective of robot arm 2 with target and anti-targets, and form analogously the Liapunov function V^2, where

$$V^2 = V_0^2 + \sum_{i=1}^{8} \beta_2[i]/V_i^2 ,$$

where $\beta_2[i]$ $i = 1,\ldots,8$ are also positive constants of our choice.

4. COORDINATION CONTROL ALGORITHMS

To analyse coordination control we separate out two cases: when both arms are not in the common workspace at the same time, and when they are. The latter case shall be formally examined here. The control algorithms for the other case can be obtained by setting $\beta_1[8] = \beta_2[8] = 0$. In the numerical simulations, program control switches to the appropriate control algorithms dependent upon the positioning of the robot arms. Let us assume therefore that both arms lie initially within the common workspace.

To establish a coordination control algorithm, we reason as follows. Consider the state equation for robot arm 1 integrated over the interval $[t , t+\delta t]$.

Information on states $\underset{\sim}{x}$ and $\underset{\sim}{y}$ are assumed known at time t and state $\underset{\sim}{y}(t)$ is assumed constant over this interval. Provided δt is sufficiently small this last assumption is not unrealistic if the systems are relatively slow moving.

In this interval we select control strategy $\underset{\sim}{u}^1$ such that

$$\dot{V}^1 \le -p_1\gamma_2x_2^2 - p_1\gamma_4x_4^2 ,$$

where $p_1\gamma_2$ and $p_1\gamma_4$ are positive constants of our choice.

Arranging terms appropriately, this can be achieved with the following selection of control

$$u_1^1 = -f_1^1 + R_1^1I_1^1/D_1^1$$

where

$$R_1^1 = -p_1\gamma_2x_2^2 - [\nabla V_0^1[1] - \beta_1[7]\ \nabla V_7^1[1]/(V_7^1)^2 - \beta_1[8]\ \nabla V_8^1[1]/(V_8^1)^2]\,x_2$$

$$D_1^1 = \nabla V_0^1[2] - \beta_1[1]\ \nabla V_1^1[2]/(V_1^1)^2 - \beta_1[2]\ \nabla V_2^1[2]/(V_2^1)^2$$

and

$$u_2^1 = -f_2^1 + R_2^1I_2^1/D_2^1$$

where

$$R_2^1 = -p_1\gamma_4x_4^2 - [\nabla V_0^1[3] - \beta_1[3]\ \nabla V_3^1[3]/(V_3^1)^2 - \beta_1[7]\ \nabla V_7^1[3]/(V_7^1)^2$$
$$- \beta_1[8]\ \nabla V_8^1[3]/(V_8^1)^2]\,x_4$$

$$D_2^1 = \nabla V_0^1[4] - \beta_1[5]\ \nabla V_5^1[4]/(V_5^1)^2 - \beta_1[6]\ \nabla V_6^1[4]/(V_6^1)^2 .$$

The notation $\nabla V_j^1[k]$ used above, define the kth coordinate of the vector ∇V_j^1.

Given now the value of the state $\underset{\sim}{x}(t+\delta t)$ at the end of this interval, it is used in a similar way to develop a strategy $\underset{\sim}{u}^2$ to obtain $\underset{\sim}{y}(t+\delta t)$ from $\underset{\sim}{y}(t)$ with integration of the state equations of robot arm 2 over the same interval $[t, t+\delta t]$.

We determine that

$$\dot{V}^2 \le -p_2\gamma_2y_2^2 - p_2\gamma_4y_4^2$$

with strategy selection

$$u_1^2 = -f_1^2 + R_1^2I_1^2/D_1^2$$

where

$$R_1^2 = -p_2\gamma_2y_2^2 - [\nabla V_0^2[1] - \beta_2[7]\ \nabla V_7^2[1]/(V_7^2)^2 - \beta_2[8]\ \nabla V_8^2[1]/(V_8^2)^2]\,y_2$$

$$D_1^2 = \nabla V_0^2[2] - \beta_2[1]\ \nabla V_1^2[2]/(V_1^2)^2 - \beta_2[2]\ \nabla V_2^2[2]/(V_2^2)^2$$

and

$$u_2^2 = -f_2^2 + R_2^2I_2^2/D_2^2$$

where

$$R_2^2 = -p_2\gamma_4 y_4^2 - [\nabla V_0^2[3] - \beta_2[3]\,\nabla V_3^2[3]/(V_3^2)^2 - \beta_2[4]\,\nabla V_4^2[3]/(V_4^2)^2$$
$$- \beta_2[7]\,\nabla V_7^2[3]/(V_7^2)^2 - \beta_2[8]\,\nabla V_8^2[3]/(V_8^2)^2]\,y_4$$
$$D_2^2 = \nabla V_0^2[4] - \beta_2[5]\,\nabla V_5^2[4]/(V_5^2)^2 - \beta_2[6]\,\nabla V_6^2[4]/(V_6^2)^2 .$$

Positive parameters $p_2\gamma_2$ and $p_2\gamma_4$ are of our choices to strengthen convergence (discussed later).

With these control strategies we observe that for all

$$\underset{\sim}{x} \in R^4 \setminus (T_1 \cup AT_1^1 \cup \cdots \cup AT_8^1)\ ,$$
$$V^1 > 0 \quad \text{and} \quad \dot{V}^1 \le 0\ ,$$

for all

$$\underset{\sim}{y} \in R^4 \setminus (T_2 \cup AT_1^2 \cup \cdots \cup AT_8^2)$$
$$V^2 > 0 \quad \text{and} \quad \dot{V}^2 \le 0\ ,$$

in the interval $[t\,,\,t+\delta t]$. Liapunov conditions for Lagrange stability of both state equations hold.

Moreover, on substitution of these strategies into the state equations, we see that any equilibrium state for either robot arm 1 or 2 lies within T_1 or T_2 respectively. So the conditions described above are sufficient for asymptotic stability to respective targets, provided the $\beta_1[i]$ and $\beta_2[i]$ are chosen sufficiently small. These constants may be referred to as avoidance parameters in the algorithm as they are associated with the avoidance of the constructed antitargets. They have been chosen as positive constants non-time varying within the duration of motion. Clearly this is a limitation, and the algorithm may be improved by defining the $\beta_i[j]$ as functions of the 'distance' of the arm from the various antitargets.

Positive constants $p_i\gamma_j$ may be referred to as convergence parameters associated with capture in the target sets. They determine within the algorithm the rate of attraction or convergence to the target sets.

Selection of these positive constants within the control algorithm is discussed in the next section on numerical simulation.

To incorporate bounding constraints on the accelerations, we define control laws u_1^1 and u_2^1 by:

If $\dot{x}_2 > 1.0$ then $u_1^1 = -f_1^1 + I_1^1$

If $\dot{x}_2 < -1.0$ then $u_1^1 = -f_1^1 - I_1^1$

Else u_1^1 determined as above,

If $\dot{x}_4 > 1.0$ then $u_2^1 = -f_2^1 + I_2^1$

If $\dot{x}_4 < -1.0$ then $u_2^1 = -f_2^1 - I_2^1$

Else u_2^1 determined as above,

to be calculated and used in the next interval of integration.

5. NUMERICAL SIMULATIONS

Numerical simulations were performed for various selections of convergence and avoidance parameters, initial states, and target/antitarget constants. A number are presented here with both targets T_1 and T_2 placed within the common workspace.

EXAMPLE 1

Consider the defined problem with values:

RK4 : step size = 0.02

T_1 : center = (1,0.9) radius = 0.05

T_2 : center = (1,-0.5) radius = 0.05

Tolerances: $\varepsilon_1 = \varepsilon_2 = 0.05$

Initial states: $\underset{\sim}{x} = [5.5 \quad 0.2 \quad 1.3 \quad -0.2]^T$

$\underset{\sim}{y} = [2.0 \quad 0.1 \quad 1.3 \quad 0.1]^T$

Avoidance and convergence parameters:

$\underset{\sim}{\beta}_1 = [1 \quad 1 \quad 0.05 \quad 0.001 \quad 1 \quad 1 \quad 0.2 \quad 0.05]^T$

$\underset{\sim}{\beta}_2 = [1 \quad 1 \quad 0.05 \quad 0.001 \quad 1 \quad 1 \quad 0.2 \quad 0.05]^T$

$\beta_1\gamma_2 = 20$, $p_1\gamma_4 = 10$, $p_2\gamma_2 = 20$, $p_2\gamma_4 = 10$.

Target T_1 is placed close to the outer boundary for arm 1. Arm 2 is initially placed outside the common workspace. The path of the gripper of each arm is shown in Figure 5.1.

By $t_f = 79.0$, both target sets were reached. The final states and position coordinates were:

Arm 1:

$\underset{\sim}{x}_f = [7.0526 \quad -0.00266 \quad 1.3580 \quad 0.0005]^T$

Coordinate position (0.9763 , 0.9439)

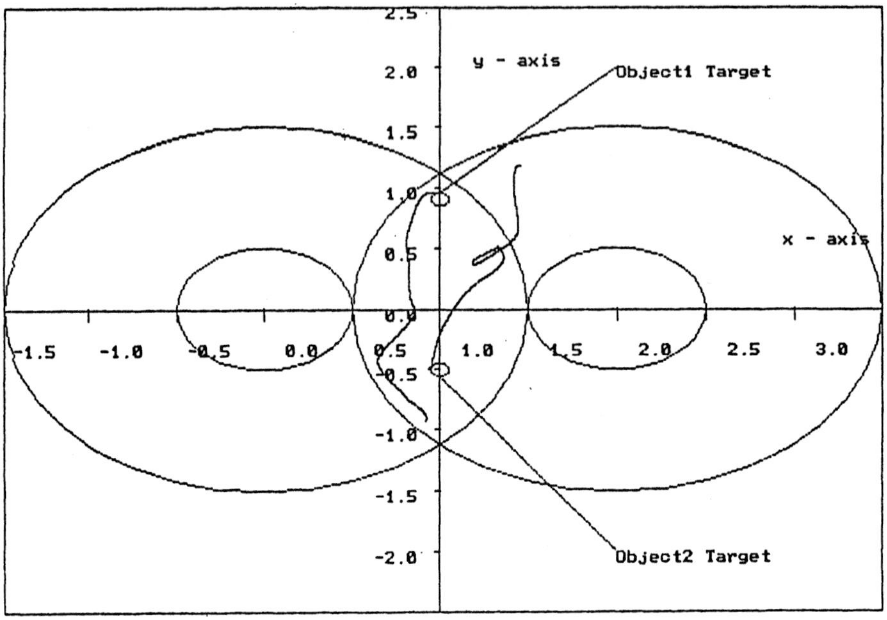

FIGURE 5.1

Arm 2:

$\underset{\sim}{y}_f = [3.5771 \quad 0.0022 \quad 1.1722 \quad 0.0069]^T$

Coordinate position (0.937 , -0.49) .

These parameter settings yield a smooth path for the arm avoiding T_2 and arriving at T_1. Both arms arrive at the targets with near zero speed, but t_f is considerably large. Final t_f may be reduced by a change of parameter to smaller $\beta_i[j]$, but this invariably requires a change also to the convergence parameters. Arm 2, however, is repelled back on its path. This situation is improved considerably by making a small change to the parameters, namely:

$\underset{\sim}{\beta}_1 = [1 \quad 1 \quad 0.05 \quad 0.001 \quad 1 \quad 1 \quad 0.2 \quad 0.2]^T$

$\underset{\sim}{\beta}_2 = [1 \quad 1 \quad 0.02 \quad 0.001 \quad 1 \quad 1 \quad 0.2 \quad 0.005]^T$

$p_1\gamma_2 = 20$, $p_1\gamma_4 = 10$, $p_2\gamma_2 = 30$ and $p_2\gamma_4 = 10$.

The resulting trajectories are shown in Figure 5.2 below. Final t_f is increased to 90 with (to four decimal places):

Arm 1:

$\underset{\sim}{x}_f = [7.0241 \quad 0.0000 \quad 1.3609 \quad 0.0000]^T$

Coordinate position (1.0046 , 0.9182)

Arm 2:

$\underset{\sim}{y}_f = [3.5279 \quad 0.0036 \quad 1.1517 \quad 0.0007]^T$

Coordinate position (0.9332 , -0.4340) .

Example 2

Consider the defined problem with values:

RK4 : step size 0.02

T_1 : center = (1 , 0.9) radius = 0.05

T_2 : center = (1 , -0.5) radius = 0.05

Tolerances: $\varepsilon_1 = \varepsilon_2 = 0.05$

Initial states: $\underset{\sim}{x} = [0.2 \quad 0.01 \quad 0.8 \quad -0.2]^T$

$\underset{\sim}{y} = [2.7 \quad 0.1 \quad 1.3 \quad -0.1]^T$

Avoidance and convergence parameters:

$\underset{\sim}{\beta}_1 = [1 \quad 1 \quad 0.05 \quad 0.005 \quad 1 \quad 1 \quad 0.02 \quad 0.0005]^T$

$\underset{\sim}{\beta}_2 = [1 \quad 1 \quad 0.02 \quad 0.001 \quad 1 \quad 1 \quad 0.02 \quad 0.005]^T$

$p_1\gamma_2 = 30$, $p_1\gamma_4 = 30$, $p_2\gamma_2 = 30$, $p_2\gamma_4 = 10$.

In this illustration the arms have been initially placed in positions where they are required to be manoeuvered around each other to their respective targets. The above selection of avoidance and convergence parameters works well as shown in Figure 5.3. Values of $\underset{\sim}{\beta}_i[3]$, $\underset{\sim}{\beta}_i[4]$ and $\underset{\sim}{\beta}_i[8]$ required careful selection in relation to the convergence parameters. By the time $t_f = 85$, both targets were reached. The final states and position coordinates were:

Arm 1:

$\underset{\sim}{x}_f = [0.7361 \quad 0.0003 \quad 1.2976 \quad 0.0001]^T$

Coordinate position (0.9616 , 0.8712)

Arm 2:

$\underset{\sim}{y}_f = [3.5779 \quad 0.0014 \quad 1.1572 \quad 0.0003]^T$

Coordinate position (0.9512 , -0.4891) .

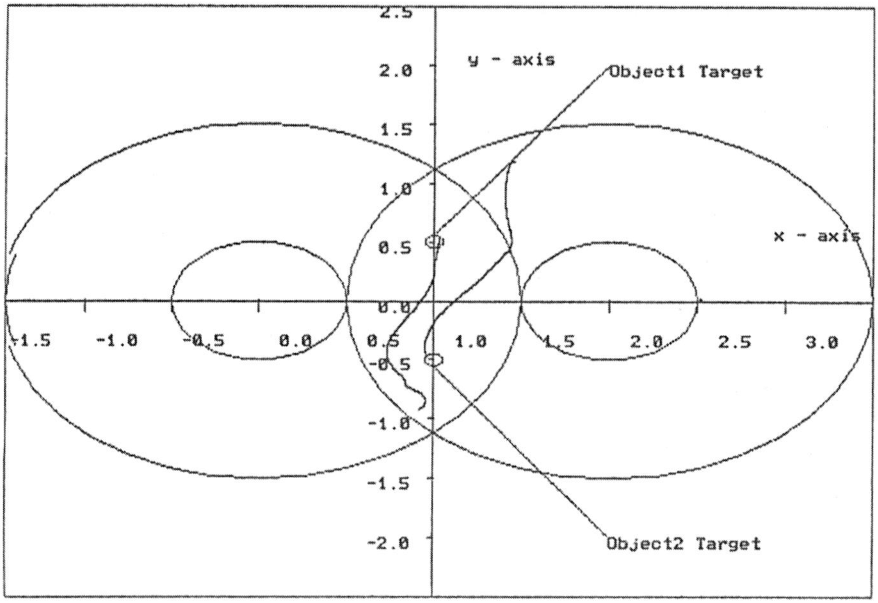

FIGURE 5.2

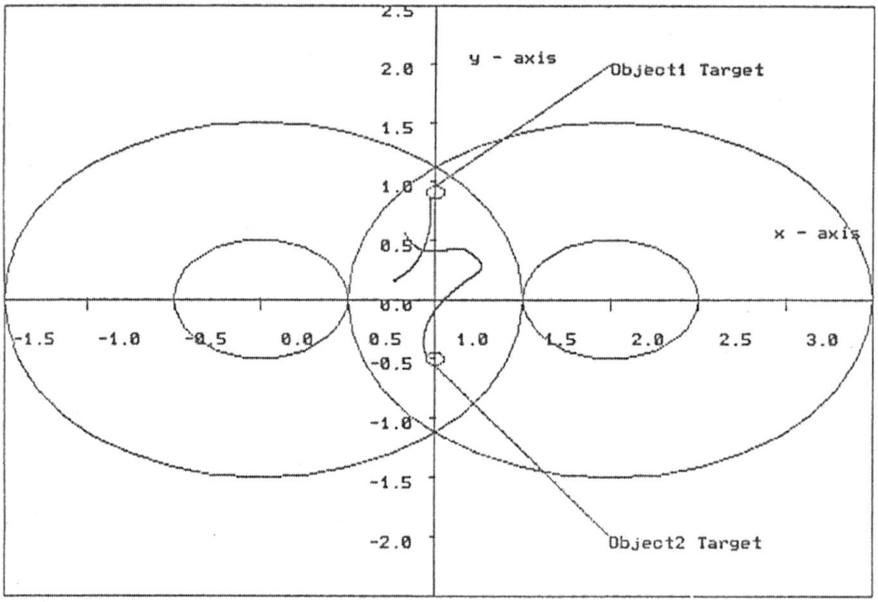

FIGURE 5.3

Example 3

Consider the problem with values:

RK4 : step size 0.01

T_1 : center = (1,0.5) radius = 0.05

T_2 : center = (1,-0.5) radius = 0.05

Tolerances: $\varepsilon_1 = \varepsilon_2 = 0.5$

Initial states: $\underset{\sim}{x} = [5.5 \quad 0.2 \quad 1.3 \quad 0.5]^T$

$\underset{\sim}{y} = [2.0 \quad 0.1 \quad 1.3 \quad 0.1]^T$.

Avoidance and convergence parameters:

$\underset{\sim}{\beta}_1 = [0.5 \quad 0.5 \quad 0.05 \quad 0.01 \quad 0.5 \quad 0.5 \quad 0.1 \quad 0.0005]^T$

$\underset{\sim}{\beta}_2 = [0.5 \quad 0.5 \quad 0.02 \quad 0.001 \quad 0.5 \quad 0.5 \quad 0.02 \quad 0.0005]^T$

$p_1\gamma_2 = 8$, $p_1\gamma_4 = 10$, $p_2\gamma_2 = 15$, $p_2\gamma_4 = 8$.

The scenario is as in Example 1, but parameters were sought to reduce the time taken for each arm to enter its respective target. With the above selection of parameters, arm 1 enters T_1 at $t = 24.24$ and arm 2 enters T_2 at $t = 47.1$. The final state and position coordinates were:

Arm 1:

$\underset{\sim}{x}_f = [6.7601 \quad 0.0001 \quad 1.1521 \quad 0.0000]^T$

Coordinate position (1.0237 , 0.5284)

Arm 2:

$\underset{\sim}{y}_f = [3.5678 \quad 0.0061 \quad 1.1444 \quad 0.0017]^T$

Coordinate position (0.95799 , -0.4731) .

The trajectories are shown in Figure 5.4 below.

Example 4

As a final example, consider the problem with values:

RK4 : step size = 0.02

T_1 : center = (1,0.5) radius = 0.05

T_2 : center = (1,-0.5) radius = 0.05

Tolerances: $\varepsilon_1 = \varepsilon_2 = 0.05$

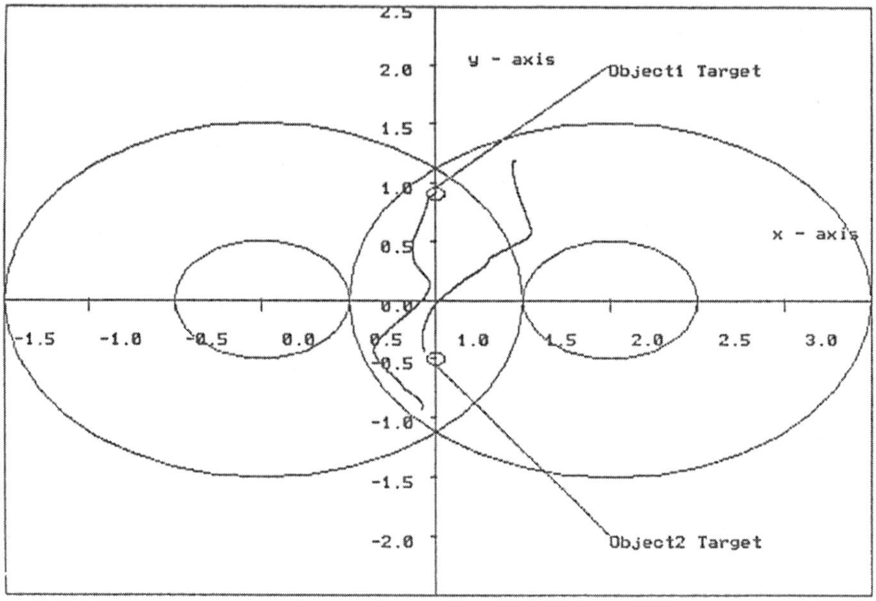

FIGURE 5.4

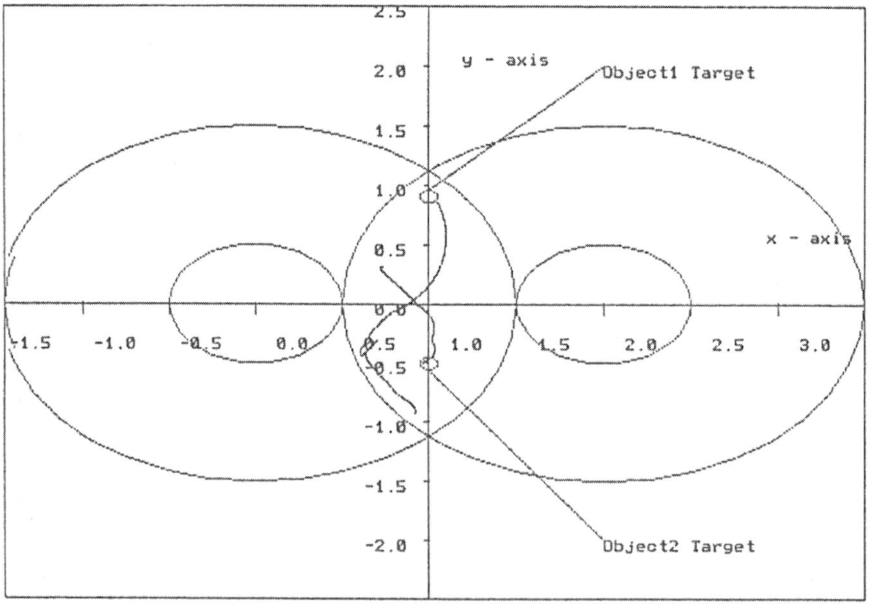

FIGURE 5.5

Initial states: $\underset{\sim}{x} = [5.5 \quad 0.2 \quad 1.5 \quad -0.2]^T$

$\underset{\sim}{y} = [2.9 \quad 0.1 \quad 1.3 \quad 0.1]^T$

Avoidance and control parameters:

$\underset{\sim}{\beta}_1 = [1 \quad 1 \quad 0.05 \quad 0.001 \quad 1 \quad 1 \quad 0.02 \quad 0.02]^T$

$\underset{\sim}{\beta}_2 = [1 \quad 1 \quad 0.02 \quad 0.001 \quad 1 \quad 1 \quad 0.02 \quad 0.005]^T$

$p_1\gamma_2 = 20$, $p_1\gamma_4 = 10$, $p_2\gamma_2 = 30$, $p_2\gamma_4 = 10$.

In this illustration arm 2 is placed centrally so that it is not required to manoeuver around T_1, whilst arm 1 must still avoid target T_2. As can be seen in Figure 5.5 above, arm 1 is repelled by arm 2 but still reaches the target around $t_f = 90$. Arm 2 reaches its target much earlier at $t = 58$. The final states and position coordinates were:

Arm 1:

$\underset{\sim}{x}_f = [6.9696 \quad 0.0051 \quad 1.3511 \quad 0.0017]^T$

Coordinate position (1.04549 , 0.8558)

Arm 2:

$\underset{\sim}{y}_f = [3.6015 \quad 0.0004 \quad 1.1593 \quad 0.0001]^T$

Coordinate position (0.9611 , -0.5145) .

A smooth solution to the problem was sought which would include the removal of the 'wobble' in arm 2 trajectory as it moves to its target. A minimal change to the parameters was required, a reduction of $\beta_1[3]$ allowing closer approach to the inner radial constraint and a strengthening of the convergence parameters. The following selection of parameters

$\underset{\sim}{\beta}_1 = [1 \quad 1 \quad 0.01 \quad 0.001 \quad 1 \quad 1 \quad 0.02 \quad 0.005]^T$

$\underset{\sim}{\beta}_2 = [1 \quad 1 \quad 0.02 \quad 0.001 \quad 1 \quad 1 \quad 0.02 \quad 0.005]^T$

$p_1\gamma_2 = 30$, $p_1\gamma_4 = 15$, $p_2\gamma_2 = 30$, $p_2\gamma_4 = 10$

yielded the trajectories shown in Figure 5.6 with final state and position coordinates:

Arm 1:

$\underset{\sim}{x}_f = [6.9859 \quad 0.0023 \quad 1.3135 \quad 0.0012]^T$

Coordinate position (1.0026 , 0.8485)

Arm 2:

$\underset{\sim}{y}_f = [3.6055 \quad 0.0002 \quad 1.1605 \quad -0.0000]^T$
Coordinate position (0.9621 , -0.5193) .

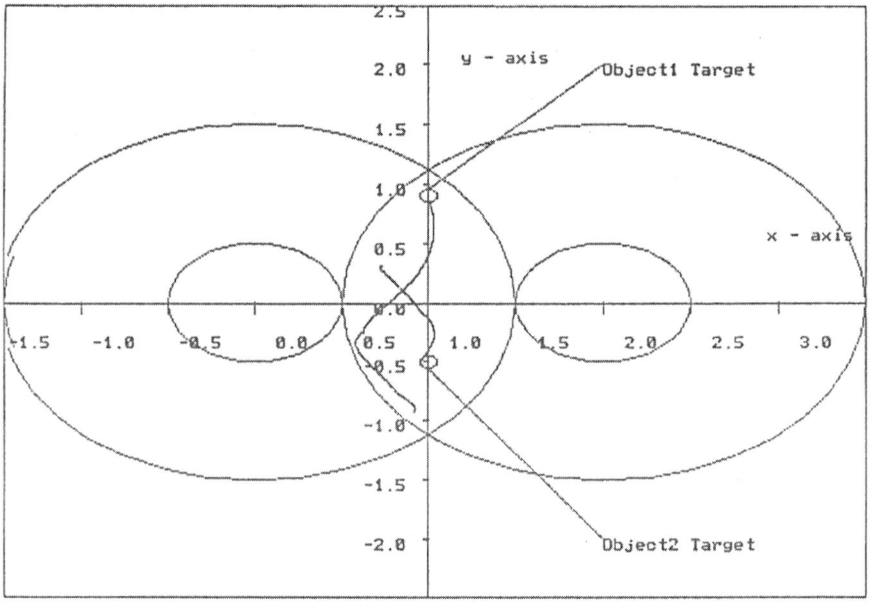

FIGURE 5.6

6. CONCLUSION

In this paper we have illustrated that the Liapunov method introduced in [1] may be applied to the collision-avoidance problem associated with planar movement of two PR robot arms operating within a common workspace. Physical state constraints are included as antitargets or obstacles in the analysis. The illustrations show that with respect to simulation for the simple robot system, constant avoidance and collision parameters can be found so that acceptable qualitative behavior of each arm is obtained in moving to respective targets. No analysis of control histories was undertaken at the time of presentation of this paper.

There exists concern over the length of time taken for the arms to reach the target sets. This is due in part to the parameter selection and selection of control. It should be noted that the method does not achieve the fastest decrease of V as would be the case associated with a 'steepest descent' type selection for the control algorithms.

The depth of constant parameter selection becomes more obvious as we move to more complicated systems.

The possibility of determing time varying $\underset{\sim}{\beta}_i$ with a possible adaptive mechanism is a direction of further research. There is also the incorporation of imposing a quantitative objective such as the minimisation of a specified performance index upon this qualitative analysis.

REFERENCES

[1] R.J. Stonier, *Use of Liapunov Techniques for Collision and Avoidance of Robot Arms*, Control & Dynamic Systems, Vol. 35 (1990), pp. 185-214.

[2] A.J. Koivo, *Fundamentals for Control of Robot Manipulators*, John Wiley & Sons, 1989.

APPENDIX

We consider the calculation of $D(\underset{\sim}{x},\underset{\sim}{y})$ given state $\underset{\sim}{y}$, the reverse case follows a similar analysis. Suppose the gripper of arm 2 is placed at (a,b) relative to the $x_1 - x_2$ coordinate system as shown in Figure A.1 below.

Let the gripper of arm 1 be positioned at (e,f), and construct lines L, L_1, L_2, L_3 and L_4 as shown in the diagram. We define the distance function D as follows.

If (e,f) lies in the region $L_1 > 0$ then

$D(\underset{\sim}{x},\underset{\sim}{y})$ = distance from (a,b) to line L.

If (e,f) lies in the region $L_1 < 0$ and $L_2 > 0$ then

$D(\underset{\sim}{x},\underset{\sim}{y})$ = distance from (a,b) to (e,f).

If (e,f) lies in the region $L_1 < 0$ and $L_2 < 0$ then

$D(\underset{\sim}{x},\underset{\sim}{y})$ = distance from (e,f) to line L_1.

If (e,f) lies in the region $L_4 < 0$ then

$D(\underset{\sim}{x},\underset{\sim}{y})$ = distance from (e,f) to line L.

If (a,b) is positioned below the x_1 axis, the definition of D is defined analogous to the above.

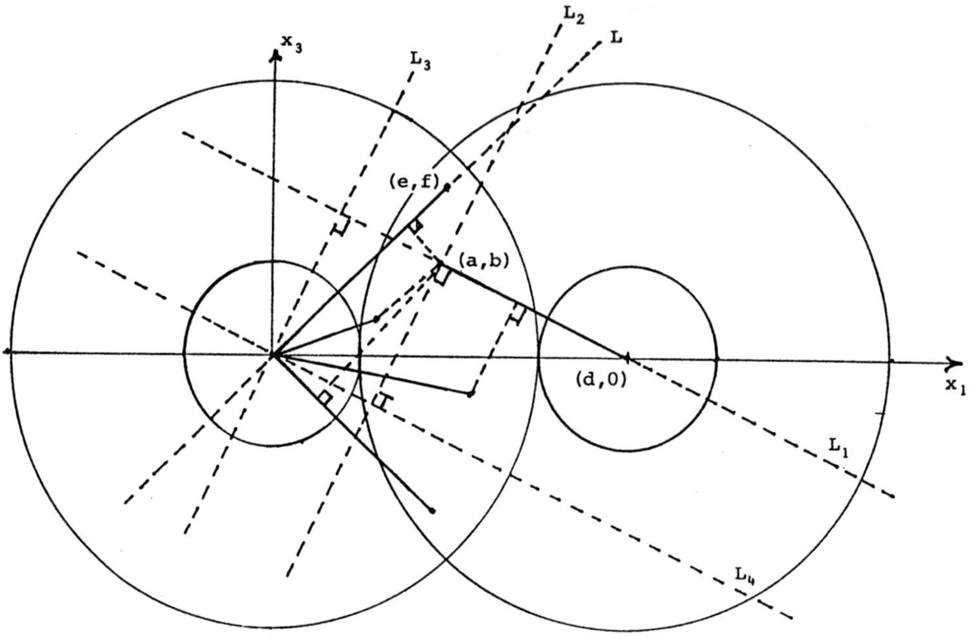

FIGURE A.1

REGIONS OF ATTRACTION FOR TWO DIMENSIONAL UNCERTAIN DYNAMIC SYSTEMS

Thomas L. Vincent and Yeong Ching Lin

Department of Aerospace and Mechanical Engineering

University of Arizona

Tucson, AZ 85721

1. INTRODUCTION

Consider a two dimensional system of the form

$$\begin{aligned} \dot{x}_1 &= f_1(x_1, x_2, u, v) \\ \dot{x}_2 &= f_2(x_1, x_2, u, v) \end{aligned} \tag{1}$$

where f_1 and f_2 are C^1 functions of the state x_1, x_2, control u, and uncertain input v. The dot denotes differentiation with respect to time. The control is assumed to be bounded for all time according to

$$|u| \leq u_m \tag{2}$$

and the uncertain input is assumed to be bounded for all time by

$$|v| \leq v_m \tag{3}$$

where u_m and v_m are both positive numbers and $v_m < u_m$. Furthermore, assume that when $u = v = 0$, the system has a single equilibrium point at the origin.

The control u and uncertain input v may, in general, be specified functions of the state and/or time. Such functions will be called control laws and a given control law for u and v will be said to be admissible if it satisfies (2) and (3) and when substituted into (1) yields a unique solution for $x_1(t)$ and $x_2(t)$. There are several "controllability" type of sets in the x_1, x_2 state space which are of interest in problems of this type.

Controllable Set: A given point in state space is said to be controllable to the origin if under the admissible control law $v(t) = 0$, there exists an admissible control law for u such that the system can be driven from the given point to an arbitrary small neighborhood of the origin in finite time. The controllable set is all points controllable to the origin.

Domain of Attraction: A given point in state space is said to be attracted to the origin if under the admissible control law $v(t) = 0$, and a given specified admissible control law for u, the system will be driven from the given point to an arbitrary small neighborhood of the origin in finite time. The domain of attraction is all points attracted to the origin.

Guaranteed Controlled Set (or Playable Set): A given point in state space is said to be guaranteed controllable to the origin if in spite of any admissible control law for v to the contrary, there exists an admissible control law for u such that the system can be driven from the given point to an arbitrary small neighborhood of the origin in finite time. The guaranteed controllable set is all points guaranteed controllable to the origin.

V − Reachable Set: A given point in state space is said to be v-reachable from the origin if there exists an admissible control law for v such that under a given admissible control law for u, the system can be driven from the origin to the given point in finite point. The v-reachable set is all v-reachable points. (Vincent, 1980, Gayek and Vincent, 1985)

V – Attractive Set: A given point in state space, not in the v-reachable set, is said to be v-attractive if in spite of any admissible control law for v to the contrary, under a given admissible control law for u, the system will be driven from the given point to an arbitrary small neighborhood of the v-reachable set in finite time. The v-attractive set is all v-attractive points.

Since the system may be unstable and because of the limited control available, the controllable and playable sets may be subsets of the state space. Our design objective here is to determine a state feedback control law for u such that the system will be driven to the smallest possible v-reachable set from the largest possible v-attractive set.

For a given admissible stabilizing feedback control law $u(x_1, x_2)$, (1) may be rewritten as

$$\begin{aligned} \dot{x}_1 &= g_1(x_1, x_2, v) \\ \dot{x}_2 &= g_2(x_1, x_2, v) \end{aligned} \tag{4}$$

where now with $v = 0$, this system should have a domain of attraction which is some subset of the controllable set.

For two dimensional systems, the v-reachable set can readily be determined using methods of qualitative control theory. For example, in (4) the boundary of the v-reachable set may be found by satisfying the necessary conditions for a control that will drive the system along the boundary of the v-reachable set (Grantham and Vincent, 1975). In particular, along the boundary of the v-reachable set the input v must maximize (Grantham and Vincent, 1975)

$$H = \lambda_1 g_1(x_1, x_2, v) + \lambda_2 g_2(x_1, x_2, v) \tag{5}$$

such that the maximum value is zero. The $\lambda's$ are adjoint to the system perturbation equations. For those situations where the v-reachable set boundary control is bang-bang, the adjoint equations used to determine the $\lambda's$ may be dispensed with, since a switching curve for determining the control may be obtained from the conditions

$$\begin{aligned} \mathrm{H} &= 0 \\ \partial \mathrm{H}/\partial \mathrm{v} &= 0. \end{aligned} \tag{6}$$

Hopefully, the v-reachable set will be a small subset of the v-attractive set.

The v-attractive set under a given control law for u, subject to the uncertainty v, is donut like in shape with an inner hole defined by the v-reachable set. The actual shape and size of both the v-reachable set and the v-attractive set depend on the control law specified for u. While there have been several methods proposed for dealing with system uncertainty, few studies calculate the v-reachable set or the v-attractive set and compare the efficiency of a given method with respect to others based on this information.

2. METHODS USED TO COPE WITH UNCERTAIN SYSTEMS

Classical control theory has always been concerned with uncertain inputs. However, this concern has generally been implicit rather than explicit. The idea of "turning up the gain" of an output feedback system to improve "system robustness" is an implicit way of handling system uncertainty (Chen, 1987). Indeed, it works quite well in some cases. We call this approach method 1.

Modern control theory, with its focus on pole placement, implicitly buffers the system from uncertain inputs by placing the system poles well to the left of the imaginary axis. For controllable systems, full state feedback allows for arbitrary placement of the controlled system poles. Alternately, the feedback gain may be determined using LQ design by solving the matrix Riccati equation (Hollot and Barmish, 1980). We will use the latter method here (method 2) since it will allow us a direct comparison with method 3 to follow.

An explicit account of uncertainty has its roots in game theory. Indeed the game theoretic approach would be the logical way to deal with system uncertainty if it were not so difficult (or impossible) to solve the Issacs equations (Issacs, 1965) for closed-loop control. There is one exception to this however, with linear system (with unbounded control and uncertainty). Lee and Bryson, 1989; Athans and Falb, 1966 has proposed using linear game theory to design controllers subject to uncertainty. Again the matrix Riccati equation is used, but now, with account taken of the uncertainty in the quadratic cost criterion. This method has the advantage of simplicity, however it still does not explicitly deal with for the bounds on u or v. We call this approach Method 3.

The "second method of Lyapunov" has been used to provide an alternate method for the design and analysis of control systems (Kalman and Bertram, 1960). Leitmann (1983) further developed this type of analysis for use with systems subject to uncertainty (Gutman and Leitmann, 1976; Corless and Leitmann, 1983; Leitmann, 1989). A recent version of this approach (Soldatos, Corless, and Leitmann, 1990) uses the matrix Riccati equation to solve for the feedback gains for the linear portion of the controller, just as in method 2. The same gains are then used to define a switching surface for the bang-bang part of the controller. The total controller is composed of the linear and bang-bang parts. We call this approach method 4.

The final method uses the qualitative nonlinear game theoretic necessary conditions to generate feedback strategies for u. This approach has the advantage of being applicable to nonlinear systems. Its disadvantage is that it can only be applied to one and two dimensional problems. We call this approach method 5.

3. EXAMPLE SYSTEM

We will design controllers for the linear system

$$\begin{aligned} \dot{x}_1 &= x_2 \\ \dot{x}_2 &= a_1 x_1 + a_2 x_2 + u + v \end{aligned} \tag{7}$$

with the nonlinear constraints given by (2) and (3) with $v_m < u_m$. In output format, with $y = x_1$, (7) can be written as

$$\ddot{y} - a_2 \dot{y} - a_1 y = u + v \tag{8}$$

In matrix format (7) is of the form

$$\dot{\mathbf{x}} = \mathbf{A}\mathbf{x} + \mathbf{B}u + \mathbf{B}v \tag{9}$$

where

$$\mathbf{x} = \begin{bmatrix} x_1 \\ x_2 \end{bmatrix},$$

$$\mathbf{A} = \begin{bmatrix} 0 & 1 \\ a_1 & a_2 \end{bmatrix}, \tag{10}$$

$$\mathbf{B} = \begin{bmatrix} 0 \\ 1 \end{bmatrix}. \tag{11}$$

3.1 Method 1 - Output Feedback

The transfer function for the system (8) is given by

$$G_p(s) = \frac{1}{s^2 - a_2 s - a_1}. \tag{12}$$

Consider placing a phase lead compensator

$$H(s) = \frac{1 + \tau_e s}{1 + \tau_a s} \tag{13}$$

with $\tau_e \gg \tau_a$ in a control feedback loop as illustrated in Figure 1. Let the command input $R(s) = 0$. The Laplace transform of the output is related to the Laplace transform of the uncertain input $V(s)$ by

$$Y(s) = \frac{G_p(s)}{1 + KG_p(s)H(s)} V(s) \tag{14}$$

which is equivalent to

$$Y(s) = \frac{1 + \tau_a s}{\tau_a s^3 + (1 - \tau_a a_2)s^2 + (K\tau_e - a_2 - a_1\tau_a)s + (K - a_1)} V(s). \tag{15}$$

Equation (15) is equivalent to the following system

$$\tau_a \dddot{y} + (1 - \tau_a a_2)\ddot{y} + (K\tau_e - a_2 - a_1\tau_a)\dot{y} + (K - a_1)y = v + \tau_a \dot{v}. \tag{16}$$

For a constant input, $v = \bar{v}$, we have the steady state solution

$$y = \frac{\bar{v}}{K - a_1}. \tag{17}$$

It follows that for high gain, the steady state error due to a constant uncertainty can be made small by making the gain large.

For $\tau_e \gg \tau_a$, the two dominant roots to the characteristic equation are approximated by

$$\lambda^2 + (K\tau_e - a_2)\lambda + (K - a_1) = 0. \tag{18}$$

For a given choice of K and τ_e, this system will be equivalent to a second order system with the characteristic equation

$$\lambda^2 + 2\xi\omega_n\lambda + \omega_n^2 = 0 \tag{19}$$

The nominal control design may be obtained by specifying the damping ratio ξ and undamped natural frequency ω_n. The controller for this system is given by

$$U(s) = -K[\frac{1+\tau_e s}{1+\tau_a s} Y(s)] \tag{20}$$

which for small τ_a reduces to

$$U(s) = -K[(1+\tau_e s)Y(s)] \tag{21}$$

or equivalently in the time domain

$$u = -K(y + \tau_e \dot{y}). \tag{22}$$

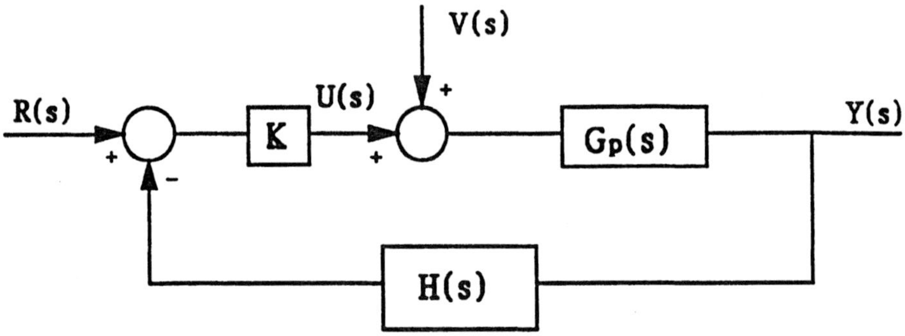

Figure 1 Block diagram for method 1.

In what follows we will designate those curves in state space along which the control is constant by σ. For this output feedback design

$$\sigma = Kx_1 + K\tau_e x_2 \tag{23}$$

In this case the slope of the σ curve is given by $-\frac{1}{\tau_e}$. Whenever $|\sigma| > u_m$ a control determined by (22) will violate (2). In order to satisfy the control constraint we modify the controller to allow for control saturation. Specifically the control law we will use under method 1 is

$$\text{IF } |\sigma| \leq u_m \text{ THEN } u = -\sigma \text{ ELSE } u = -u_m \text{SGN}(\sigma) \tag{24}$$

where σ is determined from (23). Note that SGN(σ) is the sign function defined by

$$\text{IF } \sigma \neq 0 \text{ THEN } \text{SGN}(\sigma) = \frac{\sigma}{|\sigma|} \text{ ELSE } \text{SGN}(\sigma) = 0.$$

3.2 Method 2 - LQ Design

For the system in the form of (9)-(11) the control u is choosen to minimize the cost criterion:

$$J = \int_0^{t_f} (\mathbf{x}^T \mathbf{Q} \mathbf{x} + u^T R u) dt, \tag{25}$$

where $\mathbf{Q}$ is an arbitrary positive semidefinite matrix, and R is an arbitrary constant. The optimal control has a feedback form

$$u = -\mathbf{B}^T \mathbf{S} \mathbf{x} \tag{26}$$

where S is determined from the matrix Riccati equation

$$\mathbf{Q} + \mathbf{S}\mathbf{A} + \mathbf{A}^T \mathbf{S}^T - \mathbf{S}\mathbf{B}R^{-1}\mathbf{B}^T\mathbf{S} = 0. \tag{27}$$

For the class of problems considered here the control will be of the form

$$u = -s_{12}x_1 - s_{22}x_2 \tag{28}$$

where s_{12} and s_{22} are the elements of the last column of the S matrix. Thus

$$\sigma = s_{12}x_1 + s_{22}x_2 \tag{29}$$

with the slope of the σ curve is given by $-s_{12}/s_{22}$. In order to satisfy (2) the control law we will use under Method 2 is again given by (24) with σ determined by (29).

3.3 Method 3 - Linear Game Theory

The system is again given by (9)-(11). Only now,the oontrol u is choosen to minimize the cost criterion:

$$J = \int_0^{t_f} (\mathrm{x}^T \mathbf{Q}\mathrm{x} + \mathrm{u}^T \mathrm{R}\mathrm{u} - \mathrm{v}^T \hat{\mathrm{R}}\mathrm{v})\,dt, \tag{30}$$

where $\mathbf{Q}$ is an arbitrary positive semidefinite matrix, and R and $\hat{\mathrm{R}}$ are arbitrary constants provided $\hat{\mathrm{R}} > \mathrm{R}$. The optimal control is again of the form (26) where $\mathbf{S}$ is determined from

$$\mathbf{Q} + \mathbf{SA} + \mathbf{A}^T\mathbf{S}^T - \mathbf{SB}[\mathrm{R}^{-1} - \hat{\mathrm{R}}^{-1}]\mathbf{B}^T\mathbf{S} = 0. \tag{31}$$

For the class of problems considered here, the control will be of the form of (28). In order to satisfy (2) the control law we will use under method 3 is again given by (24) with σ determined by (29).

3.4 Method 4 - Lyapunov Approach

Following Sodatos, Corless, and Leitmann (1990) choosing Q_R any symmetric positive definite matrix and R = 1, the Riccati eqation (27) is of the form

$$\mathbf{Q_R} + \mathbf{SA} + \mathbf{A}^T\mathbf{S} - \mathbf{SBB}^T\mathbf{S} = 0 \tag{32}$$

which has a unique positive definite solution $\mathbf{S}$. For any $\varepsilon > 0$, the proposed controller is giving by

$$u = -\mathbf{B}^T\mathbf{S}\mathbf{x} - v_m \mathrm{SAT}(\varepsilon^{-1}\mathbf{B}^T\mathbf{S}\mathbf{x}), \tag{33}$$

Where SAT(y) is a saturation fanction defined by

$$\text{IF } |y| \leq 1 \text{ THEN } \mathrm{SAT}(y) = y \text{ ELSE } \mathrm{SAT}(y) = \mathrm{SGN}(y). \tag{34}$$

For the class of problems considered here

$$\mathbf{B}^T\mathbf{S}\mathbf{x} = \sigma \tag{35}$$

with σ defined by (29). Thus the control may be expressed as

$$u = -\sigma - v_m \mathrm{SAT}(\sigma/\epsilon). \tag{36}$$

In order to satisfy (2) the control law we will use under Method (4) will be of the form

$$\text{IF } |\hat{\sigma}| \leq u_m \text{ THEN } u = -\hat{\sigma} \text{ ELSE } u = -u_m \mathrm{SGN}(\hat{\sigma}) \tag{37}$$

where

$$\hat{\sigma} = \sigma + v_m \mathrm{SAT}(\sigma/\epsilon). \tag{38}$$

The first term of (36) is linear state variable feedback which will stablize the system in the LQ method. The second term of (36) is a bang-bang type control which is unique to Leitmann's method.

3.5 Method 5 - Qualitative Nonlinear Game Theory

Necessary conditions for determining controls u and v which will drive the system (7) along the boundary of the playable set are obtained from a Min-Max principle. In particular if a control law for u and v exists which will drive the system along the boundary of the playable set, then u must minimize and v must maximize the function (Isaacs, 1965)

$$H = \lambda_1 x_2 + \lambda_2(a_1 x_1 + a_2 x_2 + u + v) \tag{39}$$

such that the min-max value is zero. Again, the $\lambda's$ are adjoint to the system perturbation equations. From the necessary conditions, it may be shown that the boundary controls are bang-bang. The state space switching function which divides the state space into regions of maximum control and minimum control is given by

$$sw = x_2 \tag{40}$$

which is obtained from the conditions

$$\begin{aligned} &H = 0 \\ &\partial H/\partial u = \partial H/\partial v = \lambda_2 = 0. \end{aligned} \tag{41}$$

Boundary control is of the form

$$\text{IF} \;\; x_2 > 0 \;\; u = -u_m \;\; \text{ELSE} \;\; u = u_m, \tag{42}$$

$$\text{IF} \;\; x_2 > 0 \;\; v = v_m \;\; \text{ELSE} \;\; v = -v_m. \tag{43}$$

In order to have a guaranteed domain of attraction approach the size of the playable set, the controller for u should be of the form of (42) and (43) near the boundary of playable set. However using the x_1 axis as a switching function for any points off of the boundary of the playable set will not work since the system moves perpendicular to the x_1 axis as it crosses it. Because of this difficulty and because we wish to avoid chatter, the following control law is proposed. We first define a curve of constant control effort

$$\begin{aligned} &\text{IF} \;\; x_1 < 0 \;\; \text{THEN} \;\; \sigma = K(x_2 - h) \\ &\text{IF} \;\; x_1 > 0 \;\; \text{THEN} \;\; \sigma = K(x_2 + h) \\ &\text{IF} \;\; |x_1| < h \;\; \text{THEN} \;\; \sigma = K(x_1 + x_2) \end{aligned} \tag{44}$$

where K and h represent a "gain" and a displacement from the x_2 axis. The control is given by

$$u = -\sigma. \tag{45}$$

In order to satisfy (2) the control law we will use under Method 5 is again by (24) with σ determined by (44).

4. INVERTED PENDULUM

Consider now the problem used by Solatos, Corless, and Leitmann (1990) in their recent paper on uncertain systems. The system is an inverted pendulum in which the nonlinear gravitational term has been replaced by an uncertain input. This results in a system of the form

$$\begin{aligned} \dot{x}_1 &= x_2 \\ \dot{x}_2 &= u + v \end{aligned} \tag{46}$$

which is also equivalent to the one dimensional motion of a mass acted on only by forces u and v. This system is of the form of (7) with $a_1 = a_2 = 0$. If we set $v = 0$ and seek a minimum time control to the origin, this reduces to Bushaw's problem (Bushaw, 1958). The minimum time controller is bang-bang with at most one switch. The system may be driven in minimum time from every point in state space to the origin. Thus, the controllable set must be the entire state space. The playable set to the origin is also the entire state space. This follows from the fact that the minimum time game theoretic control for v is opposite of that for u. Thus, an equivalent system under game theoretic control is obtained by setting $r = u + v$, $|r| \le u_m - v_m$. All of the different control approaches may now be compared with this result. Namely, the v-reachable set can be made arbitrary small and the playable set is the entire state space. For this example we choose $u_m = 2$ and $v_m = 1$.

4.1 Method 1 - Output Feedback

If we choose $K = 10$ and $\tau_e = 0.3162$ then the controlled system under output feedback will be approximatly equivalent to a second order system with a characteristic equation given by (19) with $\xi = 0.5$ and $\omega_n = \sqrt{10}$.

Figure 2 illustrates both the v-reachable set and the domain of attraction obtained by solving (46) with the control u determined by (23) and (24) and the control for v determined by (43). The v-reachable sets (the smaller regions) are obtained by integrating (46) forward in time from an initial point near the origin. After a sufficiently long time the system trajectories approach the boundaries of the v-reachable set for values of $K = 10$ and 20 illustrated. As expected,the v-reachable is reduced in size by increasing K.

Figure 2 also shows the v-attractive set obtained by integrating (46) backward in time with u and v again determined by (23), (24) and (43) but now with an initial point outside the v-reachable set. The v-attractive set is seen to increase with an increase in K. However, increasing K much beyond 20 has little effect on the v-attractive set. The size of the v-attractive set is much more greatly effected by the value chosen for τ_e as we will demonstrate later. Indeed much larger, but finite v-attractive sets are possible. Any point outside of the v-attractive set cannot be driven to the origin under the control law specified by (23) and (24). Any point inside the v-attractive set can be guaranteed to be driven to be the boundary of the v-reachable set, but not necessarily to the origin.

4.2 Method 2 - LQ Design

If we choose

$$\mathbf{Q} = \begin{bmatrix} 1 & 0 \\ 0 & 1 \end{bmatrix}, \quad \mathrm{R} = 1. \tag{47}$$

then the solution to the Riccati equation (27) yields

$$\mathrm{S} = \begin{bmatrix} 1.7321 & 1.0 \\ 1.0 & 1.7321 \end{bmatrix}.$$

Thus

$$\sigma = 1.0\mathrm{x}_1 + 1.7321\mathrm{x}_2. \tag{48}$$

The system (46) is integrated using (24) and (48) for u and (43) for v. By integrating forward in time starting near the origin we obtain the v-reachable set as shown in Figure 3. By integrating backward in time from the outside the v-reachable set we obtain the v-attractive set as shown in Figure 4. The shape and size of both the v-reachable set and the v-attractive set depend on Q and R and different results may be obtained than illustrated here.

4.3 Method 3 - Linear Game Theory

If we choose

$$\mathbf{Q} = \begin{bmatrix} 1 & 0 \\ 0 & 1 \end{bmatrix}, \quad \mathrm{R} = 1, \quad \hat{\mathrm{R}} = 2. \tag{49}$$

Solving the Riccati equation (31) yields

$$\mathrm{S} = \begin{bmatrix} 1.9566 & 1.4142 \\ 1.4142 & 2.7671 \end{bmatrix}$$

Thus

$$\sigma = 1.4142\mathrm{x}_1 + 2.7671\mathrm{x}_2. \tag{50}$$

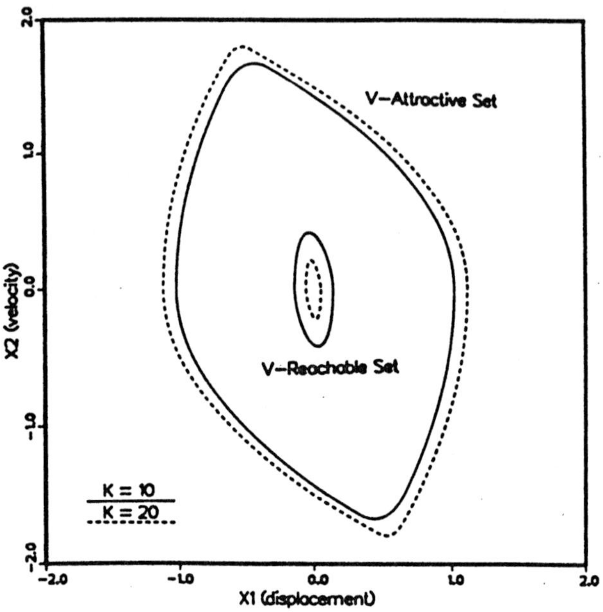

Figure 2 V-reachable and v-attractive sets for problem 1 using a method 1 controller with two different values of K ($\tau_e = .3162$).

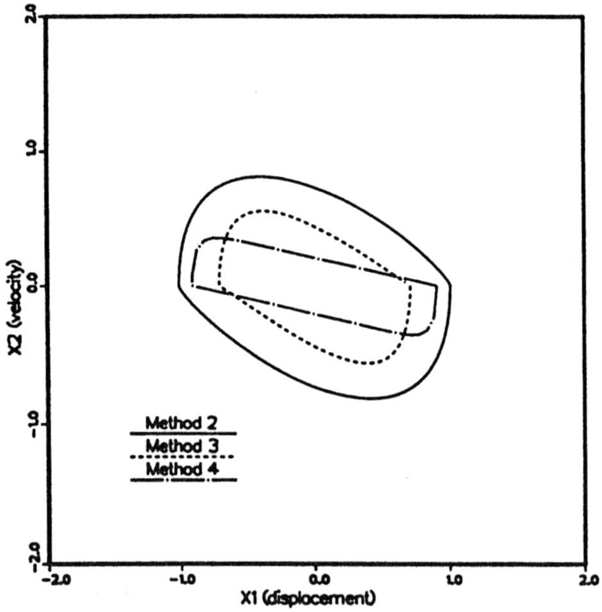

Figure 3 V-reachable sets for problem 1 using controllers based on methods 2, 3, and 4.

We integrate (46) using (24) and (50) for u and (43) for v. By integrating forward in time starting near the origin we obtain the v-reachable set as shown in Figure 3. By integrating backward in time from the outside the v-reachable set we obtain the domain of attraction as shown in Figure 4.

4.4 Method 4 - Lyapunov Approach

As in Soldatos, Corless, and Leitmann (1990) we choose

$$\mathbf{Q_R} = \begin{bmatrix} \alpha^2 & 0 \\ 0 & \alpha^2 \end{bmatrix}. \tag{51}$$

Solving the Riccati equation (32) yields

$$s_{11} = \alpha(\alpha^2 + 2\alpha)^{1/2}$$

$$s_{12} = s_{21} = \alpha$$

$$s_{22} = (\alpha^2 + 2\alpha)^{1/2}$$

Thus,

$$\sigma = \alpha x_1 + (\alpha^2 + 2\alpha)^{1/2} x_2. \tag{52}$$

For this particular example, we choose $\varepsilon = .1$.

We integrate (46) under the controller u obtained using (37)-(38) and (52) with $\alpha = 0.1$ and with v determined by (43). Again, starting near the origin and integrating forward in time we obtain the v-reachable set illustrated in Figure 3. Figure 4 shows the v-attractive set. For a given ε the v-attractive set and the v-reachable set depends on the value of α as illustrated in Figures 5 and 6. Decreasing α increases both the v-reachable set and the v-attractive set. Note however decreasing ε can further decrease the v-reachable set but at the risk of introducing chatter.

4.5 Method 5 - Qualitative Nonlinear Game Theory

The controller for u in this case is given by (24) with σ determined by (44). The system (46) is integrated with u under this control and v determined by (43). By choosing $K = 25$ and $h = 1$ we obtain the v-reachable set illustrated in Figure 7. This is an order of magnitude smaller than the v-reachable sets obtained with the other methods. In addition the v-attractive set is equal to the playable set, that is, it is equal to the entire state space.

4.6 Comparison of Methods as Applied to the First Problem

One might incorrectly conclude from these figures that the effectiveness of the methods in terms of the v-attractive set increase in the order given. Actually the differences between the methods in terms of v-reachable and v-attractive sets can be made small. For example, if using method 1, we choose $K = 1.4$ and $\tau_e = 6.4$ we obtain the results shown in Figures 8a-8b which compares very favorably with Method 4. Figures 8a-8b were obtained by using the same slope as method 4 ($\sigma = 0.05$) with sufficiently high gain so as to closedly match results. It is interesting to note that among the first four methods, method 1 produced the smallest v-reachable set. In this regard, its performance is comparitable to method 5 which, in turn, produced a somewhat smaller v-reachable set.

The first 4 methods use $\sigma = 0$ as a curve which separates positive control from negative control. The slope of this curve for methods 1-4 used to produce Figures 2-6 was:

$$-3.1626$$

$$-0.5773$$

$$-0.5111$$

$$-0.1562 \ (\alpha = 0.05)$$

Each increase in slope (approaching zero) represented an increase in performance. The ultimate performance would appear to be the non-linear game theoretic solution of zero slope. It should be pointed out however that $\sigma = 0$ is not a good choice. As σ approaches zero the time it takes for the system to approach the v-reachable set increases. This is illustrated in Figures 9a-9c for methods 1, 3, and 4 ($\alpha = 0.1$, $\epsilon = 0.1$). The initial condition for each case is the same. As $\sigma \to 0$ the time required to reach the v-reachable set is seen to increase. In fact when $\sigma = 0$ infinite time would be required. Thus none of the methods can use $\sigma = 0$ and as a result the v-attractive set for first four methods will always be finite. This is because for the first four methods no matter how close σ is to zero, a long way away from the origin, the control law for u will not correspond to the non-linear game solution.

By avoiding a straight x_2 curve, method 5 is able to have x_2 "close" to $x_2 = 0$ a long way from the origin (producing an unbounded v-attractive set) and yet be similar to the traditional method 1 controller with high gain near the origin (producing a small v-reachable set). This is accomplished without undue sacrifice in performance as illustrated in Figure 10, where the quick time response and the small v-reachable set may be noted.

All methods are prone to chattering under selection of certain parameters. Method 1 and 5 will chatter if the gain K is to large. Method 2 and 3 will chatter if s_{12} is too large and method 4 will chatter if ϵ is too small. Parameters were choosen in all the examples to avoid chattering.

Only method 4 explicity deals with the bound on the uncertainty v_m. It is implicit in the other methods. There is a theoretical advantage to having v_m explicite in method 4 in that exact statements about the procees can be made.

However the other methods have the pratical advantage of not having to deal with it.

5. UNSTABLE SECOND ORDER SYSTEM

Consider now a problem which differs from the previous one in that the playable set is bounded. The system is of the form

$$\begin{aligned} \dot{x}_1 &= x_2 \\ \dot{x}_2 &= -x_1 + 2x_2 + u + v \\ |u| &\leq 2 \\ |v| &\leq 1 \end{aligned} \tag{53}$$

which is equivalent to a spring-mass system with negative damping. If we set $v = 0$ then (53) is of the form

$$\dot{x} = \mathbf{A}x + \mathbf{B}u. \tag{54}$$

it can be shown (Gayek and Vincent, 1985) that if all the eigenvalues of the A matrix (any dimension) have positive real parts, and if there exists an input u for the system (54) which will generate a trajectory which remains on the boundary of the reachable set for the retro system

$$\dot{x} = -\mathbf{A}x - \mathbf{B}u \tag{55}$$

[which is indeed the controllable set for (54)] for all time $t > 0$, then this same control law will drive the system to the boundary of the reachable set from any point within the reachable set. Furthermore the existance of such a control law is guaranteed for single input, controllable, two dimensional systems of the form given here (Gayek and Vincent, 1988). The input u which will drive the system along the boundary of the reachable set for the retro system may be easily determined by

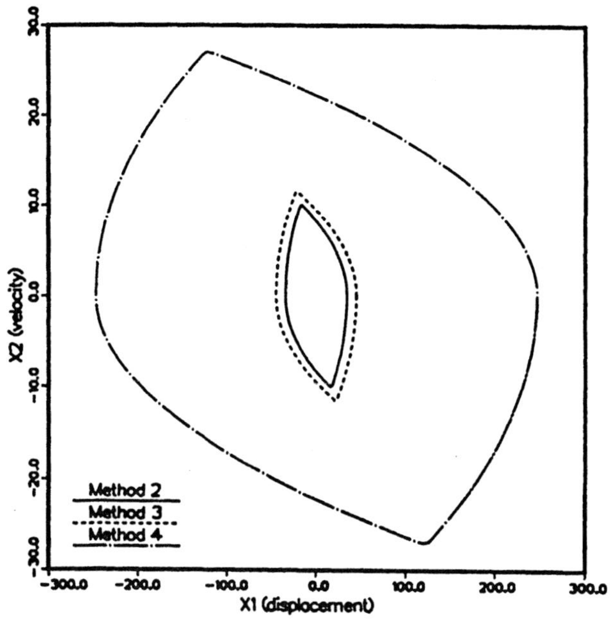

Figure 4 V-attractive sets for problem 1 using controllers based on methods 2, 3, and 4.

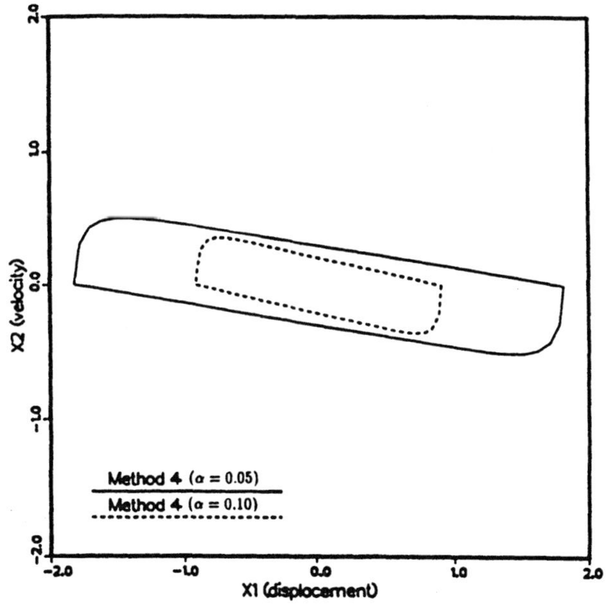

Figure 5 Effect of varying α on the v-reachable sets for problem 1 using a method 4 controller.

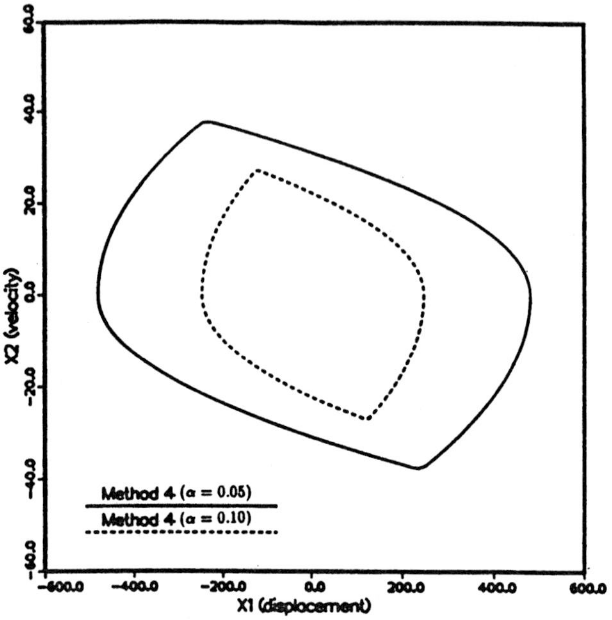

Figure 6 Effect of varying α on the v-attractive sets for problem 1 using a method 4 controller.

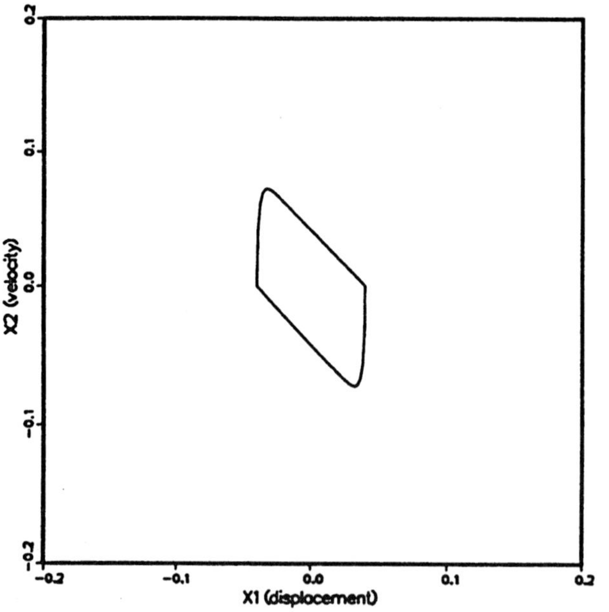

Figure 7 V-reachable set for problem 1 using a method 5 controller.

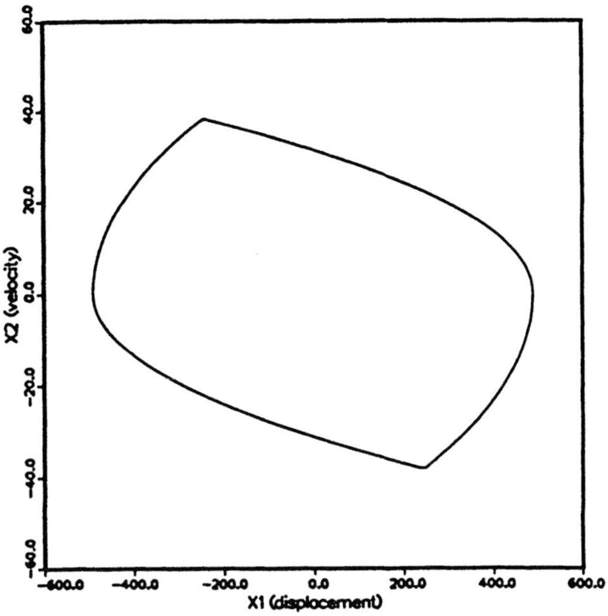

Figure 8a V-attractive set for problem 1 using a method 1 controller (K = 1.4, τ_e = 6.4).

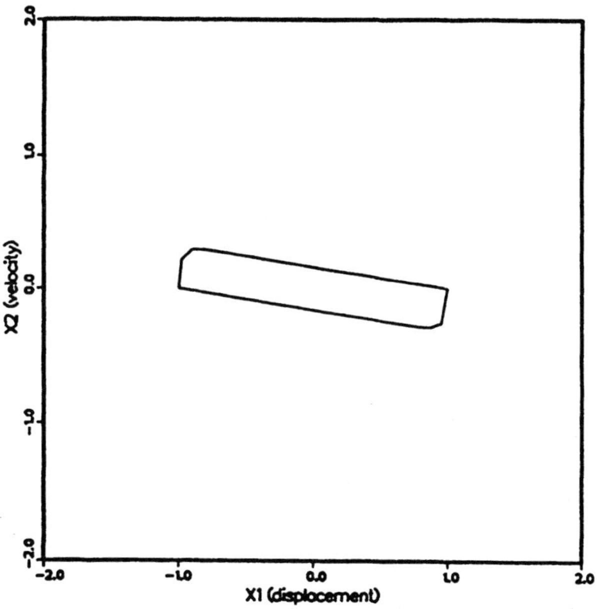

Figure 8b V-reachable set for problem 1 using a method 1 controller (K = 1.4, τ_e = 6.4).

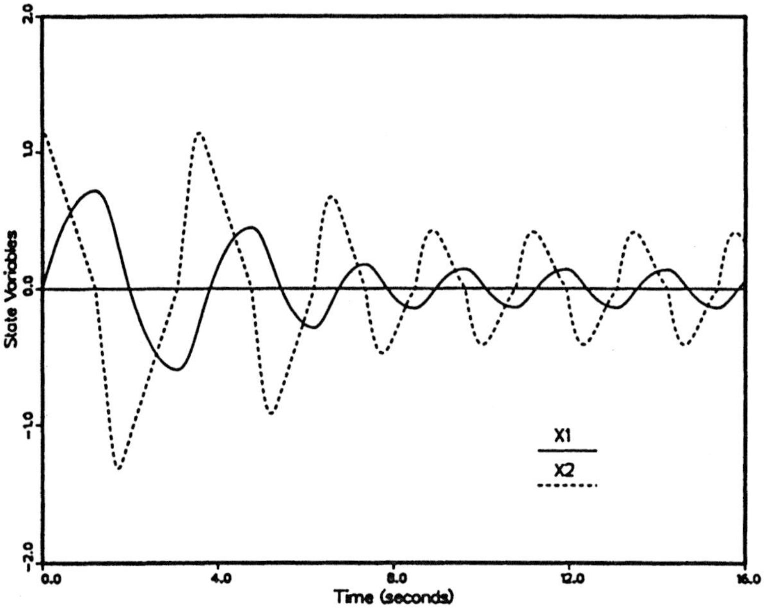

Figure 9a Time response for inverted pendulum using the method 1 controller of figure 2.

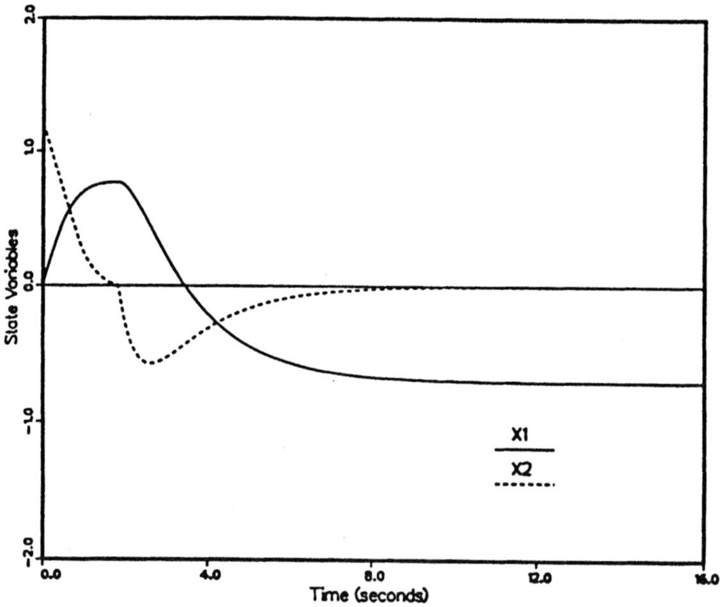

Figure 9b Time response for inverted pendulum using the method 3 controller of figure 3.

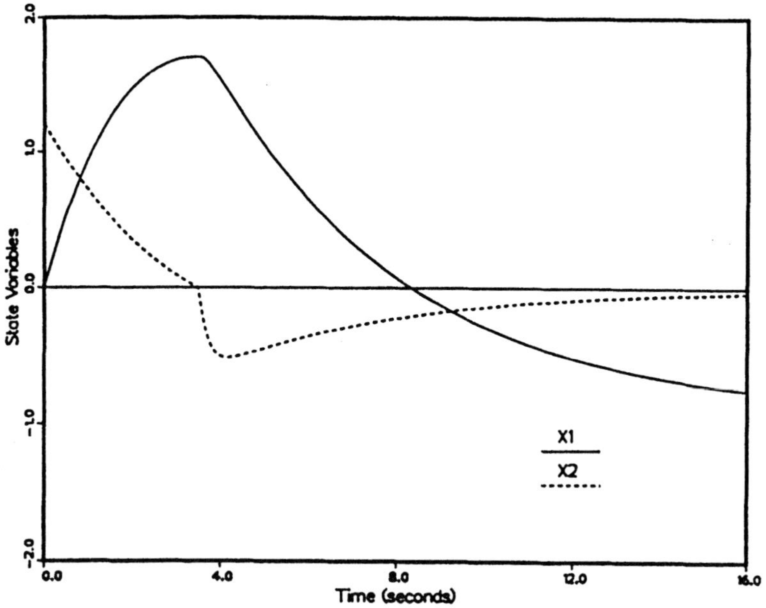

Figure 9c Time response for inverted pendulum using the method 4 controller of figure 3.

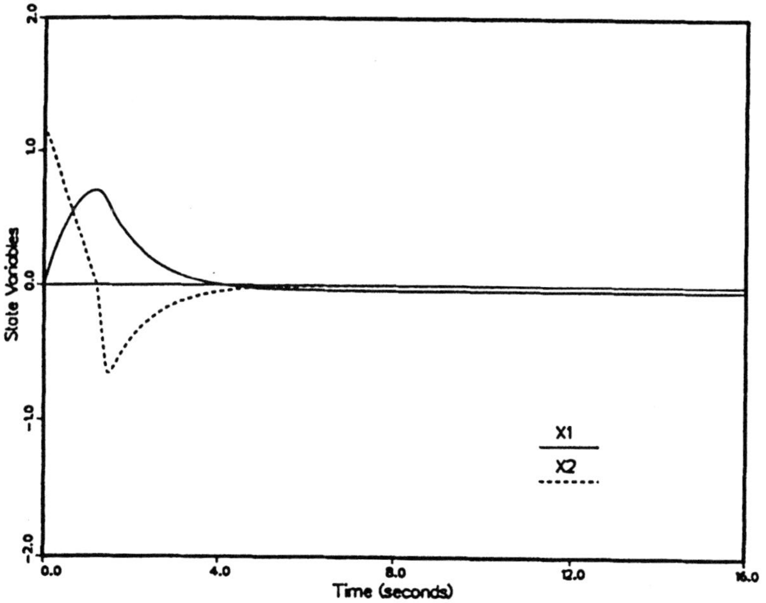

Figure 10 Time response for inverted pendulum using the method 5 controller of figure 7.

applying the controllability maximum principle (Grantham and Vincent, 1975). In particular application of this principle to (53) yields

$$u = \begin{cases} 2SGN(x_2) & \text{if } x_2 \neq 0; \\ 2 \text{ or } -2 & \text{if } x_2 = 0. \end{cases} \tag{56}$$

It may easily be shown that equilibrium solutions corresponding to u = 2 or u = -2 will always lie on the boundary of the reachable set for the retro system to (53). Thus one may start at either equilibrium point and apply (56) to generate a trajectory to the opposite equilibrium point. This trajectory will lie on the boundary of the reachable set for the retro system which is the same as the boundary of the controllable set for the forward system (53). The controllable set for (53) illustrated in Figure 11 was obtained in this way.

The playable set, also illustrated in Figure 11, was obtained in a similar fashion. Qualitative game theory yields the result that on the boundary of the playable set the v control is always opposite the u control. The playable set boundary is generated by starting at the equilibrium points corresponding to u = +2, v = -1 and u = -2, v = +1, switching sign on both u and v and integrating the system to the opposite equilibrium point.

The controllable set represents the largest domain of attraction a system can have when $v \equiv 0$ and the playable set represents the largest v-attractive set a system can have when $v \not\equiv 0$. Under a given control law for u there will also exist a v-reachable set (generally non empty) so that a "hole" will appear in the v-attractive set to form a "donut" shape. We will now compare the size and shape of the v-reachable set and the v-attractive set using the various methods.

5.1 Method 1 - Output Feedback

It follows from (18) that the dominant roots to the characteristic equation are given by

$$\lambda^2 + (K\tau_e - 2)\lambda + (K+1) = 0.$$

Clearly we must choose $K\tau_e > 2$. The controller for this system is again given by (23) and as before we will use (43) for v. We start with the same nominal design as the previous example with $\zeta = 0.5$ and $\omega_n = \sqrt{10}$, thus

$$K\tau_e - 2 = 2\zeta\omega_n = \sqrt{10}$$

$$K + 1 = \omega_n^2 = 10$$

which yields $K = 9$, $\tau_e = .5736$. Under this nominal design the v-reachable set intersects the playable set of Figure 11. Thus this design is unusable as the uncertainty can always destablize the system. However by turning up the gain a usable solution is obtained as illustrated in Figure 12. Both the v-reachable set and the v-attractive set are obtained by solving (53) with the control for u determined by (23) and (24) with $K = 19$, $\tau_e = .5736$ and the control for v by (43).

Further increasing the gain K will shrink the v-reachable set, but will have little efffect on the v-attractive set as illustrated in Figure 13 where K has been increased to $K = 29$ ($\tau_e = .5736$).

5.2 Method 2 - LQ Design

The system (53) is in the form of (9) where

$$\mathbf{x} = \begin{bmatrix} x_1 \\ x_2 \end{bmatrix}, \mathbf{A} = \begin{bmatrix} 0 & 1 \\ -1 & 2 \end{bmatrix}, \mathbf{B} = \begin{bmatrix} 0 \\ 1 \end{bmatrix}. \tag{57}$$

As in Section 4.2 if we choose **Q** and R according to (47) we obtain

$$\mathbf{S} = \begin{bmatrix} 1.4142 & 0.4142 \\ 0.4142 & 4.4142 \end{bmatrix}$$

which yields

$$\sigma = 0.4142\mathrm{x}_1 + 4.4142\mathrm{x}_2. \tag{58}$$

Applied the saturated control law (24) with σ defined by (58) and v determined by (43) results in a v-reachable set that intersects the playable set. Thus this control law is unusable.

If instead we choose

$$\mathbf{Q} = \begin{bmatrix} 5 & 0 \\ 0 & 5 \end{bmatrix}, \quad \mathrm{R} = 1.$$

Then

$$\mathbf{S} = \begin{bmatrix} 2.4495 & 1.4495 \\ 1.4495 & 5.4495 \end{bmatrix}$$

and from (29) we obtain

$$\sigma = 1.4495\mathrm{x}_1 + 5.4495\mathrm{x}_2. \tag{59}$$

Applying the saturated control law (24) with σ defined by (59) and v determined by (43) to system (53) results in the v-reachable set illustrated in Figure 14. Integrating backward in time under the same control laws starting outside the v-reachable set yields the v-attractive set also illustrated in Figure 14. The control law for u is only marginally usable. Clearly only a small portion of the v-attractive set results in guaranteed controllability to a rather large v-reachable set.

5.3 Method 3 - Linear Game Theory

The system is again in the form of (9) with **x**, **A**, and **B** given by (57). The control u is choosen to minimize the cost criterion given by (30) with the feedback gains determined from (31). If we choose

$$\mathbf{Q} = \begin{bmatrix} 5 & 0 \\ 0 & 5 \end{bmatrix}, \quad \mathrm{R} = 1, \quad \hat{\mathrm{R}} = 2.$$

Then (31) yields

$$\mathbf{S} = \begin{bmatrix} 9.4495 & 0.4495 \\ 0.4495 & 8.4495 \end{bmatrix}$$

and from (29) we obtain

$$\sigma = 0.4495\mathrm{x}_1 + 8.4495\mathrm{x}_2. \tag{60}$$

Applying the saturated control law (24) with σ defined by (60) and v determined by (43) to system (53) gives the results illustrated in Figure 15. The v-reachable set is obtained by integrating forward in time from the origin and the v-attractive set is obtained by integrating backward in time from a point outside the v-reachable set. This method provides a larger v-attractive set than method 2. However a good portion of the playable set is still not within the domain of attraction. What is more critical however is the fact that the v-reachable set is just inside the domain of attraction. The uncertain input is dangerously close to being able to drive the system unstable.

5.4 Lyapunov Approach

We again choose $\mathbf{Q_R}$ according to (51). Solving the Riccati equation (32) for

$$\mathbf{S} = \begin{bmatrix} s_{11} & s_{12} \\ s_{21} & s_{22} \end{bmatrix}$$

yields

$$\begin{aligned} \mathrm{s}_{11} &= 2 + \sqrt{\alpha^2+1}\sqrt{2+\alpha^2+2\sqrt{\alpha^2+1}} \\ \mathrm{s}_{12} &= s_{21} = -1 + \sqrt{\alpha^2+1} \\ \mathrm{s}_{22} &= 2 + \sqrt{2+\alpha^2+2\sqrt{\alpha^2+1}} \end{aligned}$$

The control law for u is given by (37)-(38) where

$$\sigma = s_{12}x_1 + s_{22}x_2. \tag{61}$$

Applying this control law for u ($\alpha = 1, \epsilon = 0.1$) with v determined by (43) to system (53), we ontain the results illustrated in Figure 16. The v-reachable set and the v-attractive set are obtained as with the other methods. Here we see a substantial reduction in the v-reachable set with a small increase in the v-attractive set when compared with method 3.

5.5 Qualitative Game Theory

Integrating (53) under the u control law given by (24) with σ determined by (44) with K=50, h=0.05, and the v control law determined by (43) we obtain the v-reachable set and v-attractive set illustrated in Figure 17. This method provides a further reduction in the v-reachable set along with an increase in the v- attractive set.

5.6 Comparison of Methods as Applied to the Second Problem

The results presented here represent what might be a reasonable first look at this problem from five different points of view. Considerably different results can be obtained from those presented here. For example, if using method 1, we choose K=2 and τ_e =14.1, the result illustrated in Figure 18 are obtained. This controller has a v-attractive set which compares favorably with method 5, however the v-reachable set is perhaps unacceptably large.

Again the best results are obtained when the σ curve is of small slope and a relatively high gain is used (providing control saturation near the $\sigma = 0$ curve). Method 5 attempts to mimic the game theoretic solution near the boundary of the playable set and mimic a traditional controller with high gain near the origin. This approach produces good results for this problem.

6. DISCUSSION

Depending on the choice of parameters, the first three methods can yield identical results and hence do not differ except in philosophical approach. Under bounded control, they all use the same basic control law as given by (24) with σ defined either by

$$\sigma = \mathrm{K}\mathrm{x}_1 + \mathrm{K}\tau_e\mathrm{x}_2$$

or

$$\sigma = s_{12}\mathrm{x}_1 + s_{22}\mathrm{x}_2$$

Clearly K and τ_e in method 1 or **Q** and R in method 2 or **Q**, R, and $\hat{\mathrm{R}}$ in method 3 which in turn determine s_{12} and s_{22} can be chosen to yield the same result.

The first four methods have the advantage of being generally applicable to problems of any dimension whereas the applicability of method 5 to higher dimensional problems is uncertain. Method 4 has a strong theoretical background which makes it more reliable as a direct tool. The other methods rely more on experimenting. It would appear from problem 1 that method 5 will produce large v-attractive sets when the playable set is also large. However in problem 2 where the playable set is relatively small, the advantage of using method 5 is not so pronounced. That is all methods are capable of producing a control which approximates boundary control when the system is near the boundary.

From the examples presented here it would appear that method 5 has the advantage of being able to achieve small v-reachable sets and large v-attractive sets while maintaining a good time response. It would be of interest to extend

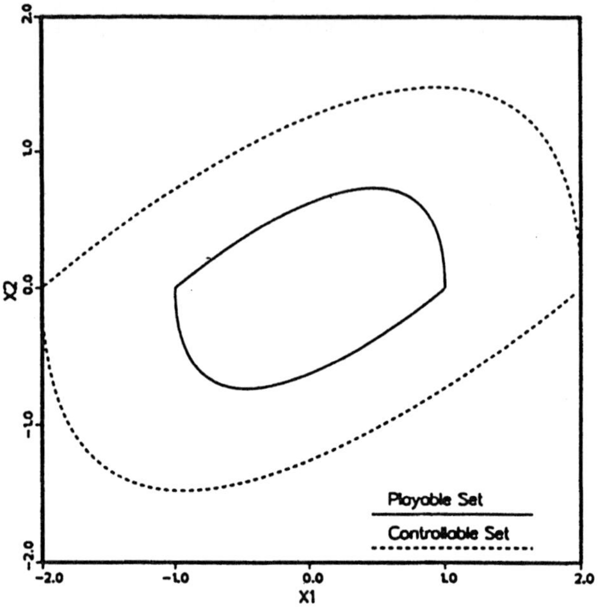

Figure 11 Controllable and playable sets for problem 2.

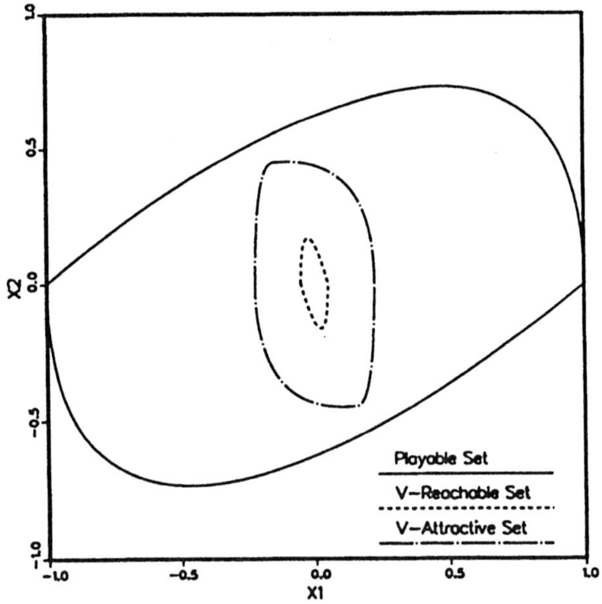

Figure 12 V-reachable and v-attractive sets for problem 2 using method 1 (K = 19, τ_e = .5736)

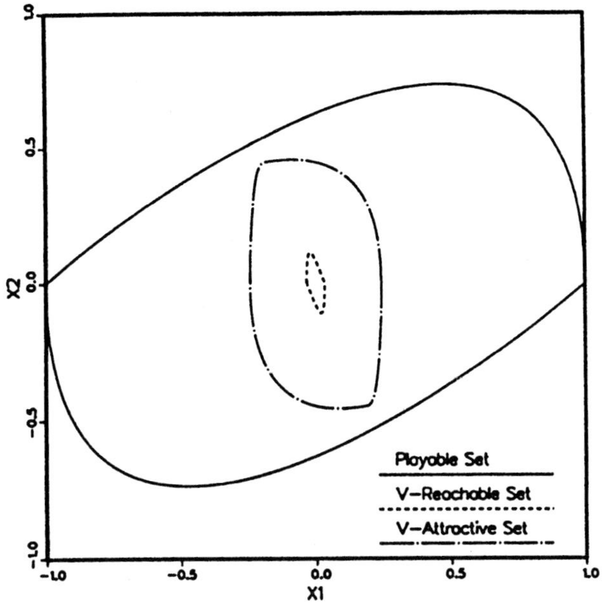

Figure 13 V-reachable and v-attractive sets for problem 2 using method 1 (K = 29, τ_e = .5736)

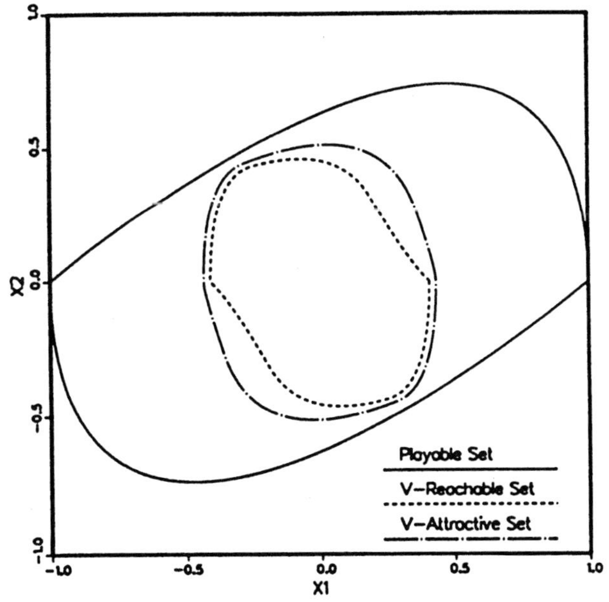

Figure 14 V-reachable and v-attractive sets for problem 2 using method 2.

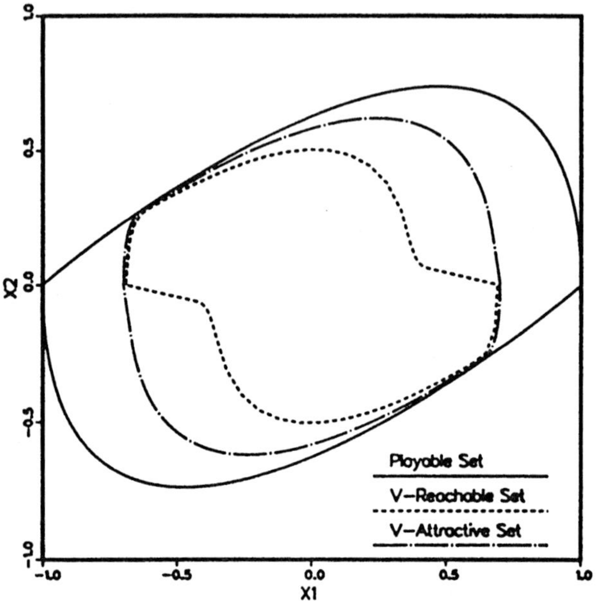

Figure 15 V-reachable and v-attractive sets for problem 2 using method 3.

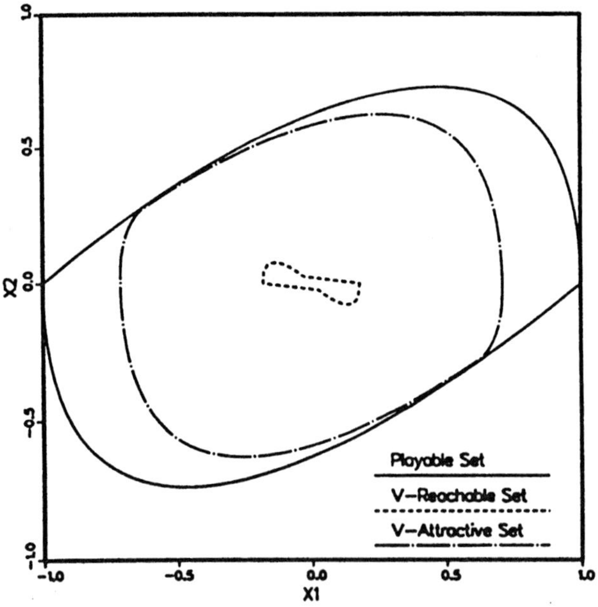

Figure 16 V-reachable and v-attractive sets for problem 2 using method 4.

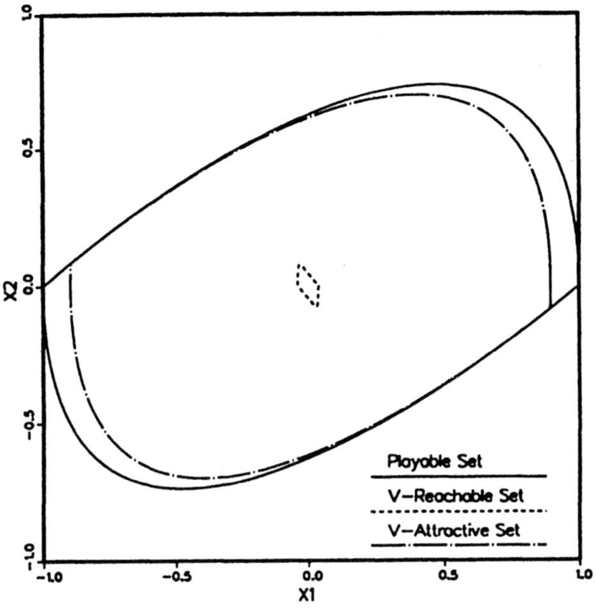

Figure 17 V-reachable and v-attractive sets for problem 2 using method 5.

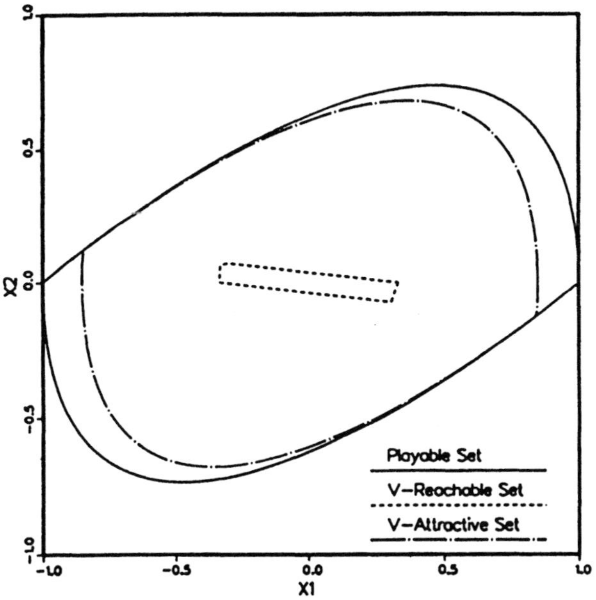

Figure 18 V-reachable and v-attractive sets for problem 2 using method 1 (K = 1, τ_e = 14.1)

this method for use with higher dimensional problems. Whether it would compare favorably for such systems remains to be seen.

No matter which method is used, for at lease two dimensional systems, the qualitative methods presented here represent an effective tool for adjusting parameters to achieve a good control design.

7. REFERENCES

1. Gayek, J. H., and Vincent, T. L., "On the Asymptotic Stability of Boundary Trajectories," *International Journal of Control*, Vol. 41, No. 4, pp. 1077-1086, 1985.
2. Grathan, W. J. and Vincent, T. L., "A Controllability Minimum Principle," *Journal of Optimization Theory and Applications*, Vol. 17, No. 5-6, pp. 523-543, 1975.
3. Chen, Y. H., "Deterministic Control for a New Class of Uncertain Dynamic systems," *IEEE Trans. Automatic Control*, AC-32, No. 73, 1987.
4. Hollot, C. V., and Barmish, R. B.,"Optimal Quadratic stabilizability of Uncertain Linear Systems,"*Proceeding 18th Allerton Conf. Communications, Control*, 1980.
5. Isaacs, R., Differential Games, Addison-Wesley Publishing Company, New York, 1965.
6. Lee, A. Y. and Bryson, A. E., "Neighbouring Extremals of Dynamic Optimization Problems with Parameter Variations," *Journal of Optimal control Applications and Methods*", Vol. 10, No. 1, Jan.-Mar., pp. 39-52, 1989.
7. Athans, M., and Falb, P. L., Optimal Control, McGraw-Hill, New York, 1966.

8. Kalman, R. E. and Bertram, J. E., "Control System Analysis and Design Via the Second Method of Lyapunov," *Journal of Basic Engineering*", Vol. 82, pp. 394-400, 1960.

9. Leitmann, G.,"New Class of Stabilizing Controllers for Uncertain Dynamical Systems," *SIAM Journal of Control Optimization*, Vol. 21, No. 2, March, pp. 246-255, 1983.

10. Gutman, S., and Leitmann, G., "Stabilizing Feedback Control for Dynamic Systems with Bounded Uncertainty," *Proceedings IEEE Conf. Decision Control*, 1976.

11. Leitmann, G., "Deterministic Control of Uncertain System via a Constructive Use of Lyapunov Stability Theory," *14th IFIP Conference on System Modelling and Optimization*, Leipzig, German Democratic Republic, July 3-7, 1989.

12. Sodatos, A. G., Corless, M., and Leitmann, G.,"Stabilizing Uncertain Systems with Bounded Control," *3rd Control Mechanics Workshop*, USC, 1990.

13. Gayek, J. E. and Vincent, T. L.,"An Existence Theorem for Boundary Trajectories," *Journal of Mathematical Analysis and Applications*, Vol. 132, No. 1, pp. 290-299, 1988.

14. Vincent, T. L.,"Control Design for Magnetic Suspension," *Optimal Control Applications and Methods*, Vol. 1, pp. 41-53,1980.

Lecture Notes in Control and Information Sciences

Edited by M. Thoma and A. Wyner

Vol. 117: K.J. Hunt
Stochastic Optimal Control Theory with Application in Self-Tuning Control
X, 308 pages, 1989.

Vol. 118: L. Dai
Singular Control Systems
IX, 332 pages, 1989

Vol. 119: T. Başar, P. Bernhard
Differential Games and Applications
VII, 201 pages, 1989

Vol. 120: L. Trave, A. Titli, A. M. Tarras
Large Scale Systems: Decentralization, Structure Constraints and Fixed Modes
XIV, 384 pages, 1989

Vol. 121: A. Blaquière (Editor)
Modeling and Control of Systems in Engineering, Quantum Mechanics, Economics and Biosciences
Proceedings of the Bellman Continuum Workshop 1988, June 13–14, Sophia Antipolis, France
XXVI, 519 pages, 1989

Vol. 122: J. Descusse, M. Fliess, A. Isidori, D. Leborgne (Eds.)
New Trends in Nonlinear Control Theory
Proceedings of an International Conference on Nonlinear Systems, Nantes, France, June 13–17, 1988
VIII, 528 pages, 1989

Vol. 123: C. W. de Silva, A. G. J. MacFarlane
Knowledge-Based Control with Application to Robots
X, 196 pages, 1989

Vol. 124: A. A. Bahnasawi, M. S. Mahmoud
Control of Partially-Known Dynamical Systems
XI, 228 pages, 1989

Vol. 125: J. Simon (Ed.)
Control of Boundaries and Stabilization
Proceedings of the IFIP WG 7.2 Conference Clermont Ferrand, France, June 20–23, 1988
IX, 266 pages, 1989

Vol. 126: N. Christopeit, K. Helmes M. Kohlmann (Eds.)
Stochastic Differential Systems
Proceedings of the 4th Bad Honnef Conference June 20–24, 1988
IX, 342 pages, 1989

Vol.127: C. Heij
Deterministic Identification of Dynamical Systems
VI, 292 pages, 1989

Vol. 128: G. Einarsson, T. Ericson, I. Ingemarsson, R. Johannesson, K. Zigangirov, C.-E. Sundberg
Topics in Coding Theory
VII, 176 pages, 1989

Vol. 129: W. A.Porter, S. C. Kak (Eds.)
Advances in Communications and Signal Processing
VI, 376 pages, 1989.

Vol. 130: W. A. Porter, S. C. Kak, J. L. Aravena (Eds.)
Advances in Computing and Control
VI, 367 pages, 1989

Vol. 131: S. M. Joshi
Control of Large Flexible Space Structures
IX, 196 pages, 1989.

Vol. 132: W.-Y. Ng
Interactive Multi-Objective Programming as a Framework for Computer-Aided Control System Design
XV, 182 pages, 1989.

Vol. 133: R. P. Leland
Stochastic Models for Laser Propagation in Atmospheric Turbulence
VII, 145 pages, 1989.

Vol. 134: X. J. Zhang
Auxiliary Signal Design in Fault Detection and Diagnosis
XII, 213 pages, 1989

Vol. 135: H. Nijmeijer, J. M. Schumacher (Eds.)
Three Decades of Mathematical System Theory
A Collection of Surveys at the Occasion of the 50th Birthday of Jan C. Willems
VI, 562 pages, 1989

Vol. 136: J. Zabczyk (Ed.)
Stochastic Systems and Optimization
Proceedings of the 6th IFIP WG 7.1 Working Conference,
Warsaw, Poland, September 12-16, 1988
VI, 374 pages. 1989

Lecture Notes in Control and Information Sciences

Edited by M. Thoma and A. Wyner

Vol. 137: S. L. Shah, G. Dumont (Eds.)
Adaptive Control Strategies for Industrial Use
Proceedings of a Workshop
Kananaskis, Canada, 1988
VI, 360 pages. 1989

Vol. 138: D. C. McFarlane, K. Glover
Robust Controller Design Using Normalized Coprime Factor Plant Descriptions
X, 206 pages. 1990

Vol. 139: V. Hayward, O. Khatib (Eds.)
Experimental Robotics I
The First International Symposium
Montreal, June 19-21, 1989
XVII, 613 pages. 1990

Vol. 140: Z. Gajic, D. Petkovski, X. Shen
Singularly Perturbed and Weakly Coupled Linear Control Systems
A Recursive Approach
VII, 202 pages. 1990

Vol. 141: S. Gutman
Root Clustering in Parameter Space
VIII, 153 pages. 1990

Vol. 142: A. N. Gündeş, C. A. Desoer
Algebraic Theory of Linear Feedback Systems with Full and Decentralized Compensators
V, 176 pages. 1990

Vol. 143: H.-J. Sebastian, K. Tammer (Eds.)
System Modelling and Optimization
Proceedings of the 14th IFIP Conference
Leipzig, GDR, July 3-7, 1989
X, 960 pages. 1990

Vol. 144: A. Bensoussan, J. L. Lions (Eds.)
Analysis and Optimization of Systems
Proceedings of the 9th International Conference, Antibes, June 12-15, 1990
XII, 992 pages. 1990

Vol. 145: M. B. Subrahmanyam
Optimal Control with a Worst-Case Performance Criterion and Applications
X, 133 pages. 1990

Vol. 146: D. Mustafa, K. Glover
Minimum Entropy H_∞ Control
IX, 144 pages. 1990

Vol. 147: J. P. Zolésio (Ed.)
Stabilization of Flexible Structures
Third Working Conference, Montpellier, France, January 1989
V, 327 pages, 1990

Vol. 148: In preparation

Vol. 149: K.H. Hoffmann, W. Krabs (Eds.)
Optimal Control of Partial Differential Equations
Proceedings of the IFIP WG 7.2 International Conference, Irsee, April 9-12, 1990
VI, 245 pages. 1991

Vol. 150: L. C. G. J. M. Habets
Robust Stabilization in the Gap-topology
IX, 126 pages. 1991

Vol. 151: J. M. Skowronski, H. Flashner, R. S. Guttalu (Eds.)
Mechanics and Control
Proceedings of the 3rd Workshop on Control Mechanics in Honor of the 65th Birthday of George Leitmann
January 22-24, 1990, University of Southern California
IV, 497 pages. 1991

Vol. 152: J. D. Aplevich
Implicit Linear Systems
XI, 176 pages. 1991

Vol. 153: O. Hájek
Control Theory in the Plane
X, 272 pages. 1991

Vol. 154: A. Kurzhanski, I. Lasiecka (Eds.)
Modelling and Inverse Problems of Control for Distributed Parameter Systems
Proceedings of IFIP (W.G. 7.2)-IIASA Conference
Laxenburg, Austria, July 24-28, 1989
VII, 179 pages. 1991

Vol. 155: M. Bouvet, G. Bienvenu (Eds.)
High-Resolution Methods in Underwater Acoustics
V, 249 pages. 1991

Lecture Notes in Control and Information Sciences

Edited by M. Thoma and A. Wyner